Slope Engineering for Mountain Roads

Geological Society books refereeing procedures

The Society makes every effort to ensure that the scientific and production quality of its books matches that of its journals. Since 1997, all book proposals have been refereed by specialist reviewers as well as by the Society's Books Editorial Committee. If the referees identify weaknesses in the proposal, these must be addressed before the proposal is accepted.

Once the book is accepted, the Society Book Editors ensure that the volume editors follow strict guidelines on refereeing and quality control. We insist that individual papers can only be accepted after satisfactory review by two independent referees. The questions on the review forms are similar to those for *Journal of the Geological Society*. The referees' forms and comments must be available to the Society's Book Editors on request.

Although many of the books result from meetings, the editors are expected to commission papers that were not presented at the meeting to ensure that the book provides a balanced coverage of the subject. Being accepted for presentation at the meeting does not guarantee inclusion in the book.

More information about submitting a proposal and producing a book for the Society can be found on its web site: www.geolsoc.org.uk.

It is recommended that reference to all or part of this book should be made in one of the following ways:

Hearn, G. J. (ed.) 2011. *Slope Engineering for Mountain Roads.* Geological Society, London, Engineering Geology Special Publications, **24**.

Hearn, G. J. 2011. A3 Slope materials, landslide causes and landslide mechanisms. *In*: Hearn, G. J. (ed.) *Slope Engineering for Mountain Roads.* Geological Society, London, Engineering Geology Special Publications, **24**, 15–57.

GEOLOGICAL SOCIETY ENGINEERING GEOLOGY SPECIAL PUBLICATION NO. 24

Slope Engineering for Mountain Roads

BY

G. J. HEARN
URS Scott Wilson Ltd, UK

2011
Published by
The Geological Society
London

THE GEOLOGICAL SOCIETY

The Geological Society of London (GSL) was founded in 1807. It is the oldest national geological society in the world and the largest in Europe. It was incorporated under Royal Charter in 1825 and is Registered Charity 210161.

The Society is the UK national learned and professional society for geology with a worldwide Fellowship (FGS) of over 9000. The Society has the power to confer Chartered status on suitably qualified Fellows, and about 2000 of the Fellowship carry the title (CGeol). Chartered Geologists may also obtain the equivalent European title, European Geologist (EurGeol). One fifth of the Society's fellowship resides outside the UK. To find out more about the Society, log on to www.geolsoc.org.uk.

The Geological Society Publishing House (Bath, UK) produces the Society's international journals and books, and acts as European distributor for selected publications of the American Association of Petroleum Geologists (AAPG), the Indonesian Petroleum Association (IPA), the Geological Society of America (GSA), the Society for Sedimentary Geology (SEPM) and the Geologists' Association (GA). Joint marketing agreements ensure that GSL Fellows may purchase these societies' publications at a discount. The Society's online bookshop (accessible from www.geolsoc.org.uk) offers secure book purchasing with your credit or debit card.

To find out about joining the Society and benefiting from substantial discounts on publications of GSL and other societies worldwide, consult www.geolsoc.org.uk, or contact the Fellowship Department at: The Geological Society, Burlington House, Piccadilly, London W1J 0BG: Tel. +44 (0)20 7434 9944; Fax +44 (0)20 7439 8975; E-mail: enquiries@geolsoc.org.uk.

For information about the Society's meetings, consult *Events* on www.geolsoc.org.uk. To find out more about the Society's Corporate Affiliates Scheme, write to enquiries@geolsoc.org.uk

Published by The Geological Society from:
The Geological Society Publishing House, Unit 7, Brassmill Enterprise Centre, Brassmill Lane, Bath BA1 3JN, UK

(*Orders*: Tel. +44 (0)1225 445046, Fax +44 (0)1225 442836)
Online bookshop: www.geolsoc.org.uk/bookshop

British Library Cataloguing in Publication Data

A catalogue record for this book is available from the British Library.
ISBN 978-1-86239-331-8

Distributors
For details of international agents and distributors see:
www.geolsoc.org.uk/agentsdistributors

Typeset by Techset Composition Ltd, Salisbury, UK
Printed by Cambrian Printers, Aberystwyth, UK

Contents

Foreword

This book is special in that it has made a significant advance in coalescing engineering, geology and geomorphology into one orderly and comprehensive volume which can be read and enjoyed by an engineer with a lack of knowledge of geology and geomorphology, or a geologist with a lack of engineering and geomorphology. The book, I believe, is the first, or at least one of the first, fully cohesive engineering geology texts, unlike many predecessors which are only a partially successful integration between the disciplines of engineering and geology. The latter principally reflect the discipline of the authors and even when one is a geologist and one an engineer, the joins are often patently visible. Not so with this book.

How has this come about?

Firstly, the authors are a team of practising engineers, geologists and geomorphologists, not academics, nor a working party, nor a conference of themed papers, but employees of a major consulting practice with decades of successful hands-on experience in the subject. As such, it is not teaching geology to engineers, or engineering to geologists: it is a book that integrates the planning, design, construction and maintenance of mountain roads in wet mountainous environments – mainly the humid tropics and subtropics.

Secondly, it is the example set by the excellent manual, 'Principles of low cost road engineering in mountainous regions', Transport Research Laboratory, UK, Overseas Road Note No. 16 (1997). This also drew heavily upon the knowledge and experience of the consulting engineers, Scott Wilson UK, earlier pioneering publications on mountain roads in Nepal, and academic leaders such as Professors Brunsden, Cooke, Doornkamp and Jones who, in the 1970s and 80s, were largely responsible for creating the broader framework in Britain of modern engineering geomorphology and visualizing its power as a tool in assisting engineering.

These two factors provided knowledge and set the scene for a wider-scoped new book some fifteen years later.

The driving force for the new book, albeit with considerable help from many others (see Acknowledgments), is Dr. Gareth Hearn, one of UK's leading second generation of engineering geomorphologists who mainly work in industry and have inherited the mantle of the original academic pioneers. He has been supported in this role by Tim Hunt: a geotechnical engineer with considerable experience in mountain road engineering.

Good road engineering in wet mountains is a matter of achieving efficiently that which is practical. There are, as far as I am aware, no substantive codes yet written, especially for wet mountain road engineering. Eurocodes used or discussed in the book and the current vogue of geotechnical modelling are often not the realistic way forward for mountain slope design because of the difficulty in obtaining hard field geo-data about mountain slopes. Each situation is a risk judgement. What factor of safety should be used on a mountain where a huge lump of the landscape could fall before, during or after the engineering works? What is the limit of what can be built in mountain terrain? The book has navigated its way through these problems. I like it very much. It is heading towards developing a mountain road philosophy but there is still quite a long way to go yet to writing a Wet-Mountain Slope Code but it has made significant headway.

What else do I like about the book? From its perceptive description of mountains and landslides, mountain roads and their feasibility, planning, site investigation, detailed design and construction to subsequent road and slope management, its logical structure is well written for tropical situations and remote areas which commonly have only limited infrastructure support. There are many case studies interwoven with the text, largely drawn from the firsthand experience of the authors. References are numerous and relevant and lead to wider reading. Figures, tables and annotated photographs abound and considerably strengthen the book, especially as colour is comprehensively used and is particularly helpful in illustrating multi-coloured tropical soils. Each photograph has been carefully selected to support and illustrate the associated subject matter. Text boxes are used to supplement engineering and geological points without disturbing the theme of the main text. All in all, a readable, valuable and authoritative volume. The authors, Scott Wilson, the Department for International Development (DFID) and the Geological Society Publishing House are to be congratulated.

Professor P. G. Fookes, F. R. Eng.
Winchester, UK
August 2011

Acknowledgements

The editor would like to acknowledge and thank the following contributors:

- T. Hunt: contributions and comment throughout the document, advice and encouragement
- J. Howell: bio-engineering measures for slope protection, comments on the text in general and some of the photographs
- I. Muir: contributions to soil and rock reinforcement
- A. Hart: satellite image interpretation, Geographical Information Systems
- G. Pettifer: terrain model illustrations
- C. Massey: rock slope stability.

I. Hodgson reviewed and commented on the entire book and his contribution is gratefully acknowledged. M. Sweeney and C. Manby acted as reviewers for the Geological Society Publishing House and provided useful comments. P. G. Fookes acted in the capacity of Peer Reviewer and has written a Foreword to the document. His overall guidance and encouragement are especially gratefully acknowledged. M. Holt provided updated information on satellite imagery. All maps and drawings were prepared by K. Jones.

The following were significantly involved in one or more of the illustrated projects:

P. G. Fookes	N. Carruthers
D. Petley	P. Ward
R. Blong	J. Dowling
R. Mason	C. Lawrance
C. Massey	S. Sharma
C. Chant	N. Man Shakya
J. Henley	X. Chonephetsarath
D. Wise	S. Sysouvanthong
J. Hart	B. Siliphon
C. Chant	L. Ayalew
P. Quinlan	S. Tiwari
M. Molinari	

This book has been developed from an original document prepared by Scott Wilson Ltd (now URS Scott Wilson Ltd) for the Department for International Development (DFID) UK. D. Salter played a key role in facilitating the original document and his continued support, enthusiasm and encouragement are gratefully acknowledged.

Most of the illustrations contained herein are taken from projects carried out in the humid tropics and subtropics, although some examples are given from areas outside these zones in order to illustrate a particular aspect. The editor and URS Scott Wilson Ltd would like to thank the following for appointing them to carry out the work illustrated in this book:

- Department for International Development (DFID) UK for work carried out in Nepal, Bhutan and Laos
- Asian Development Bank for work carried out in Nepal, Sri Lanka and Tajikistan
- World Bank (IDA) for work carried out in Nepal, Philippines and Ethiopia
- International Finance Corporation (IFC) for work carried out in Tajikistan
- Islamic Development Bank for work carried out in Tajikistan
- United Nations Environment Programme (UNEP) for work carried out in China
- SNV Netherlands for work carried out in Bhutan
- Department of Public Works and Highways, Philippines
- Ministry of Transport, Tajikistan
- Department of Roads and Railways, Libya
- Azerroadservice, Azerbaijan
- Ministry of Education, Azerbaijan
- Department of Local Infrastructure Development and Agricultural Roads, Nepal
- Department of Roads, Nepal
- Department of Roads, Bhutan
- Road Development Authority, Sri Lanka
- Ministry of Public Works and Transport, Laos
- Ethiopian Roads Authority
- IKRAM, Malaysia
- Geotechnical Engineering Office, CEDD, Hong Kong
- Rio Tinto for work carried out in Guinea
- Ok Tedi Mining Ltd (Broken Hill Proprietary) and Xstrata Frieda River Ltd for work carried out in Papua New Guinea
- ArcelorMittal, Liberia Limited
- Konkan Railway Corporation Ltd, India

Figures showing details from engineering drawings have often been simplified and may only be appropriate for the examples from which they have been derived. These should not be used for design without modification to suit site-specific conditions and materials. The publisher, the editor and URS Scott Wilson Ltd cannot accept responsibility for the use of material contained in this book by a third party.

How to use this book

This book covers the design, construction and maintenance of mountain roads in the humid tropics and subtropics and focuses on slope stability aspects. It concentrates on low-cost, low-volume roads, but many of the techniques described are equally applicable to higher road classifications.

The book is split into four parts.

Part A. Landslides and mountain roads
Part B. Site investigation
Part C. Design and construction
Part D. Slope management

An Index and a Glossary of Terms are also provided. References quoted are listed at the end of each Section. Note that references are only indicative, that is, they do not represent a comprehensive listing. The reader should carry out their own literature searches if a more comprehensive bibliography is required.

Part A describes and illustrates the background to landslide and slope instability problems affecting roads in hilly and mountainous areas of the humid tropics and subtropics. Basic considerations of hazard and risk are discussed.

Part B contains a description and review of techniques of site investigation, ranging from desk study, through field mapping to ground investigation and monitoring.

Part C provides practical advice on a range of issues that relate to the design and construction of alignments, slopes, retaining structures, drainage and erosion protection works.

Part D focuses on slope inspections, works prioritization and emergency management during road maintenance and operation.

Most practitioners will probably not wish to read this book cover to cover, but prefer to focus on aspects most immediately relevant to them. Consequently, Activity Flow Charts 1–4 have been prepared. These flow charts provide summary recommendations of the activities that should be undertaken when:

- designing new roads to minimize slope instability (Flow chart 1);
- forming new slopes during road construction and road improvement (Flow chart 2);
- maintaining slopes during road operation (Flow chart 3); and
- responding to slope and retaining wall failures that occur during road operation (Flow chart 4).

The relevant sections of this book, where each activity is described, are indicated in the flow charts. Project phasing and construction procurement are not referred to in the flow charts, and are discussed in Section A2.

Each of the disciplines of geology and civil engineering, including their various specializations and subdivisions such as geomorphology, engineering geology, geotechnical engineering and hydrology, for example, offers techniques and skills that can contribute variously to the design, construction and maintenance of mountain roads. Multi-disciplinary teams are most common on large and complex construction projects but guidance may be required in compiling these teams, or in seeking the advice of a specialist following an instability event during road operation for example. Consequently, Table 1 shows the broad range of tasks that each of these specialists might ordinarily undertake. However, there will be much variation and many exceptions according to training and experience, and each situation will require careful team selection. The definitions of the various specialists listed in Table 1 are provided in the Glossary.

Flow chart 1: Design of new roads to minimise slope instability

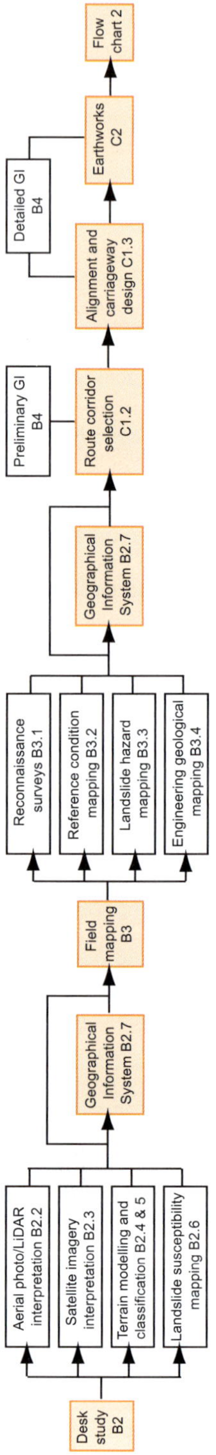

Flow chart 2: Design and construction of new slopes (for new road construction and road improvement)

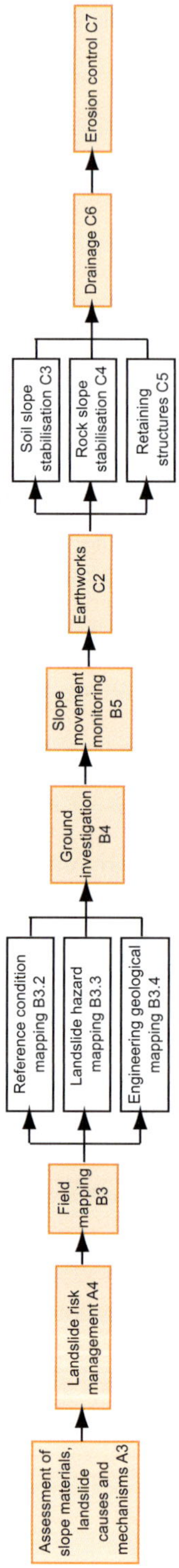

Flow chart 3: Maintenance of existing slopes

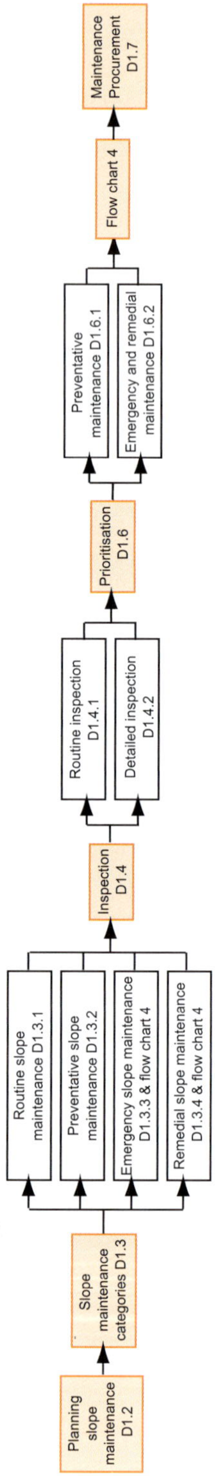

Flow chart 4: Slope or retaining wall failure during road operation

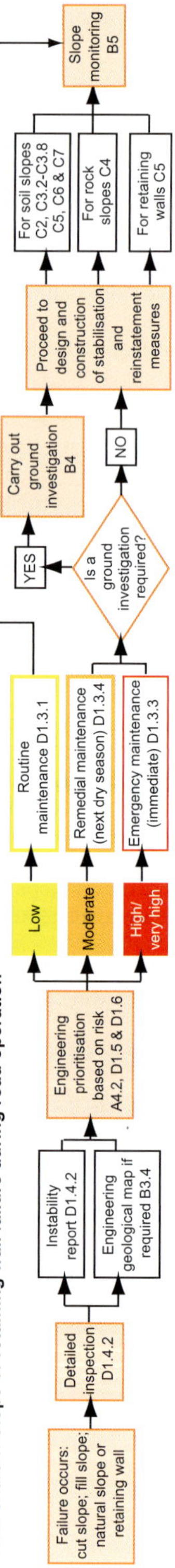

Table 1. *Common specialist skill sets for the assessment of terrain and slope stability and the design of mountain roads*

Specialist	*Terrain classification (B2.4 & 2.5)*	*Landslide mapping*		*Identifying areas of future instability*	*Ground investigation (B4)*			*Slope stability assessment & analysis*		*Design of engineering works*						
		From remote sensing (B2.2 & B2.3)	*From field observation (B3.4)*	*From remote sensing & field observation (B2.6 & B3.3)*	*Planning*	*Supervising*	*Interpreting*	*Soil slopes (C3.2)*	*Rock slopes (C4.2)*	*Alignment (C1)*	*Earthworks (C2)*	*Soil slope stabilisation (C3.3–C3.8)*	*Rock slope stabilisation & protection (C4.3 – C4.5)*	*Retaining walls (C5)*	*Drainage (C6)*	*Erosion protection (C7)*
Geologist	Some skills likely	Some skills possibly	Some skills possibly	Some skills possibly	Some skills possibly	Some skills possibly	Some skills possibly	Skills unlikely	Some skills likely	Some skills possibly	Some skills possibly	Skills unlikely	Skills unlikely	Skills unlikely	Skills unlikely	Skills unlikely
Geomorphologist	Main skill fields	Main skill fields	Main skill fields	Main skill fields	Some skills likely	Some skills possibly	Some skills possibly	Some skills possibly	Some skills possibly	Some skills likely	Skills unlikely	Skills unlikely	Skills unlikely	Skills unlikely	Some skills possibly	Skills unlikely
Engineering geologist	Some skills likely	Some skills likely	Main skill fields	Main skill fields	Main skill fields	Main skill fields	Main skill fields	Main skill fields	Main skill fields	Some skills likely	Some skills likely	Some skills likely	Main skill fields	Some skills likely	Some skills possibly	Some skills possibly
Geotechnical engineer	Skills unlikely	Skills unlikely	Some skills possibly	Some skills possibly	Some skills likely	Some skills likely	Main skill fields	Main skill fields	Some skills possibly	Some skills possibly	Main skill fields	Main skill fields	Some skills possibly	Main skill fields	Some skills likely	Some skills possibly
Civil engineer (roads & structures)	Skills unlikely	Skills unlikely	Skills unlikely	Skills unlikely	Skills unlikely	Skills unlikely	Skills unlikely	Skills unlikely	Skills unlikely	Main skill fields	Some skills likely	Some skills likely	Some skills likely	Some skills likely	Some skills likely	Some skills likely
Drainage engineer	Skills unlikely	Skills unlikely	Skills unlikely	Skills unlikely	Skills unlikely	Skills unlikely	Skills unlikely	Skills unlikely	Skills unlikely	Skills unlikely	Skills unlikely	Skills unlikely	Skills unlikely	Skills unlikely	Main skill fields	Main skill fields
Bio-engineer/forester	Skills unlikely	Skills unlikely	Some skills possibly	Skills unlikely	Skills unlikely	Skills unlikely	Skills unlikely	Some skills likely	Skills unlikely	Skills unlikely	Skills unlikely	Skills unlikely	Skills unlikely	Skills unlikely	Some skills likely	Main skill fields

Key: Main skill fields · Some skills likely · Some skills possibly · Skills unlikely

Part A: Landslides and Mountain Roads

A1 Introduction

G. J. Hearn* & T. Hunt

URS Scott Wilson Ltd, Scott House, Alençon Link, Basingstoke, Hampshire RG21 7PP, UK
**Corresponding author (e-mail: gareth.hearn@scottwilson.com; garethhearn@talktalk.net)*

A1.1 Purpose

This book deals with landslides, earthworks (cut and fill slopes), retaining structures and erosion protection on mountain roads and embraces planning, feasibility study, investigation, design, construction, improvement and maintenance. Non-arterial roads constructed in hilly and mountainous areas are usually characterized by low traffic volume, and are low-cost in the approach adopted in their design, construction and maintenance. This book focuses on these roads but many of the techniques described are equally relevant to more highly trafficked roads and high-investment infrastructure including railways and pipelines. The reason for this is that the techniques of geomorphology and engineering geology, which constitute much of the discussion and illustration contained herein, are among the most valuable tools applicable to any linear infrastructure project in complex and unstable terrain. This is especially true for remote locations where information is frequently lacking on ground conditions.

This book also focuses on the humid tropics and subtropics where heavy seasonal rainfall is responsible for a high incidence of slope instability. Nevertheless, large parts of this book will also be of interest to practitioners working in higher latitudes. Environmental issues of mountain road construction and maintenance are not addressed *per se* although many of the engineering considerations relating to land use and vegetation cover, earthworks stability, spoil disposal, drainage and erosion are also highly relevant to environmental protection. TRL (1997) discusses environmental and social impact considerations of mountain roads and further review is given in, for example, Corbett & Gaviria (2003); Dhakal *et al.* (2010) and Campos *et al.* (2010).

A1.2 Low-volume and low-cost roads

Traffic volumes are usually measured as the AADT for motorized vehicles (Average Annual Daily Traffic). The term *low-volume roads* is used to describe roads with traffic volumes of up to 400 AADT (Keller & Sherar 2003). These authors also suggest a maximum design speed of 80 kph as another defining parameter, but this would not apply in hilly or mountainous areas where design speeds associated with steep terrain and difficult alignment geometry might be expected to be lower (Section C1). A low-cost road is one that is constructed and maintained with the minimum required investment in earthworks and structures to provide the required access serviceability.

The terms *low-volume* and *low-cost* would normally be considered to apply to rural roads and 'feeder' roads, and not to more heavily trafficked roads that form important links in a country's transport network. However, due to limited budgets, even the arterial road networks in many mountainous countries within the humid tropics and subtropics (Section A1.3) are constructed and managed within a low-cost framework despite the fact that many of these roads have AADTs in excess of 1000 (or even as much as 3000 in some cases). While the budgets for the construction and maintenance of these arterial roads are usually significantly higher than for low-volume roads *per se*, they still remain essentially low-cost by comparison with many other parts of the world. However, even in the advanced economies, available budgets may be insufficient to counter landslide and slope instability hazards, and temporary road closures and road damage can be frequent (e.g. Winter *et al.* 2009).

The term *low-cost* is therefore preferred to *low-volume* in the context of this book because it encapsulates, to varying degrees, low-volume/low-cost as well as high-volume/low-budget situations.

Although there have been several key texts published in recent decades concerning landslide hazard and risk assessment for engineering projects (e.g. Fookes *et al.* 1985; Turner & Schuster 1996; Fookes 1997; Griffiths 2001; Fookes *et al.* 2005; Glade *et al.* 2005; Waltham 2009), there have been few that focus on the assessment and management of landslides in the context of road design, construction and maintenance. The publication by Fookes *et al.* (1985) was probably the first detailed account of the use of engineering geology and geomorphology in the design and construction of low-cost mountain roads to appear in the international literature. The United States Federal Highway Authority (FHWA 1988) published a manual of landslide management on federal highways in

From: Hearn, G. J. (ed.) *Slope Engineering for Mountain Roads*. Geological Society, London, Engineering Geology Special Publications, **24**, 1–6.
DOI: 10.1144/EGSP24.1 0267-9914/11/$15.00 © The Geological Society of London 2011

six states, but this focused primarily on embankment failures. The publications by TRL (1997) and Keller & Sherar (2003) provide greater engineering detail and advice for mountain road construction and maintenance than Fookes *et al.* (1985); however, the former focuses to a large extent on Nepal while the latter does not cover landslide management and slope engineering in any great detail. Documents produced by Dhital *et al.* (1991) and Hearn *et al.* (2003) provide useful guidelines, but the former is all-encompassing and does not have a slope engineering focus, while the latter focuses on remote sensing and mapping and contains limited details on slope engineering. Finally, the Department of Roads, Nepal (2003) has published a useful guide to slope protection on low-cost mountain roads, but it is aimed principally at management level practitioners and provides only limited engineering detail.

Given the rate at which mountain road construction is taking place in many countries, and the need to maintain existing infrastructure for strategic, economic and community access purposes, an update and expansion of the work of Fookes *et al.* (1985) and TRL (1997) is required for geologists, geotechnical engineers and civil engineers responsible for the design, construction and maintenance of mountain roads. Accordingly, this book describes, illustrates and advises on:

- the geological, geomorphological and engineering context of landslide and slope instability impacts on mountain roads, with particular reference to the humid tropics and subtropics (Part A);
- the techniques available to identify, define and assess landslides and unstable slope conditions for design purposes (Part B);
- the design, construction and improvement of mountain roads with regard to topography, ground conditions and slope stability considerations in particular (Part C); and
- the maintenance of mountain roads with regard to slope stability (Part D).

A1.3 Geographical coverage

This book is aimed principally at those regions of the world where many or all of the following factors combine to create serious slope instability constraints to the construction and maintenance of low-cost roads:

- mountainous or hilly terrain;
- slopes that are composed of tectonically disturbed rocks that are prone to landsliding;
- a current or past climate that encourages the development of deep weathering profiles resulting in residual soils and weathered rocks that are prone to landsliding;
- a climate that comprises heavy rainfall leading to high water tables and surface soil saturation, promoting conditions for slope instability (it is often the seasonality of rainfall in areas that experience monsoonal climates that creates the highest intensities);
- river regimes that are characterized by frequent flooding, shifting channels and scour of adjacent slopes;
- ongoing tectonic activity that continues to weaken surface rock masses and results in earthquakes that trigger landslides and earthworks failures and damage engineering structures, including retaining walls;
- limited available information concerning all aspects of geology, ground conditions and the distribution of landslides and landslide-prone slopes;
- limited economic and technical resources to derive the geological and geotechnical data required to evaluate ground conditions fully for design purposes; and
- limited capital resources with which to mitigate landslide hazards and adverse ground conditions during road construction and maintenance.

According to the Koppen-Geiger Climate Classification (Kottek *et al.* 2006), the climate conditions outlined above are mainly represented by the following climate zones:

- Af – equatorial, fully humid;
- Am – equatorial, monsoonal;
- Aw – equatorial, winter dry season;
- Cwa – warm temperate, winter dry, hot, wet summer; and
- Cwb – warm temperate, winter dry, warm, wet summer.

The distribution of these climate zones is shown on Figure A1.1. They fall principally within the latitudes defined by the tropics of Capricorn (23.5°S) and Cancer (23.5°N). However, the Cwa and Cwb zones extend outside this band into parts of the subtropics (23.5–40°N and 23.5–40°S), and it is the seasonality of rainfall that is particularly relevant in these areas. The Cwa/Cwb zones are dominated by monsoonal climates; rainfall intensities are high and the winter dry season causes vegetation to die back and some surface soils to desiccate, rendering slopes vulnerable to erosion and failure during early summer rains.

The Cfa climate zones (warm temperate, fully humid, hot summer) occupy SE continental areas of the United States, South America, Africa (principally parts of South Africa and Madagascar), Europe (principally NE Mediterranean and Black Sea borders), Asia (principally China and Japan) and eastern Australia. These areas may experience heavy rainfall even though it is not strictly seasonal. The Cfa climate zone is therefore also considered to be relevant to the climatic focus of this book.

Figure A1.1 also shows the distribution of relative altitude derived from 2003 SRTM (Shuttle Radar Topographic Mission, Section B2.3) data to approximate the hilly and mountainous areas of the Earth's surface.

The geographical focus of this book is where these climate zones and mountainous areas coincide; Table A1.1 lists those countries considered to be broadly representative of these combined conditions. These are the *typical* countries that are the focus of this book, and from which most of the illustrations and case histories are taken. However, this is a very approximate distribution and will include some areas that are not relevant. Conversely, there will be many other parts of the world where much of this book might

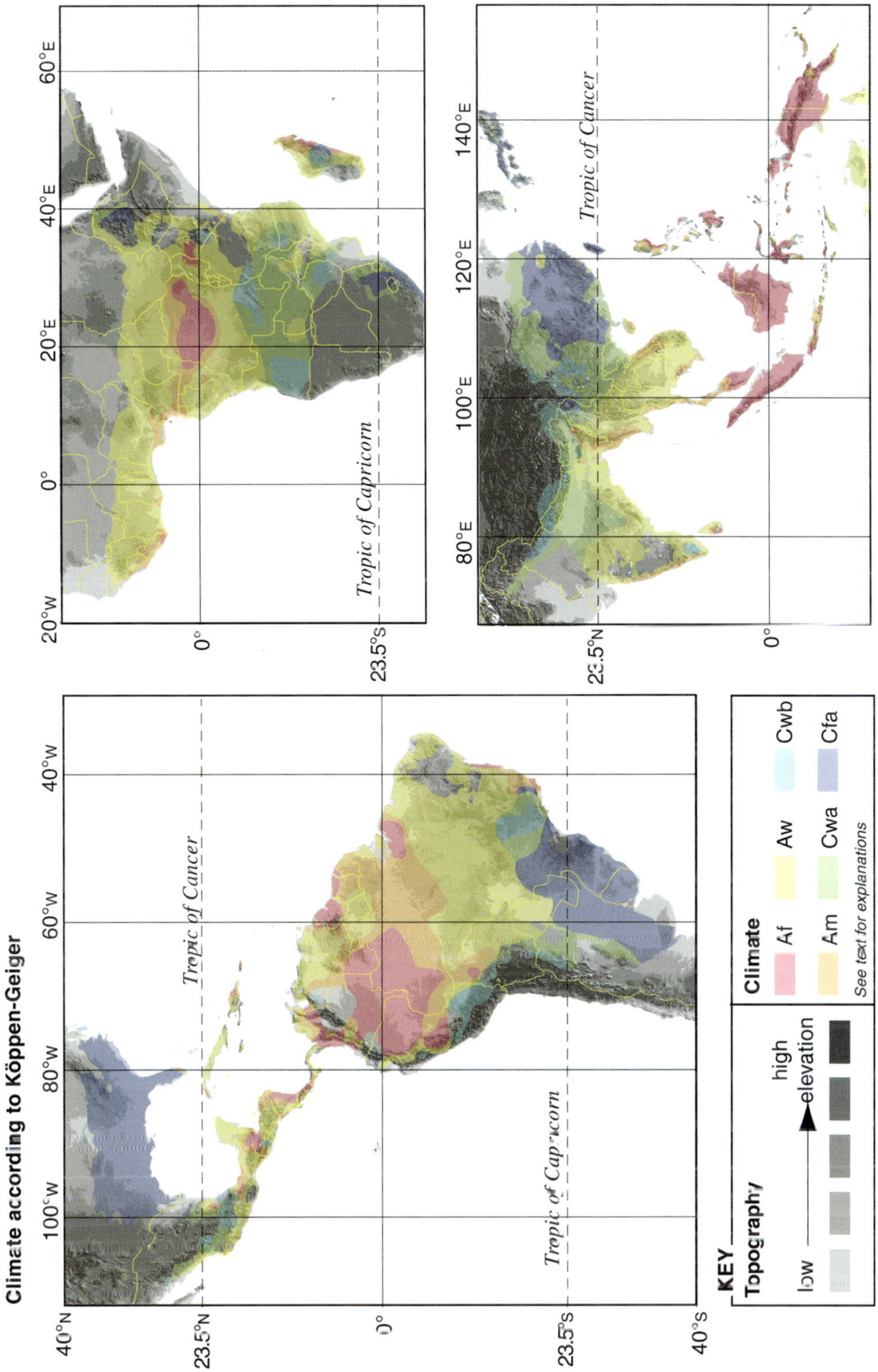

Fig. A1.1. Broad distribution of climate and topography in the humid tropics and subtropics.

Table A1.1. *Countries most relevant to this book*

Continent	*Region*	*Principal locations with hilly or mountainous terrain and annual rainfall ≥1500 mm*	*Principal locations with seasonal annual rainfall ≥1000 mm*
Asia	South Asia	Nepal, Bhutan, NE Pakistan, NW and NE India (Cwa/b), W Ghats India and Sri Lanka (Aw/Am)	
	South East Asia	Laos, Burma, Philippines, Thailand, Cambodia, Vietnam, Malaysia, Indonesia (Aw, Am, Af)	
	East Asia	Parts of SE China (Yunnan, Guangxi, Guangdong (including Hong Kong), W Sichuan, Shaanxi, Henan, Shandong, Jiangsu, Anhui) (Cwa/b)	Japan, Parts of SE China (Guizhou, E Sichuan, Hunan, Hubei, Jiangxi, Zhejiang, Fujian), Taiwan (Cfa)
Australasia	Australasia	Papua New Guinea, East Timor, South Pacific Islands (North of 20°S) (Af)	
Africa	Equatorial Africa (*c.* 10°N–10°S)	Interior Guinea, interior Liberia, NW Ivory Coast, interior Sierra Leone, Cameroon, NE Nigeria, Central African Republic, Gabon, E Zaire, Uganda, Kenya, Ethiopia (Am, Aw, Cwa/b)	N Angola, Tanzania, S Sudan (Aw),
	Southern Africa (SADC* definition)	Zambia, Malawi (Cwa/b) Mozambique, Madagascar (Aw, Af)	NE Zimbabwe (Cwa)
South America	South America	Guyana, Surinam, French Guinea, Venezuela, Columbia, Equador, N interior Peru (Am, Aw, locally Af)	S Brazil (Cfa), southern interior Peru (Am), S Bolivia (Cwa)
	Central America	Haiti, Dominican Republic, Puerto Rico, Jamaica, Honduras, Guatemala, El Salvador, Nicaragua, S Mexico, Costa Rica, Panama most of the Caribbean Islands (Aw, Am, Af)	
Europe	Southern Europe		Eastern peninsular Italy, Croatia, Albania, Serbia, northern Greece, Georgia, northern Turkey (Europe border) (Cfa)

*SADC, Southern Africa Development Community.

equally apply due to the occurrence of intense and prolonged rainfall on steep slopes. A minimum average annual rainfall of 1500 mm is probably appropriate for the creation of significant and recurrent landslide hazards in hilly and mountainous areas, although this figure might reduce to 1000 mm if rainfall is very seasonal. Higher latitude regions of the world will experience landslides that are triggered as a result of snowmelt, for example in northern and alpine Europe, parts of the Former Soviet Union and parts of North and South America. Some of the techniques described in this book might also apply to these areas.

The worst-case conditions with respect to slope instability hazards impacting low-cost roads can be expected to occur where all or most of the factors bullet-pointed at the start of this section coincide. Unfortunately, the majority of countries identified in Table A1.1 experience this combination of adverse climate, topography, geology, lack of available information and limited resources with which to combat slope instability.

Earthquakes are also responsible for many landslides in all climate zones located in seismically active, hilly and mountainous areas. No attempt has been made to delineate earthquake-prone areas on Figure A1.1. Information on seismic hazard zonation is available, for example, from the Global Hazard Assessment Program website (http://geology.about.com).

A1.4 Landslide hazards and mountain roads

Road construction within the humid tropics and subtropics is taking place at a rapid rate, and often in areas where existing geological and geotechnical information is insufficient to make informed decisions regarding the choice of alignments and their design. Key to the success of any project is the early assessment, in the planning stage, of the risk posed by potential geohazards and adverse ground conditions (e.g. Ho & Lau 2010). Geohazards can be due solely to natural processes or they can be caused or accelerated by man as a result of agricultural practices, construction activities and

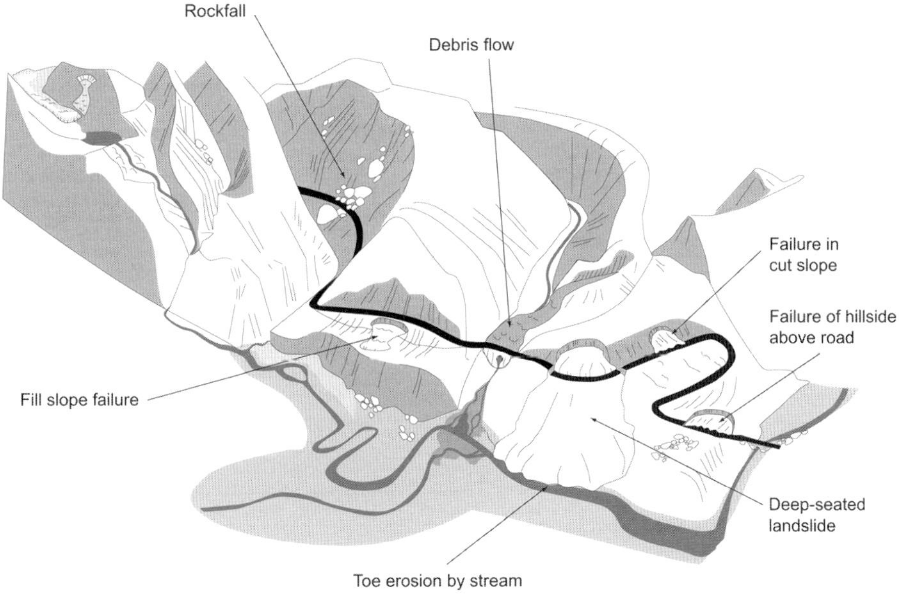

Fig. A1.2. Typical landslide hazards affecting mountain roads.

road maintenance operations (e.g. Slaymaker 2010). Effective route selection requires the identification of hazard areas, including the locations of existing landslides and those slopes that could become problematic during construction and maintenance. Even if the majority of hazard areas can be identified and avoided through route selection, there is usually a degree of residual risk that remains, especially in relation to the stability of earthworks slopes and the management of drainage from the road during heavy rain. The design and construction should strive to minimize these hazards to the greatest extent possible and maximize the protection of roadside slopes. However, there is clearly a limit to what is achievable within a low-cost framework. If there is no option other than to cross large, active landslides, it is usual to find that the affordable engineering solutions are capable of only short-term and superficial effect and do not prevent longer term movements and road damage from taking place. This outcome is frustrating to many road authorities who do not have the resources to deal with these large instability problems, but it may be the most practical solution. By contrast, route selection for higher cost roads is often less influenced by geohazard locations due to the importance of other factors, principally shortest distance, maximum connection to road users and the need to avoid high-value land uses such as urban and commercial areas. In these circumstances, geohazards are removed, stabilized or otherwise mitigated by sometimes quite costly design and construction solutions.

During road operation and maintenance, slope instability can result in significant and recurrent economic losses associated with damage and repairs to engineering structures and traffic delays caused by road blockages. These losses can be attributed to one or a number of the following:

- inadequate assessment of hazard areas during route selection, either through lack of awareness or lack of information;
- inadequate assessment of ground conditions, leading to a design that does not suit the actual ground conditions;
- lack of opportunity during route selection to avoid identified hazard areas, often as a result of cost considerations for alternatives;
- inadequate provision for permanent slope protection and drainage works during design and temporary drainage during construction, due to either a lack of awareness or a limited construction budget;
- construction and maintenance practices that are adverse to stability, particularly spoil disposal and drainage management;
- inadequate or poor supervision during construction;
- the occurrence of rainstorms, floods and earthquakes during road maintenance that are of a severity that could not have been economically accommodated within the original design.

Slope instability problems that commonly affect low-cost roads can be categorized in a variety of ways, but the most important subdivision is between shallow failures and deep-seated failures. Shallow failures involving surface soils may pose little more than nuisance effects; deep-seated failures however, often involving weathered rock, can result in significant damage and road loss. Another important distinction to be made is whether slope failure has taken place in natural ground or in fill, as a result of inherent geological instability or the effect of road construction itself or a combination of both. The risk posed by slope failures is also determined by their configuration in relation to road alignments, and the following categories can be broadly identified:

- those that occur on the slopes below a road and result in the loss of all or part of the carriageway (these can be very difficult and expensive to repair);
- those that result in road blockage from above, usually in association with failures of cut slopes (these can usually be remedied by debris clearance, road repairs and slope works);
- landslides and debris flows that originate on the natural slopes or in drainage channels above a road, leading to road blockage and scour of structures and the carriageway; and
- fill slope and retained-fill slope failures that occur due to inadequate design or poor construction.

Figure A1.2 depicts a typical mountain landscape and shows the common types of slope instability that regularly impact upon roads.

References

CAMPOS, S. J. A. M., GALLARDO, A. L. C. F. *ET AL.* 2010. Technological alternatives for erosion control in road construction. *In*: WILLIAMS, A. L., PINCHES, G. M., CHIN, C. Y., MCMORRAN, T. J. & MASSEY, C. I. (eds) *Geologically Active. Proceedings of the 11th IAEG Congress, Auckland, New Zealand, 5–10 September, 2010*, Taylor and Francis Group, London, 4005–4012.

CORBETT, M. & GAVIRIA, J. 2003. Integration of comprehensive environmental and social measures into low-volume road design for the hills of Nepal. *Transportation Research Record*, **1819**, 118–126.

DEPARTMENT OF ROADS, NEPAL 2003. *Guide to Road Slope Protection Works*. Department of Roads, Ministry of Physical Planning and Works, His Majesty's Government of Nepal, Kathmandu.

DHAKAL, I. S., BARAL, S. R. & HOWELL, J. H. 2010. What makes a mountain road environmentally sound? Experience from Nepal. *In*: WILLIAMS, A. L., PINCHES, G. M., CHIN, C. Y., MCMORRAN, T. J. & MASSEY, C. I. (eds) *Geologically Active. Proceedings of the 11th IAEG Congress, Auckland, New Zealand, 5–10 September, 2010*, Taylor and Francis Group, London, 4103–4112.

DHITAL, M., DEOJA, B., THAPA, B. & WAGNER, A. 1991. *Mountain Risk Engineering Handbook*. Volume I Subject Background and Volume II Applications. International Centre for Integrated Mountain Development, Kathmandu.

FHWA 1988. *Slope Maintenance and Slide Restoration*. Report No FHWA-RT-88-040. Prepared by HOPKINS, T. C., ALLEN, D. L., DEEN, R. C. & GRAYSON, C. G. Virginia, USA.

FOOKES, P. G. 1997. Geology for engineers: the geological model, prediction and performance. The First Glossop Lecture. *Quarterly Journal of Engineering Geology*, **30**, 293–424.

FOOKES, P. G., LEE, E. M. & MILLIGAN, G. 2005. *Geomorphology for Engineers*. Whittles Publishing, Dunbeath, UK.

FOOKES, P. G., SWEENEY, M., MANBY, C. N. D. & MARTIN, R. P. 1985. Geological and geotechnical engineering aspects of low-cost roads in mountainous terrain. *Engineering Geology*, **21**, 1–152.

GLADE, T., ANDERSON, M. & CROZIER, M. J. (eds) 2005. *Landslide Hazard and Risk*. J. Wiley & Sons Ltd., Chichester.

GRIFFITHS, J. S. 2001. *Land Surface Evaluation in Engineering Practice*. Engineering Geology Special Publication No 18. Geological Society, London.

HEARN, G., PETLEY, D., HART, A., MASSEY, C. & CHANT, C. 2003. *Lanslide Risk Assessment in the Rural Access Sector: Guidelines on Best Practice*. Department for International Development, London.

HO, K. K. S. & LAU, J. W. C. 2010. Learning from slope failures. *Quarterly Journal of Engineering Geology and Hydrogeology*, **43**, 33–68.

KELLER, G. & SHERAR, J. 2003. *Low Volume Roads Engineering*. United States Department of Agriculture, Forest Service.

KOTTEK, M., GRIESER, J., BECK, C., RUDOLF, B. & RUBEL, F. 2006. World map of the Koppen-Geiger climate classification. *Meteorologische Zeitschrift*, **15**, 259–263.

SLAYMAKER, O. 2010. Mountain hazards. *In*: ALCANTARA-AYALA, I. & GOUDIE, A. (eds) *Geomorphological Hazards and Disaster Prevention*. Cambridge University Press, Cambridge, 33–47.

TRL 1997. *Principles of Low Cost Road Engineering in Mountainous Regions*. Overseas Road Note 16, Transport Research Laboratory, Crowthorne.

TURNER, A. K. & SCHUSTER, R. L. 1996. *Landslides: Investigation and Mitigation*. Special Report, United States National Research Council, Transportation Research Board.

WALTHAM, T. 2009. *Foundations of Engineering Geology*. 3rd edn. Spon Press, London.

WINTER, M. G., MACGREGOR, F. & SHACKMAN, L. (eds) 2009. *Scottish Road Network Landslide Study: Implementation*. Transport Scotland, Edinburgh.

A2 Project phasing and procurement in relation to slope management

G. J. Hearn*, T. Hunt & R. Weekes

URS Scott Wilson Ltd, Scott House, Alençon Link, Basingstoke, Hampshire RG21 7PP, UK
*Corresponding author (e-mail: gareth.hearn@scottwilson.com; garethhearn@talktalk.net)

A2.1 Project phasing

Road construction and improvement projects (usually comprising widening, pavement reconstruction or resurfacing and improvements to horizontal geometry) are conventionally subdivided into the following stages:

- feasibility study;
- preliminary design;
- detailed design;
- construction; and
- operation and maintenance.

Low-cost road projects located in flat or gently rolling terrain typically incur costs in the following proportions (though percentages can vary significantly from project to project):

- 1% feasibility study;
- 2% design; and
- 97% construction.

For new roads in hilly and mountainous terrain several alignment options may exist, each with its own implications for length, ease of construction, stability and cost. Decisions made over alignment selection and the choice of cross-section can have profound effects on the cost of construction and the performance of the works during operation and maintenance. Investments in desk studies and engineering geological field investigations during the feasibility study and design stages can assist this decision-making and help avoid otherwise unforeseen ground conditions and stability problems during later stages. It is recommended that the opportunity be taken during these early stages to carry out these studies, especially in difficult and complex terrain. A cost distribution between the three main project stages might then be of the order of 5%, 5% and 90% respectively. For example, in the case of the Arun III hydropower access road in Nepal, where comprehensive preparatory studies were undertaken, the combined cost of the feasibility study and design amounted to a little under 9% of the tendered construction price. Additional costs incurred in the procurement of aerial photography, detailed terrain classification and engineering geological mapping (Sections B2.5 & B3.4) and the monitoring of landslides and floods for design purposes probably raised this figure to over 10% (Hearn & Lawrance 2000). These investments enabled geohazards to be identified, assessed and compared for different corridor options and assisted in the detailed design of the selected alignment.

For large and complex schemes the feasibility study is sometimes carried out in two phases: pre-feasibility and feasibility. In such cases, an outline cost estimate is developed during the initial stage for early decision-making. This is followed by a more detailed cost estimate during the feasibility stage. A two-stage feasibility study can take considerable time to complete, sometimes involving years rather than months in extreme cases. Furthermore, the programming of a new road project can become quite protracted when non-engineering issues, such as funding provision, land acquisition and environmental and social factors, prove difficult to resolve. This can result in a multi-phased and lengthy feasibility study period with attendant additional costs.

The design itself can be conducted as either a single-stage design or a two-stage preliminary and detailed design, depending upon the procurement and funding strategy. Usually, a single-stage design is adopted. This should comprise sufficient desk study, field surveys and investigations to derive the following:

- a final alignment with horizontal and vertical alignment drawings;
- cross-sections at 10 or 20 m intervals (depending on topography) showing the extent of cut slopes, fill slopes and retaining walls;
- schedules of retaining walls and other structures required to support and protect excavations and the road formation;
- schedules of culverts and other drainage works;
- the results of ground investigations for earthworks and foundations, and materials suitability for construction;
- the results of investigations at landslide sites and other areas of difficult ground;
- standard details;
- site-specific designs, such as those required at landslide sites;
- bills of quantities; and
- design standards and specifications.

From: Hearn, G. J. (ed.) *Slope Engineering for Mountain Roads*. Geological Society, London, Engineering Geology Special Publications, **24**, 7–14.
DOI: 10.1144/EGSP24.2 0267-9914/11/$15.00

For procurement expedience, detailed design is sometimes deferred until the construction stage itself, with the contract documentation based on little more than a preliminary design. In such situations, if inadequate attention is given to ground conditions in the contract documents, this could affect the cost and constructability of the works, and the stability of slopes and retaining structures. Construction projects can soon run into technical and contractual difficulties when too much design is left to a construction supervision team who are also required to work within a bill of quantities (BoQ) that may be inadequate for the ground conditions encountered. These difficulties can be exacerbated when the supervision team does not have the required skills and experience to deal with all of the geological and geotechnical considerations that may arise. This is a fairly common outcome when the same procurement and staffing provisions for lowland road projects are applied to projects in hilly and mountainous terrain.

Some of the most costly slope stability problems are caused by lack of control of access and haul road construction, spoil disposal and road runoff during construction. This can and often does stem from inadequate staffing levels to exercise sufficient supervision of contractor operations. For this reason it is recommended that supervision costs are a minimum of 3% of the construction cost (preferably 5% in the most difficult terrain) and that adequate engineering geological provision is allowed for in the supervision team.

The construction cost itself varies according to terrain, design standard and, to an extent, the project implementation strategy adopted. The cost per kilometre for a 7 m wide formation width road constructed across the Blue Nile gorge in Ethiopia in 2010, for example, varied from approximately $300 000 on the plateau above the gorge to $900 000 within the gorge itself. Although some of the cost differential is due to the lower cost of a gravel-wearing course on the plateau compared to a sealed road surface in the gorge, the majority is attributable to the need for major earthworks and retaining structures in the steep terrain of the gorge (Figs B2.7 & B2.8). The estimated per kilometre construction cost in 2010 for a hydropower access road of similar width in the Pamir mountains of eastern Tajikistan was $1 200 000. The higher estimated costs in the Pamir case were due to the presence of even steeper terrain and the frequency of difficult and unstable ground conditions, including extensive talus slopes, rockfall hazards, landslides and hard rock excavations.

In the case of road improvement, the cost of widening sections of hill road in Sri Lanka in 2010 from an average of 5 to 10 m formation width was *c.* $1 000 000 per kilometre. In comparison, the cost of improvement and widening to sections of road in the lowlands to the same design and specification was *c.* $500 000 per kilometre. Taking these figures at face value, the costs of road construction and road improvement in hilly and mountainous terrain can be between twice and three times higher than in lowland areas. This highlights the need to ensure that sufficient budgetary and technical resources are available when embarking on the construction of mountain roads, including those that are intended to be low-cost. Dahal *et al.* (2010), for example, point out that road construction at ultra-low-cost in mountain areas is often unsustainable as a result of landslides and other failures brought about by poorly planned and under-resourced engineering.

Although contractual claims might be anticipated on any construction contract for one reason or another, they may be increased significantly where design changes and delayed decision-making occur due to inadequate design in advance of construction and/or unforeseen ground conditions during construction. These outcomes can significantly increase final construction cost, depending upon the form of construction contract employed (Section A2.2).

Table A2.1 lists the principal activities usually carried out during the feasibility study, design and construction stages. The table is based on the conventional remeasurement form of contract (Section A2.2) whereby a consultant engineer (usually referred to as the Engineer once a contract is let) develops a design on behalf of a client (the Owner) and supervises the Contractor to build it. As discussed above, all design should be undertaken preferably prior to contractor procurement, but in reality this can often be the exception rather than the rule as true ground conditions are rarely known in sufficient detail prior to construction.

A2.2 Common forms of contract

Road construction and improvement projects are commonly procured using one of the following forms of contract:

- price-based
  - remeasurement
  - lump sum;
- cost-based
  - target cost
  - cost reimbursable;
- rates-only; and
- directly employed labour by force-account.

These various forms of contract are described briefly below and discussed in terms of their flexibility in dealing with unforeseen ground conditions, design change and slope management during construction. Table A2.2 lists these forms of contract in order of increasing preference from a slope management perspective and describes the main advantages and disadvantages associated with each. Further discussion can be found in CIRIA (1978, 1985) and Baynes (2010) with respect to geotechnical risk.

A2.2.1 Price-based

Price-based contracts are used most commonly on road construction and improvement projects. Among other things, the contractor is responsible for managing slopes

Table A2.1. *Common split of activities between project stages*

Activity	Owner Action	Engineer or Contractor Action[2] at Project Stage		
		Feasibility	Design	Construction
Identify project, define objectives				
Decide project procurement method[1]				
Appoint consultant				
Define required road standard				
Define geometric standards				
Carry out desk study and field reconnaissance				
Identify route corridors				
Compare ground conditions and geo-hazards				
Compare topography and river crossings				
Scope preliminary road design options				
Compare construction quantities and costs				
Compare environmental/social issues				
Produce comparative economic costs				
Select preferred corridor				
Feasibility report and cost estimate				
Decision to proceed to next stage				
Select final route				
Carry out topographic surveys				
Carry out engineering geological surveys				
Carry out environmental/social surveys				
Carry out siting and sizing of major structures				
Design/optimise alignment				
Carry out confirmatory ground investigations				
Design cross-section and earthworks				
Identify spoil disposal areas				
Design slope works, structures and drainage				
Prepare cost estimate, specifications, BoQ				
Decision to procure contractor				
Tender/procurement/award procedures				
Construction planning and method statement				
Construction implementation				
Treatment of landslides during construction				
Treatment of landslides during operation				
Performance monitoring of the constructed works				

[1]Some design activities (shaded lighter grey) extend into the construction period to address unforeseen ground conditions or to cater for construction effects. However this should be reduced to the greatest extent possible in order to keep delays and additional costs to a minimum.
[2]Depending upon form of contract.

Table A2.2. *Common forms of construction contract and their principal advantages and disadvantages*

Contract type	*Main advantages*	*Main disadvantages*
Price-based: lump sum (design and build)	1. Simple evaluation and comparison of tenders 2. Fixed price (except for inflation) 3. Contractor takes responsibility for the design and construction 4. Any unforeseen ground conditions or redesign is the contractor's responsibility	1. No basis for valuing claims and variations 2. Very limited flexibility for design changes by the owner 3. Incentive for the contractor to minimize time and costs; although favourable for project management it may not favour management of earthworks operations and slope stability 4. Owner participation excluded 5. No incentive for high quality 6. High risk for contractor, usually reflected in bid price 7. Not suited to projects of long duration
Price-based: remeasurement	1. Reasonable flexibility to accommodate variations (including reference condition approach, Section B3.2) 2. System well known and understood by all parties 3. Relatively simple evaluation and comparison of competitive tenders	1. Conflicting financial objectives for owner and contractor 2. Uncertainty of final cost 3. Requires design to be executed in advance; this may be problematic when subsurface conditions are not known 4. Design may be difficult to change during construction, though easier than for a lump sum contract
Cost-based: target cost	1. Incentive to contractor through the target mechanism 2. Greater flexibility for design changes 3. Contractor is encouraged to participate in design 4. Permits work to proceed in condition of uncertainty 5. Common financial objectives 6. Owner involvement in management 7. Full knowledge of contractor's costs 8. Relatively easy resolution of claims	1. High calibre owner management skills required 2. Difficult to evaluate tenders 3. Less certainty of final cost
Rates-only	1. Greatest flexibility in terms of unforeseen ground conditions and design changes 2. Easy owner participation 3. Contractor is able to participate in design (design responsibility is with the owner/engineer) 4. Low risk for contractor	1. Normally restricted to using the range of construction techniques and rate items listed in the tender 2. Uncertainty of final cost
Cost-based: cost reimbursable ('cost plus')	1. Greatest flexibility to design changes 2. Easy owner participation 3. Contractor is able to participate in design 4. Permits an earlier start to construction 5. Low risk for contractor 6. Full knowledge of costs 7. Design modification due to unforeseen ground conditions is readily feasible 8. Allows alternative construction techniques/rate items to be adopted during construction more easily if required	1. No incentive for contractor to reduce costs 2. Difficult to evaluate competitive tenders as no direct cost comparison can be made 3. Higher calibre of owner and contractor management skills are required 4. Uncertainty of final cost

(*Continued*)

Table A2.2. *Continued*

Contract type	*Main advantages*	*Main disadvantages*
Directly employed labour	1. Provides complete control over design changes and ground conditions 2. Provides greatest opportunity to maximize use of local labour and resources	1. Requires owner to invest considerable time and resources in the provision of project and engineering management 2. Owner is responsible for all procurement and is exposed to all risks associated with potential cost overruns 3. Best suited to labour-based rather than plant-based methods and not fast-track construction programmes, especially in steep and complex terrain

and drainage during construction but, because he is paid pre-determined prices, he will have little incentive to expend more than the minimum required contractually on temporary slope protection and drainage control. There may also be reluctance on the part of the owner (usually the national or regional road authority) to change the design to suit unforeseen ground conditions unless absolutely necessary in order to avoid potentially expensive contractual claims or cost overruns. A 'defects liability' period usually forms part of a price-based contract, particularly for remeasurement contracts, and normally applies to a 12 month period following construction completion. The contractor remains responsible during this period for the rectification of any faults or defects that are the result of his failure to carry out the construction according to the provisions in the contract. Any other defects are rectified at contract rates.

A2.2.1.1 Price-based: remeasurement

Price-based remeasurement contracts are the most commonly used contracts for low-cost road projects. Design drawings, a specification and a bill of quantities are prepared by the engineer in advance of a construction contract. Bidders prepare bids based on these documents, but the sums paid to the selected contractor are dictated by the actual quantities of work carried out during construction. If these quantities involve payment items not originally envisaged in the design, or where the quantity against an item varies significantly from that contained in the original bill of quantities, the contractor will be entitled to new or revised rates and this can lead to substantial cost overruns.

A significant proportion of the risk in this form of contract is therefore borne by the owner, and there still remains the potential for contractual difficulties if the design is changed significantly during construction to take account of unforeseen ground conditions. To reduce and manage this risk, the design is based on the known or expected ground conditions but the use of the *observational method* (Peck 1969) during construction enables contingency measures (usually through the *reference condition* approach, Section B3.2) to be employed when ground conditions and the adequacy of the design require reappraisal. The need to determine the range of likely ground conditions prior to the development of the design is, therefore, paramount.

A2.2.1.2 Price-based: lump sum

A lump sum contract is essentially 'design and build' (taken, in this instance, to include 'turnkey' contracts) whereby the contractor submits a price that covers the cost of detailed surveys, investigations, design and construction. In this case the contractor takes on most or all of the risk associated with unforeseen ground conditions and any other factors that remain unclear at the bidding stage. This is usually reflected in a high bid price to cover the risk and uncertainty. However, this risk and uncertainty (and subsequent bid price) can be reduced by the owner if an appropriately detailed pre-bid ground investigation is undertaken and this information is then made available to the bidders.

Although the selected contractor will be responsible for design and construction, it is recommended that the bid evaluation takes full consideration of the bidders' previous relevant experience, their appreciation of the site conditions, their approach to ground investigation and geotechnical design and their method statements with regards to earthworks and drainage management during construction. With this form of contract, the owner and the supervising engineer have least control on decision-making with regard to unforeseen ground conditions and slope and drainage management by the contractor. Furthermore, the contractor will wish to minimize his costs and construction programme and it will be necessary for the supervising engineer, as with a remeasurement contract, to ensure that the work is carried out to the required standard. However, the contractor will remain responsible for the design and, if this is later found to be inadequate, will bear the cost of any remedial works.

The 'design and build' approach provides the opportunity for the contractor to propose an alternative scheme or design approach as part of his bid and this could result in a more cost-effective solution. In modern 'design and build' contracts, self-certification by the contractor (with virtually no independent supervision) has become normal practice, although the use of this approach is not advocated for less engineered, less disciplined, circumstances.

DBO (design, build, operate) and DBFO (design, build, finance, operate) contracts are not suited to low-cost/low-volume road projects as the revenues to accrue from operation are usually too low or too uncertain to compensate for the risk required to be taken by the contractor.

A2.2.2 Cost-based

Cost-based forms of contract are relatively uncommon and involve reimbursing the contractor his actual costs with an agreed additional payment to cover overheads and profit. This form of contract provides a high degree of flexibility in how unforeseen ground conditions are dealt with, and allows the owner control over slope management decision-making during construction. Cost-based contracts are not generally favoured by funding agencies or owners because of the relative uncertainty in the out-turn costs.

A2.2.2.1 Cost-based: target cost
A target cost is set at the outset, usually by a bidding process, and any savings made are shared between the owner and the contractor. This is effectively an incentive for both parties to find the cheapest solutions to design and construction, and is not necessarily conducive to effective slope engineering and management. If actual cost is greater than target cost, there is a penalty on the contractor and his payment is reduced accordingly.

A2.2.2.2 Cost-based: cost reimbursable ('cost plus')
In cost-reimbursable contracts it is a requirement that the contractor's costs are fully available to the owner through 'open book' accounting. The owner pays these costs, subject to satisfactory control, and separately pays a fee to cover the contractor's overheads and profit. This form of contract allows flexibility in response to unforeseen ground conditions and enables the owner to exercise control over slope management decisions and design changes during construction.

A2.2.3 Rates-only

Rates-only forms of contract allow a contractor to be procured based on fixed rates only, that is, not necessarily with a bill of quantities. Of all forms of contract this provides greatest flexibility in the management of unforeseen ground conditions, but can result in significantly higher construction costs than might have been envisaged at the outset. These forms of contract are usually only employed where a contractor is required to mobilize immediately, for example to carry out emergency works. There is also significantly less complication in contract administration as there is no requirement to audit the contractor's costs, only his quantities.

A2.2.4 Directly employed labour

Directly employed labour is not a form of contract *per se* but a means by which the owner (who is often also his own designer, in this instance) exercises complete control over construction and is required to organize plant, labour and materials, either directly or via a managing consultant. This approach provides the greatest opportunity to cater for unforeseen ground conditions during construction without risk of contractual claims. It enables the owner to focus on slope management, but it usually requires a higher staffing and management commitment than he is able or willing to provide.

A2.3 Labour-based and local resource-based approaches

From Table A2.2 it can be seen that maximum flexibility in responding to unforeseen ground conditions is obtained when a directly employed labour (force-account) form of construction is used. Furthermore, this form of construction provides the owner with maximum control over site practices and this can have major benefits for the control of earthworks, drainage, plant, construction materials and spoil disposal. This raises the issue of whether a construction project should be approached as a labour and local resource-based exercise or as a plant-based operation. In practice, it is usual to find that a combination of plant-based and labour-based approaches are used, with plant-based operations being especially suited to projects where large earthworks quantities are involved. A labour-based approach maximizes the potential for local employment, but is only beneficial if a labour force with the required skills is available; importing labour for this purpose can lead to local conflicts and other social and health and safety issues.

During road maintenance operations, the opportunity to employ a labour-based approach is generally greater. This will apply to:

- the clearing of drains and culverts;
- the removal of small volumes of slip debris from the road;
- the maintenance of roadside vegetation; and
- the construction of retaining walls and revetments to reinstate failed slopes.

The extent to which local resource-based activities and materials can be used to combat slope instability and provide erosion control should be an important consideration for low-cost roads, and might include:

- bio-engineering works (essentially planting schemes and small-scale engineering structures) to protect cut and fill slopes from erosion and shallow failure (Section C7.2.4); and

Timber for temporary support to road fill

Bio-engineering methods of slope protection

Rock for protection at stream crossing

Soil-filled drums to support small slope failure

Retaining walls built in dry masonry

Dry masonry as scour protection

Timber crib used as minor retaining walls

Gravel-filled sand bags for erosion control

Fig. A2.1. Local resource-based approaches to slope and drainage management.

- dry masonry as retaining structures (Section C5.2.1) and as revetments and erosion control (Section C7.2.2).

Studies carried out in Nepal and Laos (e.g. Section C7.2.4) demonstrate that bio-engineering works are capable of providing protection against slope erosion and shallow (less than 0.5 m deep) slope failure, while deeper instability and failure to the road itself require a geotechnical solution. This geotechnical solution usually comprises earthworks and the greater use of retaining walls, constructed either from mortared masonry, gabion or reinforced concrete. Although these structures variously require the import of cement, steel and gabion baskets, masonry and gabion construction is essentially a labour-intensive exercise and utilizes stone and aggregate from local sources.

Figure A2.1 illustrates some low-cost, local resource-based applications for slope protection and erosion control. It is recommended that consideration be given to the use of these techniques when deciding how best to manage roadside slopes and drainage.

References

BAYNES, F. J. 2000. Sources of geotechnical risk. *Quarterly Journal of Engineering Geology and Hydrogeology*, **43**, 321–331.

CIRIA. 1978. *Tunnelling-improved contract practices.* Construction Industry Research and Information Association, Report 79.

CIRIA. 1985. *Target and cost-reimbursable construction contracts.* Construction Industry Research and Information Association, Report 85.

DAHAL, R. J., HASEGAWA, S., BHANDARY, N. P. & YATABE, R. 2010. Low cost road for the development of Nepal and its engineering geological consequences. *In*: WILLIAMS, A. L., PINCHES, G. M., CHIN, C. Y., MCMORRAN, T. J. & MASSEY, C. I. (eds) *Geologically Active. Proceedings of the 11th IAEG Congress, Auckland, New Zealand, 5–10 September, 2010*, Taylor and Francis Group, London, 4085–4094.

HEARN, G. J. & LAWRANCE, C. J. 2000. *Geomorphological Processes in Eastern Nepal as a Basis for Road Design.* Transport Research Laboratory, Project R6026, PR/INT/186/00.

PECK, R. B. 1969. Advantages and limitations of the observational method in applied soil mechanics. *Géotechnique*, **19**, 171–187.

A3 Slope materials, landslide causes and landslide mechanisms

G. J. Hearn

URS Scott Wilson Ltd, Scott House, Alençon Link, Basingstoke, Hampshire RG21 7PP, UK
(e-mail: gareth.hearn@scottwilson.com; garethhearn@talktalk.net)

A3.1 Common soil types and their influence on slope stability

Soils in hilly or mountainous areas are normally divided into two types: *in situ* weathered soils and transported soils. Table A3.1 provides a simplified classification and description of the common soils encountered in the humid tropics and subtropics and their engineering behaviour, based mainly on Fookes (1997).

A3.1.1 *In situ* weathered soils

Deep *in situ* weathered soils are often developed under humid tropical and subtropical climates (e.g. Ruxton & Berry 1957) and are the result of intense chemical decomposition and, to a lesser extent, mechanical disintegration of the parent rock. The classification of these soils is based on the degree of weathering that has occurred to the parent rock (Moye 1955; Ruxton & Berry 1957; Little 1969; Anon 1977, 1995; IAEG 1981; BSI 1999). Further discussion on weathering classifications is provided in Hencher (2008) and Norbury (2010).

Figure A3.1 shows the classification used to describe weathered rock and soil sequences in the humid tropics and subtropics by Fookes (1997). Text box A3.1 outlines some of the engineering properties associated with each of these weathering grades. Weathering grades IV, V and VI are usually classified as soil because, from an engineering perspective, they tend to behave more as soil than as rock. They are termed highly weathered rock, completely weathered rock and residual soil, respectively (BSI 1999), but because all three effectively behave as soil, they are also frequently referred to collectively as *tropical residual soil* (Fookes 1997). Due to processes of erosion and mass movement, it is usual in hilly and mountainous areas for deep residual soils to be limited in extent and confined to areas of gentle slope such as on geologically-controlled benches and broad ridge crests. Frequently, the weathering profiles found on most mountain slopes are highly variable in terms of depth, weathering grade and material/structural composition, making predictions of ground conditions very difficult without intensive investigation. Ho & Lau (2010), for example, describe this uncertainty in relation to the complex weathering profiles found in Hong Kong.

In weathering grade VI soil, all original rock material has been converted to soil and all rock fabric and mass structure has been entirely destroyed. The strength of this material is controlled by the interparticle friction and bonding of the constituent material (often associated with the deposition of iron and aluminium oxides and hydroxides as a result of chemical weathering). These soils usually comprise silt and clay particle sizes, perhaps with litho-relics (generally gravel-sized), and are often stiff to very stiff when dry. Suctions in dry soils and *in situ* grain-to-grain bonding and cementing impart a stiffness and strength that are greater than for transported soils of the same grading. Hencher & Lee (2010) consider the apparent cohesion imparted by soil suction as being an equivalent contributor to shear strength as that of interparticle bonding. Toll *et al.* (2011) consider the creation and maintenance of soil suctions to be more important than water table fluctuations in the control of slope stability.

Weathering and leaching processes can result in the precipitation of iron, aluminium, manganese, calcium carbonate, silica and gypsum compounds in the soil. Iron and aluminium precipitation imparts an orange to red-brown colour to soils. These soils are described as *latosols* in Table A3.1, although they are widely referred to incorrectly as *laterites*, and are predominantly free-draining. Given the drained topographic locations in which latosols are usually found, their short-term behaviour is often controlled by negative pore pressures (suctions or tensions between interparticle surfaces) that impart an increased apparent effective cohesion to the soil together with any incipient (i.e. developing) cementing of grains by the precipitation of the compounds referred to above. These soils often have a density and cohesion that allow them to stand vertically in cuttings until their strength is reduced by groundwater rise or surface water penetration.

Case-hardening often occurs on the surface of cut slopes formed in weathering grade V and VI materials due to the precipitation of cementing compounds, including iron and aluminium oxides and hydroxides for example. This

From: Hearn, G. J. (ed.) *Slope Engineering for Mountain Roads*. Geological Society, London, Engineering Geology Special Publications, **24**, 15–57.
DOI: 10.1144/EGSP24.3 0267-9914/11/$15.00

Table A3.1. *Common soil types found in the humid tropics and subtropics*

Material type	*Parent rock structure*	*Climate/ drainage influence*			*Engineering behaviour*	*Significance for hill or mountain roads*
		Wet/dry seasons		No dry season		
		Poorly- drained	Well-drained			
In Situ Weathered Soils (Tropical Residual Soil)						
Residual soil (WG* VI)	Absent	Smectite-rich Vertisols (commonly formed on basaltic rocks, shales, some mudstones)			Prone to shrink-swell and possible loss of inter-particle bonds as a result. High clay content, high plasticity. Can create stability problems in cuttings and increase embankment settlement. Typically on poorly-drained flat ground (eg terraces in hilly areas) but can form on slopes of up to 20° or more with attendant slope failures.	Relatively rare on hill/mountain roads, except on terraces, terrace flanks and as colluvium.
			Allophane-rich Andosols (formed on volcanic ash and fresh tuff)	Not developed	Moisture contents can be up to 250%, high liquid and plastic limits. Handling and compaction can be difficult. Low density means materials may soften during compaction. Can collapse and flow when disturbed. Can be susceptible to erosion.	Uncommon on hill/mountain roads, except in areas of volcanic soils.
			Iron/aluminium oxide-rich Latosols (formed on most rock types). Open-textured fabric can develop by leaching. Clays often leached through profile.		Collapse potential in open-textured soils. Kaolinite horizons can be poorly drained and unstable where cut. May contain corestones. Strength from bonding structure may be lost by construction reworking. May be susceptible to erosion/dispersion. Soil tests may be affected by test preparation methods. May require treatment for use as fill.	These soils are usually of limited depth on hill roads, and are uncommon on mountain roads.
		Duricrusts are common (eg calcrete alucrete, ferricrete, silcrete)		Not developed	As above. Duricrust may prove difficult to excavate.	Rarely found on hilly or mountain roads.
Completely weathered rock (WG V)	Strong micro-structural influence				Behaves homogeneously, strength controlled more by inter-particle bonds than relict structure. May contain corestones, may be susceptible to erosion.	Common on hill roads. Less common on mountain roads.
Highly weathered rock (WG IV)	Strong macro-structural influence				Behaviour commonly dominated by discontinuities. Strength, drainage and failure surfaces often controlled by relict joints. May contain corestones, may be susceptible to erosion.	Most common *in situ* weathered soil on hill and mountain roads.
Transported Soil						
Taluvium					Coarse (boulders/cobbles) and fine-grained material, usually derived from weathered or transported talus and landslide debris. Boulder content can make investigation, excavation and use as fill difficult. May prove difficult for foundations. Drainage conditions highly variable, often deep in predominantly coarse-grained deposits.	Commonly found on hill and mountain roads.
Colluvium					Predominantly fine-grained slope wash material. Can be unstable when cut. Unlikely to be suitable as fill or for foundation if silt, clay and moisture content too high.	Occasionally to commonly found on hill/mountain roads
Alluvium					Variable but usually bedded, coarse-grained material in mountain areas and may be gap-graded. May be susceptible to erosion in cuts. Possible source of fill and construction materials, silt and moisture content may be problematic.	Found in flood plains and terraces of mountain rivers

*WG,Weathering Grade. See Figure A3.1 for explanation. Most cut slopes along mountain roads will contain a range of weathering grades and transported soils.

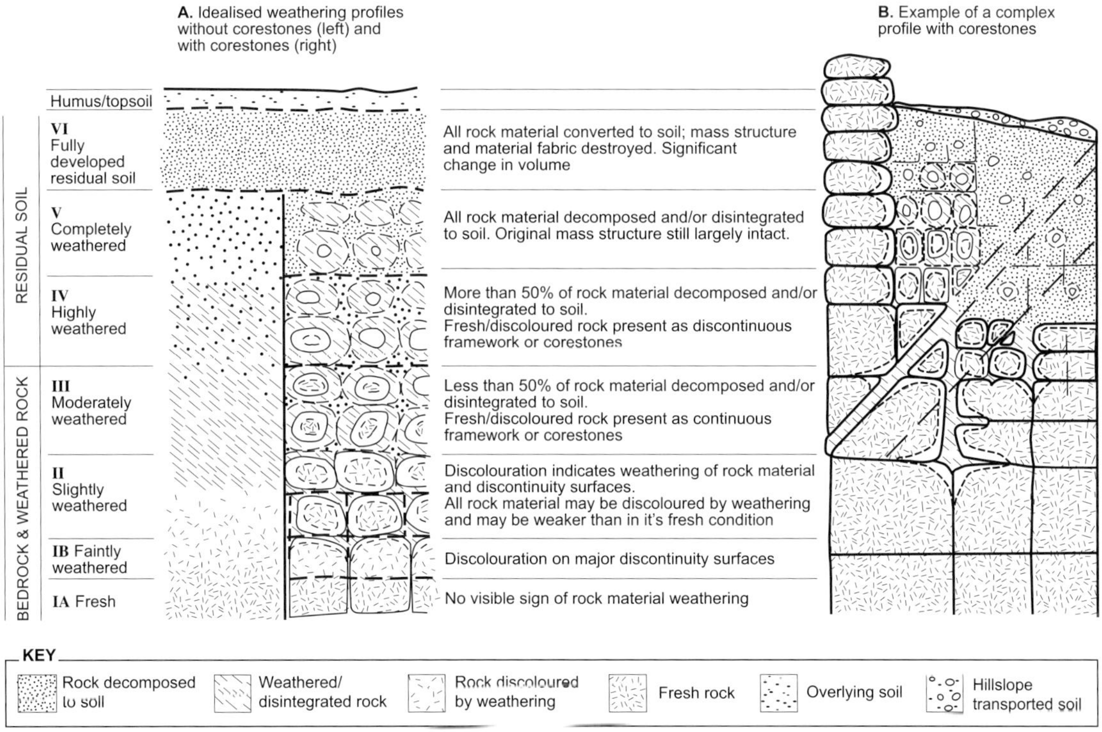

Fig. A3.1. Rock weathering grade classification (modified from Fookes 1997).

Text box A3.1. Weathering grade classification and its engineering application

Weathering Grade I

Rock is fresh with no visible signs of rock material weathering.

Weathering Grade II

Rock is slightly weathered: there has been some loss of material strength; >90% of materials remain as competent rock; <10% of materials have soil properties; more weathered, weaker materials are located along joints; joint shear strength is typically markedly lower than for joints in fresh rock; rock mechanics principles should be applied to excavation design; potential for kinematic (joint-controlled) failure may exist; blasting required for excavation (depending upon rock type and structure); excavated materials behave as clean, competent, essentially free-draining rockfill (depending upon rock type and structure).

Weathering Grade III

Rock is moderately weathered: *in situ* rock framework controls mass strength and stiffness; in excess of 50% of the material forms clasts that cannot be broken by hand but which may break down/degrade over time; shear strength along joints is typically markedly lower than for slightly weathered rock; combination of rock mechanics and soil mechanics principles to be applied to excavation design; potential for kinematic failure may exist; combination of ripping and blasting required for excavation depending on percentage of materials weathered to soils and joint pattern; when excavated described as boulders or cobbles, with some (5 to 20%) or much (20 to 50%) fines; behaves as a 'dirty' rockfill which requires careful screening of fines and moisture control during placement and compaction. Requires intensive investigation to get a clear picture of sub-surface conditions.

Weathering Grade IV

Rock is highly weathered: *in situ* rock fabric or texture contributes to mass strength; matrix or weathering products control stiffness; more than 50% of the material is decomposed or disintegrated to soil; remainder forms clasts that cannot be broken by hand and do not readily disaggregate or slake when a dry sample is immersed in water, but which may break down/degrade over time and are present as a discontinuous framework or corestones 'floating' in a soil matrix; combination of soil mechanics and rock mechanics principles to be applied to excavation and foundation design; typically rippable during excavation but potentially problematic due to presence of boulders/corestones within the soil matrix (blasting of large remnant blocks may be required to break them down to a size that can be excavated and transported); when excavated, described as fine material with some (5–20%) or many (20–50%) boulders or cobbles; may not be suitable as fill due to gap grading e.g., boulders in a fine matrix. A 'mixed fill' category might be required.

Weathering Grade V

Rock is completely weathered: all rock material is decomposed and/or disintegrated to soil; original mass structure still largely intact; considerably weakened compared to weathering grade IV material; slakes when wet; weathering products and relict structure control strength and stiffness; soil mechanics principles to be applied to excavation design, with a kinematic check required due to the relict structure (e.g. persistent at an unfavourable attitude), rippable during excavation; when excavated described as fine material; depending on soil characteristics excavated materials treated as common fill (if suitable), treated fill (where removal, mixing or blending is required to allow usage as fill) or unsuitable (cannot be used as fill due to susceptibility to erosion (unless protection is provided), too high a clay content or too low a plasticity); moisture control required during placement; potential for loss of structural strength during excavation, haulage, placement and compaction, and potential for loss of strength on wetting. High plasticity clay soils are not uncommon in tropical residual soil profiles and these will have low friction and may be subject to long-term softening as a result of loss of effective cohesion.

Weathering Grade VI

Residual soil: all rock material converted to soil; mass structure and material fabric destroyed; behaves as a soil; soil mechanics principles to be applied to excavation design; rippable during excavation; when excavated described as fine material; depending on soil characteristics excavated materials treated as common fill (if suitable), treated fill (where removal, mixing or blending is required to allow usage as fill) or unsuitable (cannot be used as fill due to susceptibility to erosion (unless protection is provided), too high a clay content or too low a plasticity); moisture control required during placement. High plasticity clay soils are not uncommon in tropical residual soil profiles and these will have low friction and may be subject to long-term softening as a result of loss of effective cohesion. Howell (2005), for example, describes failures on 10° slopes in highly plastic residual clays along hill roads in Trinidad and Tobago.

surfacing is able to provide a degree of erosion protection to the cut face. Case-hardening can be seen as the darkened cut face shown in Figure A3.2.

The progressive decrease in weathering grade with depth shown schematically in Figure A3.1 is often complicated by changes in lithology, rock structure or slope drainage and hydrogeological effects. Figure A3.3 illustrates some of these departures from the standard weathering profile that can pose difficulties for the prediction of material strength and excavatability in cut slopes.

In areas of flat to gently sloping terrain with a humid tropical climate and distinct dry season, soils can develop duricrust horizons over a long period of time in which iron or aluminium oxides and hydroxides and other compounds are precipitated to form a hardened layer. This layer can be several metres thick, and may require excavation by ripping or blasting. Text box A3.2 provides further discussion of these soils.

Corestones (Figs A3.2 & A3.4) are often found in weathering profiles developed on igneous and some sedimentary rocks with wide discontinuity spacing. These do not control stability except where they are allowed to protrude from a cut face, posing a potential rockfall hazard when undermined by erosion. However, their unpredictability can create problems when encountered during ground investigations and excavations.

The term *saprolite* is used to describe *in situ* weathered soil (less than 30% rock) that is intermediate between residual soil (weathering grade VI) and weathered bedrock (weathering grade III). Irfan (1996) and Wesley & Irfan (1997) describe

Fig. A3.2. Case-hardening on the surface of a residual soil slope formed on granite.

and classify the saprolites developed on granitic rocks in Hong Kong according to common engineering properties derived from mineralogical and structural characteristics. Weathering profiles developed on metamorphic rocks are often complex and contain the juxtaposition of weathered rock and *in situ* weathered soil (Fig. A3.5).

Weathering grade IV and V soils, by definition, contain some of the original rock structure including relict discontinuities (Text box A3.3). Geological structure, fabric and texture survive (at least in part) to the final stages of rock weathering, exerting significant control on stability (Hencher & Lee 2010). Faults, joints and intrusive igneous structures, such as dykes and sills, result in differentially weathered rock masses, complex groundwater conditions and the creation of low-strength surfaces along which failure can take place when orientated out of the slope (Fig. A3.6).

Vertisol soils can develop where rocks composed of mafic minerals weather to residual soils in areas of impeded drainage. In Ethiopia, for example, tropical weathering of the basalt rocks has created the widespread development of vertisols that are rich in smectite clay minerals and are commonly termed 'black cotton' soils (Fig. A3.7). The shrink–swell cycle caused by the seasonal drying and wetting of these soils results in settlement and heave. Wide, deep shrinkage cracks develop in the dry season and may become infilled with wind-blown dust and surface wash materials, forming a net permanent heave. Volume change under road embankments and light-weight buildings will reflect upwards and lead to deformation. In addition to creating problems for subgrade and embankment stability, these soils also pose significant slope stability hazards when exposed in deep cuttings (Table A3.1). Vertisols are widespread in Ethiopia, Kenya, India and parts of Southern Africa.

Volcanic ash (Fig. A3.8) usually has a low density and high porosity, and the constituent silt and sand particles are susceptible to crushing under load. Andosols (containing allophane clay minerals) are among the most common soil types developed on volcanic ashes. They can pose significant engineering problems, especially when used as fill, and may have extremely high water contents. Compaction can result in softening (loss of effective cohesion; Section C3.2) and they are often prone to erosion. These soils occur significantly in young volcanic terrain, including parts of Indonesia, Africa (Ethiopia, Kenya and other countries bordering the Rift Valley) and Papua New Guinea.

A3.1.2 Transported soils

Transported soils are the most commonly encountered materials on mountain slopes (Fig. A3.9). The terms taluvium and colluvium have been used (for example Reeves *et al.* 2006) to differentiate between coarse-grained and fine-grained transported soils, respectively. They are derived

Abrupt interface between WG V and WG I-II, i.e. rock that requires blasting is found directly beneath residual soil

WG II-III overlies WG V due to differential weathering between rock types. Creates weaker materials with depth

Marked lateral change from WGI on left to WG V on right with no topographic indication

Weathering grade changes laterally as well as with depth (WG decreases in direction of arrows). Note the infilled vertical joints

Fig. A3.3. Some typical departures from the standard weathering profile and their engineering implications.

from processes of mass movement and hill wash and commonly comprise the following:

- fine-grained soils accumulated over time on slopes through the progressive transport of materials downslope through gravity or hill wash (colluvium);
- rock fragments that have accumulated on the slopes below or at the base of cliffs as either a veneer or wedge of material (talus);
- soils developed through the progressive deposition of materials derived from rockfalls, other landslide mechanisms and weathered and reworked talus (these tend to be chaotic or jumbled deposits of fine-grained soils and rock fragments of varying size, i.e. taluvium);
- soils developed on hillsides derived from discrete landslide events (these can often contain chaotic deposits of large 'rafts' of rock and a range of material sizes, i.e. landslide debris); and
- debris flow and mudflow deposits in stream channels and on slopes (Section A3.4).

Transported soils usually have a variable grain size and their density depends upon their age and post-depositional history. Where a transported soil remains stable following its deposition, the high rainfall and temperatures of the humid tropics and subtropics can cause rapid weathering leading to the development of corestone erratics within a medium-dense or firm fine-grained matrix (Fig. A3.10). However, most transported soils on mountain slopes are relatively recent in age (less than 1000–10 000 years), undergo periodic movement and are loose to medium-dense. Their strength is usually derived entirely from their interparticle friction,

Text box A3.2. Fully developed residual soil (weathering grade VI)

The structure of fully developed residual soil is largely the result of the weathering processes by which it is formed. The structure frequently involves a wide range of pore sizes, some being larger than would normally be associated with the grading and grain size of the soil. There is usually some interparticle bonding in residual soils. In a fully developed residual soil this is more likely to be due to the effects of changes during weathering and mineral alteration, and to the precipitation of cementing material. In the extreme, represented by various forms of duricrust, cementation may give sufficient strength for a rock-like material to be re-formed, but in most residual soils the bonding is much weaker. It should be noted, however, that even a bond so weak that a sample can scarcely be handled still provides a component of strength and stiffness which may have a strong influence on engineering behaviour. The *in situ* stiffness and strength of a residual soil is likely to be underestimated by laboratory tests on remoulded samples. By contrast, *in situ* tests on collapsible soils, for example, may significantly overestimate their strength characteristics when disturbed, loaded or 'wetted up'.

The void ratio of fully developed residual soils may vary widely, independently of the source of rock, the type of weathering and the stress state. This may be due to variations in the amount of material that has been leached from the soil. Void ratio is a function of the weathering process and is not directly related to stress history. In a weakly bonded soil the void ratio has a strong influence on drained strength, which increases with dry density.

Modified from Fookes (1997).

and it is rare to find natural slopes in these materials at angles greater than 40° (and then only in fully drained conditions in densely packed, coarse-grained soils comprising angular cobbles and boulders). Interparticle cementation by secondary deposition of carbonate compounds for example can also cause these materials to stand at angles greater than those due to friction alone. Quite often, however, these soils have a significant volume of voids (high void ratio) and are highly porous and permeable. During heavy rain, water quickly drains through to less permeable and more clayey layers causing saturation and ground movements. Seepages are often apparent at the interface between transported soil and

Fig. A3.4. Corestones within weathering grade V granite gneiss.

Fig. A3.5. Weathering grade V–VI soil developed on folded metamorphic rock with weathering grade III rock adjacent.

Text box A3.3. Relict discontinuities in highly and completely weathered residual soils (weathering grades IV and V)

In weathering grade IV and V soils discontinuities often occur and are usually relicts from the parent rock. The low strength along these planes is due to coating of particles of low-friction clay and iron/manganese/organic compounds. Angles of shearing resistance on these surfaces may be of the order $\varphi' = 15-20^\circ$ when the seams are unsheared, dropping to about $\varphi' = 10^\circ$ when they are pre-sheared and slickensided. Low-strength discontinuities are very difficult to identify by boring or drilling. Their influences depend on their continuity, extent and the degree to which they form planar features at critical angles to the stresses imposed by slope geometry and engineering works.

Modified from Fookes (1997).

underlying rock, or between coarse-grained taluvium overlying fine-grained colluvium, and movement can take place along these boundaries.

Natural slope angles in fine-grained soils will usually be significantly lower, perhaps a maximum of 30–35° when fully-drained. Temporary soil suctions, however, may allow these soils to stand steeper in the short to medium term, depending upon the slope drainage regime. By contrast, pore-water pressures in the same material may be seasonally high, causing ongoing movements on low-angle slopes.

A3.2 Common rock types and structures and their influence on slope stability

In most mountain regions it is usual to find that tectonic processes of folding, faulting and shearing have given rise to complex geological structures (Figs A3.11–A3.13) that can significantly influence slope stability. Waltham (2009) provides an overview of the effects of tectonics on geological structure, lithology and landforms for engineers.

Fig. A3.6. Shear strength along relict joints is often the most important control on stability.

Fig. A3.7. Black cotton soil (vertisol), overlying weathered tuff.

Fig. A3.8. Volcanic ash overlying silt/clay residual soil.

Basalt talus comprising medium dense angular gravel and fine material

Basalt talus overlying weathered tuff. Note more recent soil formation above

Taluvium with long axes orientated out of the slope, potentially unstable

Multiple failure events have caused 'layering' in coarse-grained taluvium and fine-grained colluvial deposits

Colluvium derived from weathered quartzite

Entire spur formed in taluvium

Fig. A3.9. Typical transported soils found in hilly and mountainous regions.

Fig. A3.10. Corestone relics within a weathered colluvial soil.

In the Himalayas, for example, metamorphic rocks form a significant proportion of the slope materials encountered. They range in metamorphic grade from slates and phyllites, which retain some of the fabric of the original sedimentary structure of the parent rock, through to schists and gneisses, whose structure and mineralogy have been completely altered due to tectonics and heat metamorphism. The cleavage and foliation (Fig A3.11) in low- and medium-grade metamorphic rocks act as planes of weakness along which slope failure can take place if they are orientated adversely (Figs A3.14 & A3.15). The fracturing in these rocks often facilitates the ingress of groundwater into the rock mass and, as a result, the rates of weathering can be significantly higher than would otherwise be the case. Weathering often produces soils that comprise clayey silt and platy gravel particles which tend to slide and flow easily when wet. Hearn *et al.* (2008) and Hearn & Massey (2009) describe landslides in these materials along the road networks of Laos and Bhutan, respectively.

The foliation of the higher grade metamorphic rocks (schists and some gneisses) can also have a major control on slope stability. Schists tend to be less prone to weathering and slope instability than phyllites, but more prone than gneisses. Schist rocks often fail through a combination of movement along their schistosity (Fig A3.11) and failure through the rock material itself. Gneisses tend to be more massive and fail almost entirely along major joints rather than foliation. A guide to small-scale geological structures, including the fabric of metamorphic rocks, is provided by Wilson (1982) and illustrated in Figure A3.11.

Rock exposures in cut slopes along mountain roads in the humid tropics and subtropics are often extremely variable in terms of structure (Fig. A3.5), strength and weathering grade. This variability arises from lithological changes, microtectonic structure and complex slope drainage and hydrogeology, and can have a major influence on slope stability. Figure A3.16 illustrates some of the complex jointing patterns encountered in metamorphic, sedimentary and igneous rocks encountered in road excavations. The infilling of these joints with the products of weathering and fine-grained materials deposited in groundwater flow can significantly reduce the strength of a rock mass and can become the main control on slope stability (Fig. A3.17). Koor *et al.* (2000) for example note that the friction angle (φ') of these clay infills can be as low as 10°.

A3.3 Outline of the causes of landslides

Landslides occur wherever the gravitational forces acting on a slope exceed the resisting forces imparted through the strength of slope materials (Sections C3 & C4). The stability of a slope will decrease as the slope angle is increased. The presence of water in a slope will reduce stability through the development of pore pressures between soil particles

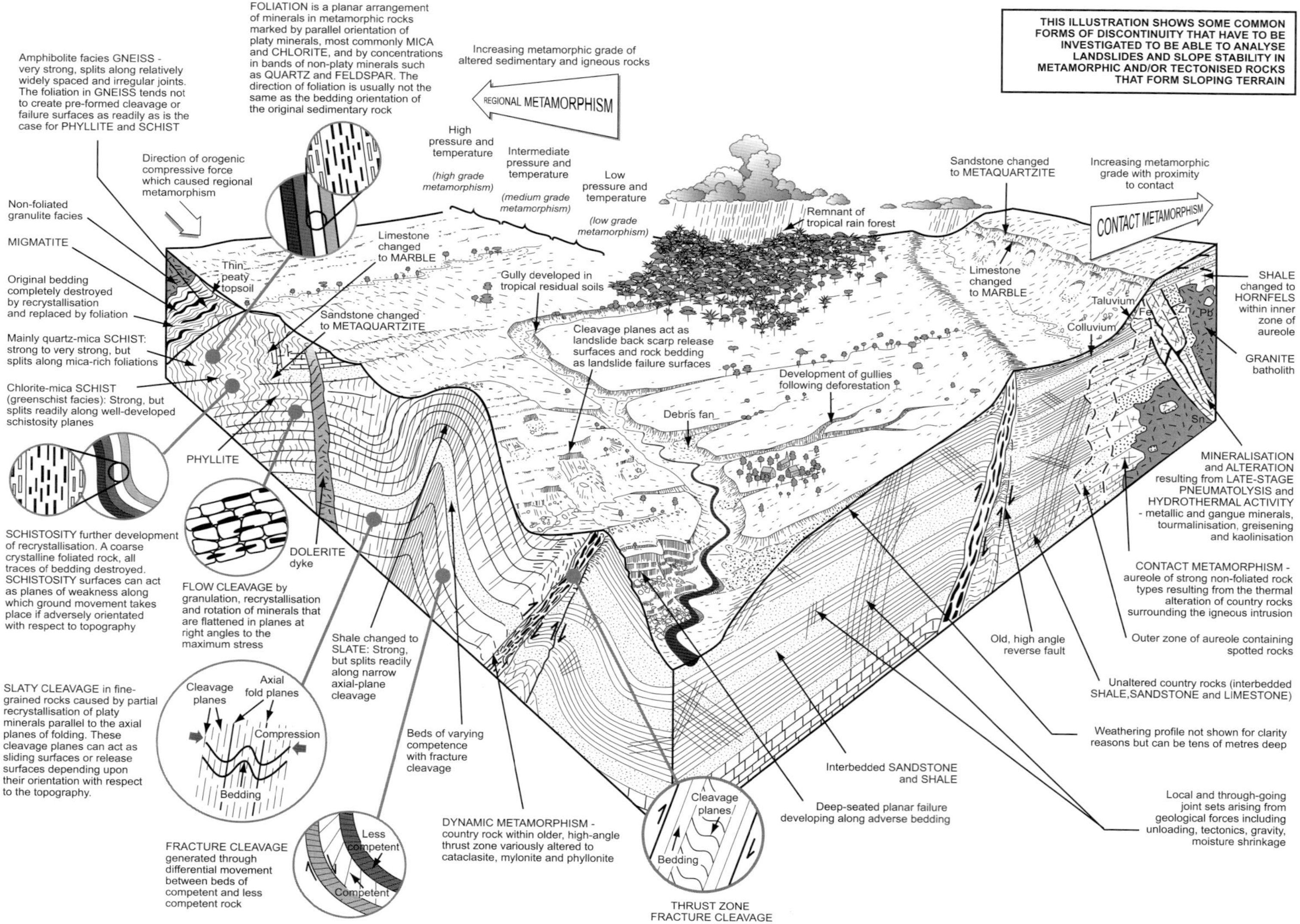

Fig. A3.11. Typical large scale and small scale geological structures encountered in metamorphic rocks and their influence on slope stability (modified from Fookes 1997).

Fig. A3.12. Folded geological structure creates complex outcrop patterns.

or between different layers or discontinuities in the soil or rock profile. This leads to a reduction in normal loads acting on existing or potential failure surfaces, thus reducing shear resistance. Loss of suctions in fine-grained tropical residual soils can also lead to slope failure.

It is important to distinguish between landslide *conditioning* and *triggering* factors when:

- assessing the stability of natural hillsides and cut slopes; and
- determining what measures are required to increase stability.

Table A3.2 and Figure A3.18 list and illustrate the most common of these factors. Heavy rain, seismicity and river erosion are among the most important triggers of slope instability in mountain areas, but trigger levels vary significantly. It is usually not only the rainfall on a given day that is important, but also the preceding rainfall (for example, Sarkar *et al.* 2011) and the extent to which this has caused groundwater and soil moisture levels to rise. The Antecedent Precipitation Index is a measure devised to take account of the moisture condition of a soil based on rainfall records prior to a given time and has been used, for example, by Lumb (1975) in Hong Kong, Soralump (2009) in Thailand and Jaiswal & van Westen (2009) in southern India to examine rainfall thresholds for landsliding. As far as 24 h rainfall is concerned, Dahal & Hasegawa (2008) describe how landslides in the Himalayas are often triggered when daily rainfalls exceed 144 mm; Hencher (2006) considers 500 mm in 24 h to be the threshold for Hong Kong.

Text box A3.4 outlines earthquake effects on slope stability. In seismically active areas, earthquakes are probably the most important cause of large landslides. Lin *et al.* (2009), for example, report in excess of 25 000 landslides triggered by the 1999 earthquake in Taiwan. Landslides also often occur where river scour becomes directed at the toe of hillside slopes, causing steepening and undercutting. These processes can be rapid, in some cases recurring over short engineering timescales. Figure A3.19 illustrates the case where slope instability is triggered by erosion on the bend of a river opposite a tributary fan inflow.

It cannot be overemphasized that, apart from earthquake effects, it is the presence of water in a slope that has the overriding influence on slope stability. It is the control of this water that will have the most beneficial effects (Sections C3 & C6).

A3.4 Landslide types and characteristics

For the purpose of this book, landslide types are usually classified according to:

- whether they are essentially slides, flows, avalanches or falls;
- whether the failure material is either soil (coarse-grained or fine-grained) or rock, or a combination of both; and
- in the case of slides, the shape of the surface(s) along which failure takes place (planar, wedge and circular, or compound).

Information on the classification of landslides can be found in, for example, Varnes (1978), Hoek & Bray (1981), Bromhead (1986), Cruden & Varnes (1996), Dikau *et al.* (1996), Wyllie & Mah (2004), Griffiths (2005) and Fell *et al.* (2007). Tables A3.3 and A3.4 summarize the common characteristics of slides, flows, avalanches and falls. These characteristics will vary widely from region to region and, consequently, these tables are intended only as a guide.

Figure A3.20 illustrates the principal failure modes of planar, wedge, circular, debris flow, avalanche and fall. Figures A3.21 and A3.22 show examples of some of these failure modes involving rock and soil in mountain regions. Many observed slope failures are compound, involving more than one mechanism and multiple material types. Many landslides also progress from one failure mode to another as they move downslope. This transition commonly applies to the development of a slide into a debris flow or a large rockfall into an avalanche, for example.

(a)

(b)

Fig. A3.13. (**a**) and (**b**) Tight folding causes fracturing of the rock mass prone to ravelling failure.

Fig. A3.14. Cleavage orientated out of the slope, adverse to stability.

Fig. A3.15. Failure along adverse jointing leading to loss of road edge.

Foliation in phyllite dips adversely to slope stability

Foliation dips into the slope (favourable to stability) but other joints dip unfavourably

Blocky rock structure; no apparent overall structural control on stability

Tectonic shearing has led to 'slickensides' along which ground movement can take place

Very different rock structure and strength due to movement or shift along a fault in meta-sediments

Pyroclastic, tuff and basalt sequences create complex groundwater patterns

Fig. A3.16. Typical small-scale rock structure variations affecting road cuts.

Fig. A3.17. Joint infill with fine-grained weathering products.

A3.4.1 Slides

A3.4.1.1 Planar failures

The simplest and by far the most common landslide types are planar failures that occur along single, approximately linear, sliding surfaces. These are differentiated here according to their constituent materials into debris slides, mudslides and rockslides. Figure A3.18 illustrates how 'daylighting' discontinuities control the stability of rock slopes in relation to planar failure.

Debris slides. This is a term often used to describe planar failures involving coarse-grained soils that occupy steep hillsides and cut slopes along mountain roads. Debris slides occur most commonly in taluvium and landslide material, that is, within soil that is not *in situ*. Failures usually take place in response to a rise in groundwater level or the saturation of the surface soil layer during heavy rain. They also occur in *in situ* weathered soils:

- through the soil mass;
- along distinct weathering grade boundaries or weak layers;
- along relict joints or existing slip surfaces (reactivation); and
- along the rock head surface (Fig. A3.23).

The term 'debris slide' is often used to describe all soil failures of shallow planar type, even when the constituent materials are fine-grained (silts and clays). Debris slides are usually up to a few metres in depth and their rate of movement varies between metres per second for first time failures in *in situ* weathered soil and metres per year for reactivations of taluvium and landslide deposits.

Table A3.2. *Main factors controlling the stability of rock and soil slopes*

Rock slopes		*Soil slopes*	
Conditioning factors	*Triggering factors*	*Conditioning factors*	*Triggering factors*
Slope angle and height Rock structure orientation, including discontinuity patterns, in relation to topography (slope direction and angle – kinematic feasibility) Rock mass strength and weathering grade* Presence of weak horizons within the rock mass, either more closely jointed or softer (more clayey) layers Presence of rock horizons/layers of varying permeability creating perched water tables	Toe erosion by streams and rivers removing lateral support, or vertical support if undercut When degree of weathering, particularly along discontinuities, reaches a critical level (strength) Earthquake acceleration, leading to increased driving forces Heavy and/or prolonged rainfall. Increased water pressure along discontinuities External influences including excavations, fills and spoil dumps, drainage changes	Slope angle and height Soil depth and the presence of any adversely orientated relict structures that are derived from the original rock fabric (if *in situ* weathered soil) or previous failure surfaces (if taluvium/colluvium) Presence of a distinct soil layer/rock head boundary along which failure takes place Soil composition and strength, a function of grain size, particle arrangement and mineralogy, density and moisture content Presence of weak horizons and permanent groundwater seepages	Prolonged/heavy rainfall leading to a rise in groundwater level and reduction in strength Intense (usually short term) rainfall leading to saturation of surface soil layers and reduction in strength Toe erosion by streams and rivers removing lateral support Earthquake acceleration, leading to increased driving forces Deforestation, and other land use changes, can lead to increased surface water runoff, erosion and slope instability External influences, including excavations, fills and spoil dumps, drainage changes

*Weathering grade obviously increases with time, but is taken here to be a constant factor over short engineering timescales.

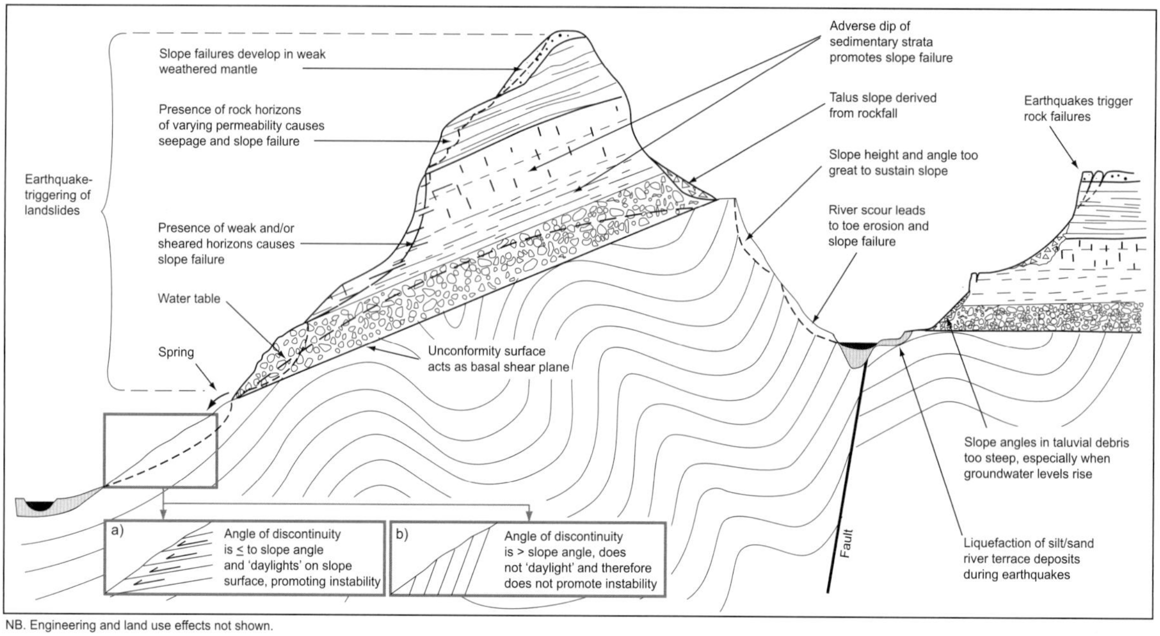

Fig. A3.18. Some factors controlling the stability of rock and soil slopes.

Text box A3.4. Earthquake effects on slope stability

Inertial forces and cyclical increases in pore-water pressure associated with ground shaking during an earthquake are often sufficient to trigger failure of slopes that are otherwise marginally to moderately stable. This is particularly the case if seismicity is combined with other climatic or land use factors which contribute to a reduction in slope stability. In addition, earthquakes usually cause rock masses to dilate (open up along joints), and the effects of this on slope stability are often not manifested until the next heavy rains when surface runoff is able to penetrate rock masses more easily than would otherwise be the case. The stability of slopes in earthquake-prone areas depends upon soil/rock properties in static and dynamic terms, hydrogeology, slope geometry, local topography and earthquake characteristics (e.g. location, duration, intensity, amplitude, frequency and cyclical effects on pore-water pressures). Further discussion on the relationship between earthquakes, rainfall and landslides is given, for example, in Chen & Hawkins (2009).

Failures in improperly constructed fill slopes (Section C2.4) sometimes take place along the interface between the fill and the underlying natural ground, and are therefore approximately planar. However, there are occasions when these failures occur within the fill layer and not at a distinct boundary. This usually follows the creation of a wetting front during heavy rain that progressively penetrates the fill until a critical depth is reached and failure occurs.

Mudslides. Mudslides (or earth slides in Cruden & Varnes 1996) are planar failures that occur in fine-grained (predominantly clayey) residual soil and weathered argillaceous rocks, such as mudstones. They are usually found on gently to moderately inclined hillsides (20–30°), and normally occur where groundwater is high. They are typically slow moving and shallow, up to a few metres in depth where they occur in soil, and potentially deeper where sliding takes place along bedding planes or other discontinuities in weathered argillaceous rocks. Rapid rises in groundwater and toe erosion by rivers can lead to significantly increased rates of movement.

Rockslides. In rock masses, planar failures occur along low-strength discontinuities. These surfaces can take the form of:

- original bedding or foliation surfaces (Figs A3.14 & A3.15) in sedimentary and metamorphic rocks that have been inclined to the horizontal through tectonic activity (folding principally);

Fig. A3.19. River scour on bend opposite an aggrading tributary fan triggers slope failure and road damage.

- discontinuities or faults formed in rock masses as a result of tectonic stresses acting upon them; and
- interlayer boundaries between successive flows of lava and pyroclastic materials in volcanic rock sequences (these sequences will have been tilted by tectonism in order for these boundaries to become inclined and adverse to stability when exposed on slopes).

For failure to occur the angle of the discontinuity must be equal to or less than the angle of slope, and the latter must be greater than the frictional angle of the discontinuity.

Rock mass deformations due to tectonic movements are usually accommodated by displacements along joints and fault surfaces, leading to slickensides and polished striated surfaces (Fig. A3.24), producing rock flour or fault gouge, with an associated reduction of friction resistance along the plane of the discontinuity. Where these surfaces are adversely orientated, they can lead to slope failure when exposed in excavations. The presence of these low strength materials derived from either weathering or tectonic displacements along discontinuities can cause slope failure to take place at low slope angles, especially if impeded drainage increases water pressures. Illustration of the relationships between geological structure, topography and rock slope stability can be found in Jaboyedoff (2011).

Discontinuities can be smooth or irregular and the failure plane itself can be continuous or stepped, as might occur for example where failure has taken place along multiple closely-spaced bedding or foliation planes.

Planar rock failures usually result in the formation of a distinct back scarp. The back scarp of a rockslide is formed either:

- along a single or a small number of discontinuities that form release surfaces (most common);
- along a number of discontinuities in the rock mass (common); or
- through the continuum of the rock (least common).

Most rockslides encountered along mountain roads are located within cut slopes and often extend into the hillside above. Although they are usually up to a few metres in depth they can cause significant damage and prolonged blockage to roads (Fig. A3.25) and, in extreme cases, they can be tens of metres deep (Fig. A3.26) and occupy large areas of natural hillside. Although the original failure will most probably have involved rapid movement (Fig. A3.26a), ongoing movements are usually intermittent and occur principally in response to wet season rise in groundwater (Fig. A3.26b). Earthquakes and toe erosion by rivers, however, can reactivate significant and rapid movements in these landslide deposits.

Deep-seated gravitational slope deformations (*sackung*) in closely jointed rock masses are described, for example, in Ambrosi & Costa (2011). These movements take place through deformation of the rock mass rather than failure along discrete discontinuities and are most prevalent in steep and high mountain slopes.

Table A3.3. *Summary characteristics of landslide types: slides*

Landslide Type / *Landslide Characteristics*	*Slide*					
	Planar			*Wedge*	*Circular*	
	Soil		*Rock*		*Soil*	*Rock*
	Debris slide	*Mudslide (earth slide)*	*Rockslide*			
Prevalence along Hill and Mountain Roads	Common	Common in hilly areas, uncommon in mountain areas	Common	Usually uncommon, but depends on geological structure	Uncommon	Rare
Form of Movement	Planar sliding surface	Planar sliding surface, sometimes with release joint(s) behind slide mass	Planar or stepped sliding surface usually with release joint(s) behind slide mass	Movement along intersecting joint surfaces	Circular or more usually curvilinear surface	
Movement Surface(s)	Weathering grade boundary or at critical depth in transported soil, sometimes at base of perched water table		Discontinuities in rock mass associated with bedding, foliation, tectonics, stress release		Usually through soil or highly shattered rock continuum, and not along discontinuities	
Common Materials	Taluvium, landslide debris, WG IV/V soil	Weathered mudstones & siltstones, WG VI soil	WG I-III rock		Fine-grained WG VI soils and WG III-V mudstones	Tectonized rock
Common Causes (see Section A3.3)	Rising groundwater and saturation of surface layer(s) due to rainfall runoff and shallow sub-surface flow		Reduced strength due to weathering along discontinuities and water pressures		Rising groundwater, weathering along discontinuities	
	Reduction in passive support by excavation or erosion in rivers/streams and earthquake loading					
Typical Locations	In cut slopes and on slopes above cut slopes, on slopes adjacent to eroding rivers, on steep slopes following heavy rain and earthquakes					
			On cliffs and other steep rock slopes			
Typical Size	100 – 10,000 m^2	1,000 – 50,000 m^2	1,000 – 100,000 m^2	100 – 1,000 m^2	1,000 – 100,000 m^2	1,000 – 250,000 m^2
Typical Depth	Up to 3m	Up to 5m	Usually 1-5m, occasionally to 10m, very infrequently to 50m		Up to 10m, very infrequently to 30m	
						Extreme cases to 50m
Typical Speed	m/s – m/a	m/a	m/s – m/a			
Typical impact on roads (see Section A3.6)	Blockage, damage to structures, loss of road edge, failure of cut or fill slope, road subsidence			Blockage, damage to structures, loss of road edge, failure of cut slope	Blockage, damage to structures, loss of road edge, failure of fill slope, road subsidence	

Table A3.4. *Summary characteristics of landslide types: flows, avalanches and falls*

Landslide Type / *Landslide Characteristics*	*Flow*					*Falls and Topples*	
	Debris flow	*Mudflow (earth flow)*	*Rock flow*	*Debris*	*Rock*	*Soil (falls only)*	*Rock*
Prevalence along Hill and Mountain Roads	Common	Common in hilly areas uncommon in mountain areas	Rare	Uncommon but locally can be common	Uncommon but locally can be common	Uncommon in hilly areas. Rare in mountain areas	Uncommon in hilly areas. Common in mountain areas
Form of Movement	Internal deformation. Fluidisation of slope or stream bed material, or downslope change in mechanism from slide to flow; Fluid flow due to high water /air content	Viscous flow due to high clay/ water content	Deep-seated rock creep by internal deformation (sackung)	Fluidisation of failing material, either by water or entrapped air		Free-fall, bounce (in case of rock)	
Movement Surface(s)	None	Internal shear surfaces may develop	Multiple fractures in rock	Intermittent contact with ground surface		Initially as shear failure through soil	Initially along dis-continuities
Common Materials	Wide range of particle size, often with large boulders	Fine-grained soils (incl WG V VI)	Rock, usually closely jointed	Soil and rock debris as initial failure material	Rock as initial failure material	Soil	Rock (WG I-III)
Common Causes (see Section A3.3)	Landslide with high water content on steep slope or landslide mixes with stream water in channel	Rising ground-water and saturation of surface layer(s) due to rainfall runoff and shallow sub-surface flow	High stresses in rock mass brought about by steep and high slopes (Jaboyedoff 2011)	Initial large volume failure from cliff/ very steep slopes often triggered by earthquakes		Water pressures in soil mass/along rock discontinuities during heavy rain, often triggered by earthquakes	
Typical Locations	In river channels draining unstable catchments	On lower valley slopes where water collects, where existing circular failures 'flow' over slopes below	Steep and high valley sides in mountain areas	In and below areas of very steep slope, on open slopes and in river channels		From over-steep cut slopes, back scarps, cliffs, river terrace edges	Cut slopes, cliffs and very steep slopes
Typical Size	30,000 – 100,000m^3	1,000 – 50,000m^3	Valley side extent (10^6-10^8m^3)	Up to 10,000m^3	Up to 10^6m^3	Up to 5,000m^3	Commonly 1-10 m^3. Often to 5,000m^3 Rarely to 10^6m^3
Typical Depth	N/A	Up to 5m	N/A				
Typical Speed	1-10m/s	m/yr	mm-cm/yr	1-10m/s, up to 70m/s in extreme cases			
Typical Impact on Roads (see Section A3.6)	Damage to culverts, bridges, walls, blockage and scour of road	Blockage, damage to structures, gradual road subsidence	Minor road deformation, but ultimately could fail totally	Blockage, damage to structures	Blockage, damage to structures, destruction of road	Limited blockage potential	Blockage, damage to structures

Fig. A3.20. Common forms of slope failure.

Fig. A3.21. Common failure mechanisms in rock.

A3.4.1.2 Wedge failures

Intersecting joints in rock masses can lead to wedge failures (Figs A3.20 & A3.21) where the line of intersection between two joint surfaces is orientated out of the slope and has an angle of inclination that is less than the slope angle and more than the friction angle. These failures are most commonly observed in steep cut slopes in jointed rock, but they are also associated with the development of large landslides on steep mountain slopes.

Planar failure in WG VI soil developed on sedimentary rock (boulders at toe are rip-rap protection)

Planar failure in taluvium derived from failed weathering profile in metamorphic rock

Planar failure of fill slope along the original ground surface

Circular (curvilinear) failure in head of eroding gully, caused by incision

Mudslide/mudflow in colluvium

Debris flows in weathered quartzite triggered by rainfall after earthquake

Fig. A3.22. Common failure mechanisms in soil/debris.

A3.4.1.3 Circular failures

Circular slope failure mechanisms occur in fine-grained (predominantly clay) soil and shattered rock, the stability of which is commonly controlled by the shear strength along intersecting multiple joints and microfractures in the rock rather than any persistent discontinuities (Section C4). In conventional soil mechanics, these failures have a slip surface that approximates to a segment of a circle and they characteristically have a back-tilted failure mass (Fig. A3.27). In practice 'circular' failures usually have

Fig. A3.23. Subtle planar failure in weathering profile overlying rockhead.

shear surfaces that are curvilinear rather than circular, and are commonly referred to by geologists as *rotational*. These failures are further subdivided into single, multiple and successive modes (Hutchinson 1968). Given the prevalence of coarse-grained soils and jointed rock masses and the approximately planar soil/rock boundary conditions found on most mountain slopes, these failures are far less common than planar failures. Where they do occur ground movements can be deep especially if the failure surface passes through rock or where the overlying soil is deep. Rates of movement will vary between rapid (metres per second) in the case of first-time failure to

Fig. A3.24. Slickensides along joint surfaces within sheared meta-sedimentary rock.

Fig. A3.25. Rockslide that caused road blockage for several days during excavations for road widening.

moderate (metres per year) where existing landslides are reactivated by a slow rise in groundwater, for example. Renewed toe erosion by rivers can result in significant and rapid displacements. This can also occur when the toe of a landslide breaks up and flows rapidly over the adjacent ground.

A3.4.2 Flows

A3.4.2.1 Mudflows

Mudflows (or earth flows in Cruden & Varnes 1996) occur in fine-grained soils, often derived from weathered mudstones, siltstones and volcanic rocks. They are usually slow-moving, and their stability is highly sensitive to changes in water table. Movements take place under high pore-water pressure conditions when effective stress is significantly reduced (Section C3). Mudflows are usually elongate in plan and fail primarily through internal deformation. They commonly occur where plane or circular failures evacuate their source areas and flow or spread over the slopes below (Fig. A3.28). Where mudflows enter stream channels, their speed of travel can increase significantly when they mix with water.

A3.4.2.2 Debris flows

Debris flows are composed of coarse-grained material (ranging from sand and gravel-sized material up to boulders) and travel at high speeds on steep slopes or in river channels. Where they flow across hillsides formed in taluvium or highly weathered rock they can erode deep channels, thereby increasing in volume and momentum. Where they become channelized into streams and rivers, their speed and travel distance increase significantly due to lateral confinement. Velocities may be enhanced when debris flows mix with river water, although this is not a prerequisite as Lin *et al.* (2009) report that the volume of debris can be as high as ten times the volume of contained water. Furthermore, debris flows can undergo significant volume change due to sediment entrainment during travel in stream channels, thus increasing momentum and travel distance (see Fannin & Bowman, 2010). Jones *et al.* (1983) describe debris flows in the Karakorams where velocities reach almost 20 m/s; Lin *et al.* (2009) quote velocities of 13 m/s for debris flows triggered by the 1999 Chi Chi earthquake in Taiwan. More extremely, Plafker & Ericksen (1978) quote avalanche flow velocities of up to 70 m/s from the Peruvian Andes. Even at the lower velocities, debris flows can be highly destructive and may be the dominant mechanism of flow in river channels during moderate to heavy rainfall in some mountain areas. They typically comprise a 'core' of debris flow travelling within a central channel with levees deposited on either side (Fig. A3.29). Debris fans (Fig. A3.30) are usually deposited on river floodplains and terraces, where debris flows generated on valley side slopes and in tributary valleys become unconfined and velocities reduce significantly.

The example shown in Figure A3.31 is taken from a steep, 10 m wide Himalayan stream where debris flow

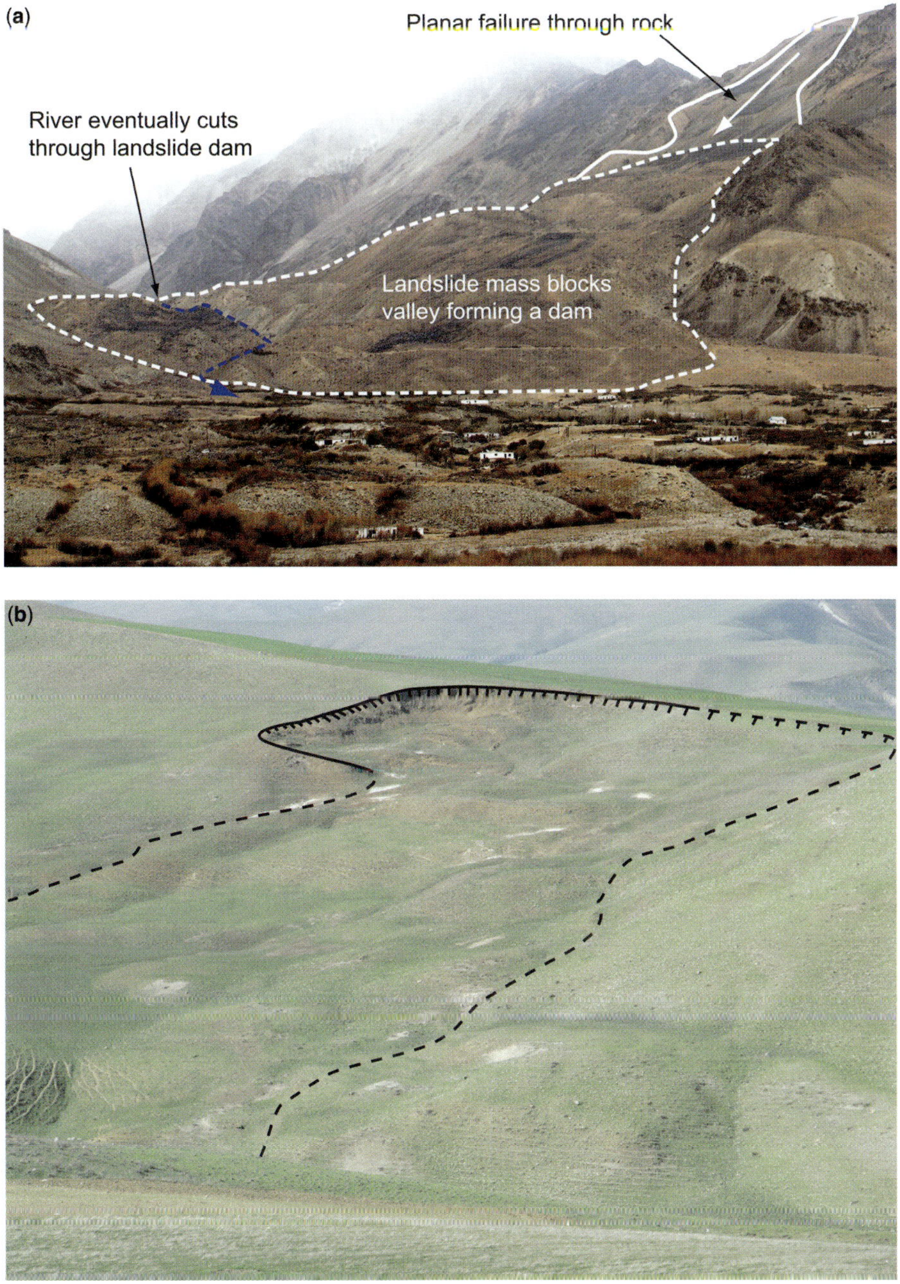

Fig. A3.26. (**a**) Deep-seated planar failure in strong igneous rock transforms to an avalanche and blocks a valley. (**b**) Deep-seated planar failure in weak sedimentary rock transforms to a flow occupying the entire valley side.

velocities ranged from 3 to 6 m/s and where up to 50 000 m^3 of debris at a time was transported during only modest rainfall. On average, between five and six of these flows were recorded in this stream during each wet season. Debris flows were generated once water levels within the bed material had risen to a level sufficient to mobilize the fine-grained matrix within the debris.

Fig. A3.27. Circular (rotational) failures in fine-grained soil and weathered rock.

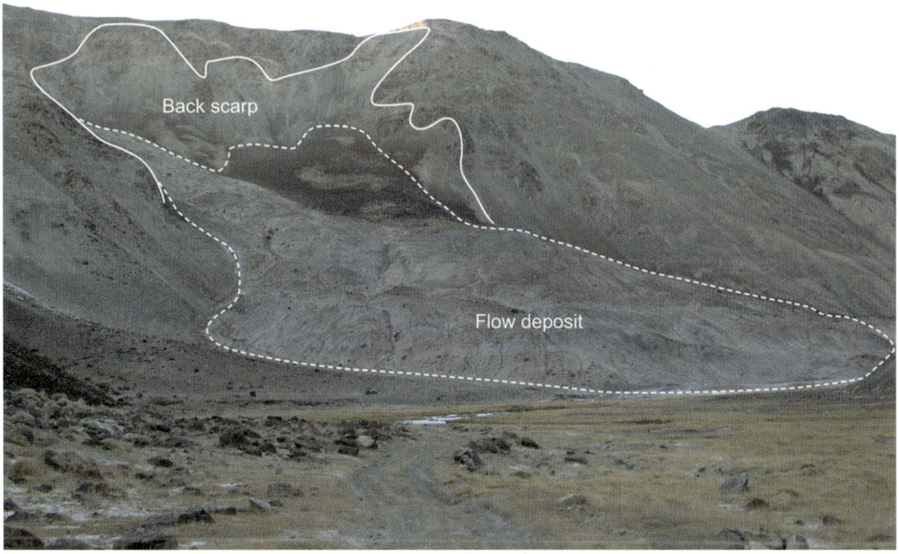

Fig. A3.28. Mudflow, probably triggered by seismicity.

Fig. A3.29. Levees deposited either side of a debris flow channel.

Fig. A3.30. Debris source, eroded flow track and depositional fan.

A3.4.3 Avalanches

Debris avalanches can occur where debris slides or soil falls fail onto steep slopes and the constituent material travels over long distances as a 'trail' of debris. They are usually generated either as a result of heavy rain or earthquake acceleration (Fig. A3.32) and are most prevalent on slopes with sparse vegetation cover.

Large rockfalls and rockslides that travel long distances are sometimes referred to as *rock avalanches*. Travel distance is usually increased considerably where compressed air within the failing mass causes the material to 'flow'. Figure A3.33 shows the scar produced by a catastrophic rock avalanche in Papua New Guinea. This failure took place when an adversely orientated fault surface was exposed by rapid stream incision. The removal of forest vegetation on the left bank by high-velocity wind caused by the passage of the flow is visible in the photograph.

Figure A3.34 illustrates a rock avalanche in the Pamirs. In this case, the downstream travel distance was relatively minor and the majority of failed material filled the valley and travelled up the opposite side, creating a temporary lake. Contemporary talus now masks much of the original scarp in the source area.

A3.4.4 Falls and topples

Falls occur in soil, debris and rock, though failures in rock are the more frequent along mountain roads (Fig. A3.35). Strictly speaking, falls and topples are forms of rockslide as they involve detachment along one or more joint planes

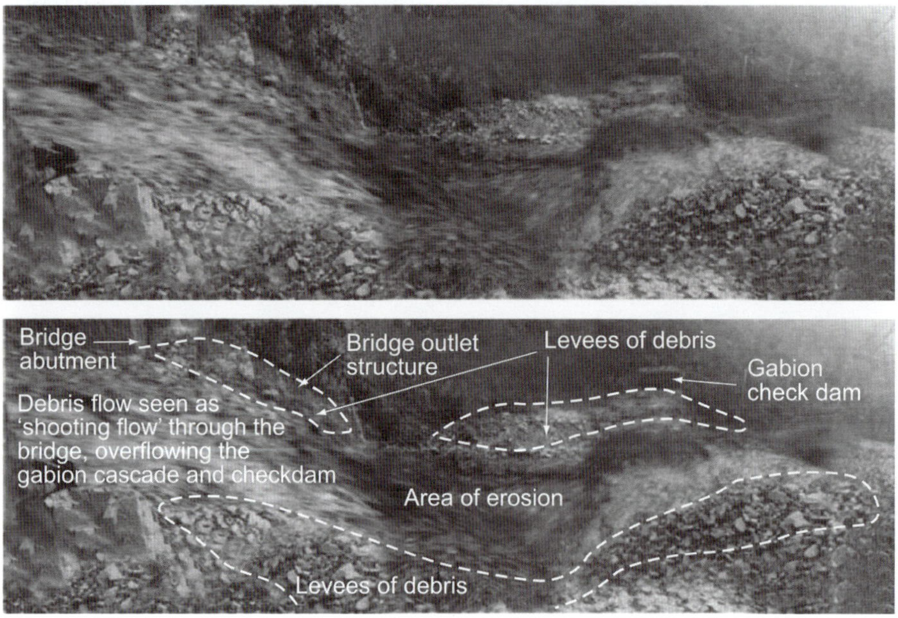

Fig. A3.31. Typical debris flow in mountain terrain.

Fig. A3.32. Debris avalanches arising from the 2008 Sichuan earthquake, China.

Fig. A3.33. Rock avalanche flow track.

Fig. A3.34. Landslide dam formed by a rock avalanche.

(Figs A3.36 & A3.37). However, this explanation is not particularly helpful in defining the hazard they pose along mountain roads, as their motion once detached is usually one of free fall or bounce. Usually, rockfalls occur as individual rock blocks from the upper portion of rock cut slopes and natural cliffs and typically amount to little more than a few cubic metres at a time. Rockfalls can be frequent in steeply cut columnar-jointed basalt, for example, where they are initiated either through wedge, topple or slide mechanisms depending on the orientation of discontinuities in relation to cut slope geometry (Fig. A3.38). Generally, rockfalls from natural slopes above or below mountain roads are less common than they are in cut slopes. They do occur, however, usually in response to intense rainfall or earthquake acceleration. Amplification of seismic acceleration often takes place along ridge lines, cliff tops and other pronounced convexities or promontories in the landscape creating large-volume rockfalls.

Progressive rockfall, which accompanies the gradual retreat of cliff faces, creates talus slopes as failed material is deposited on the slopes below (Fig. A3.39). These talus slopes are most prevalent in high-altitude mountains due to high rates of mechanical weathering (most notably freeze–thaw in cold countries or very high mountains in the tropics and subtropics), although they do develop below cliffs formed in closely jointed and fragmented rock elsewhere, including at lower elevations in the humid tropics and subtropics.

A3.5 Landslide displacement

In mountain areas all landslides have the potential to runout over long distances. The risk that such landslides pose to housing, land use and infrastructure can therefore extend over large areas, with the added disadvantage that the source of hazard and its timing may be largely unknown and unpredictable. Landslide hazard mapping and runout modelling (see Section B2.6, Hungr 1995; Legros 2002; Hungr *et al.* 2005) can provide some indication of potential landslide source areas and travel distances, but there remains considerable uncertainty given the usual lack of geotechnical and historical event data and the complex mechanisms involved.

The following four factors appear to have the most influence on the mobility of landslide debris:

- the failure mechanism of the landslide;
- the source volume (the initial failure volume) of the landslide;
- the steepness of the downslope topography and the channelization potential; and

Fig. A3.35. Rockfalls are frequent hazards along mountain roads.

- the fluidization of the failing mass by entrained water or entrapped air.

A3.6 Landslide impacts on roads

Figure A3.40 shows the typical range of potential landslide shear surface configurations in relation to a road constructed across a hill slope on side-long ground. By far the most common landslide impacts are those that relate to the partial or complete blockage of roads as a result of cut slope failures. It is not unusual to find (Hearn *et al.* 2008) that up to 70% of slope failures affecting any given mountain road are due to relatively shallow instability in cut slopes that poses recurrent though minor hazard to road operation.

The impacts of landslide hazards can be separated into engineering, commercial and social costs.

Engineering costs include:

- the cost of landslide debris clearance and slope reinstatement;
- the cost of repairs to roadside structures, such as retaining walls and side drains; and
- the cost of repairs arising from damage that might occur to the carriageway, either by the slope failure itself or by machine excavation of the slipped debris.

These costs can form a significant proportion of annual expenditure on road maintenance and can severely deplete contingency and emergency funds during periods of major landslide activity. In Laos, for example, between 50 and 80% of emergency road repair works are spent annually on landslide-related damage (Hearn *et al.* 2008). Further discussion on the impact of landslides on low-cost mountain roads can be found in Dahal *et al* (2010).

The greatest landslide impacts are usually those where slope failure leads to partial or complete loss of the carriageway. In the most extreme and least common of cases, this loss occurs instantaneously, cutting access for considerable periods of time until the road can be safely reinstated or realigned, either temporarily or permanently. More often, however, these failures occur progressively with gradual displacements to carriageways, thus allowing access to be maintained while options for remedial works are considered. Figure A3.41 illustrates this with reference to mudslide and mudflow failure mechanisms. Landslide number 1 has caused gradual displacement of the road constructed across it. Landslide number 2 poses a lower level of hazard through the deposition of debris onto the road surface and less severe road displacements.

In the Laos study, only 3% of recorded slope failures along the hill road network resulted in the displacement of the road carriageway. By contrast in the Philippines, for example, landslides affecting the Halsema Highway during the 1990 earthquake and the typhoons that followed resulted in the loss of road formation in a total of almost 40 locations over an alignment length of *c.* 100 km. In some locations, the mountain sides had retreated by as much as 50 m over a 5–6 year period as a result of slope failure and subsequent erosion. Prior to the implementation of externally-funded road improvement works (Hart *et al.* 2002), this rate of hillside retreat continually thwarted the attempts of the road authority to reinstate access along the original alignment; excavation into the hillside above was frequently the only emergency option that could be taken.

Commercial costs relate principally to the effects of delays, and can be measured as Vehicle Operating Costs (VOCs) and Value of Time (VoT). The Laos study showed that these costs increase exponentially with time, and led to the conclusion that prevented access for periods longer than 6 hours on roads with 300 AADT or more could result in significant economic losses that justified investment in low-cost slope improvement measures. Although essentially a low-cost road, the Halsema Highway experienced high traffic volumes (AADT 1300 along the mountain section). The economic losses associated with road closures due to landslides were consequently very high, justifying significant investment in slope and road reinstatement.

Fig. A3.36. Catastrophic rockfall along vertical joints in sandstone.

Social costs associated with landslides along mountain roads include:

- injury or fatality associated with landslides impacting pedestrians, vehicles and occupied buildings;
- reduced livelihood brought about by the loss of cultivated land and other land use resources;
- prevented or disrupted access to places of work, schools and health facilities;
- interruption of trade; and
- disruption to water supply.

These impacts vary widely from country to country. In Laos, for example, they are considered to be relatively small due to the low population density and the lack of cultivation within mountain road corridors. In Nepal, by contrast, the socio-economic impact of landslides in road corridors is much higher due to higher rural populations and intense agricultural land use (Petley *et al.* 2005).

Slope erosion is common on cut and fill slopes along roads in hilly and mountainous areas. Grace (1999) estimates that up to 60% of sediment generated by erosion on forest roads can originate from cut and fill slopes, and forest roads in general can contribute up to 90% of sediment from forested catchments (quoted from Rivas 2003). Erosion occurs where water runoff is capable of removing particles of soil or fragments of loose, fractured rock from slope surfaces. Silty and sandy soils are typically most prone to erosion, especially where the protection afforded by vegetation is removed by earthworks or land use change. Slope runoff, capable of causing erosion, is usually generated from the following sources:

- direct rainfall onto slopes;
- discharges from irrigation or as a result of land use change, for example forest clearance for permanent or shifting cultivation (slash and burn);
- spring seepages; or
- broken drains and water supply pipes.

Slope erosion commences with sheet erosion if runoff is distributed evenly across a slope. Rills develop where slope microtopography concentrates this runoff, and this can develop rapidly into deep gullies that ultimately may initiate slope failure. Erosion can be rapid where uncontrolled road or rainfall runoff occurs on unprotected fill slopes (Fig. A3.42) and can lead to the undermining of adjacent retaining wall foundations.

A3.7 Case studies

A3.7.1 Ethiopia

In Ethiopia many slopes comprise successive deposits of basalt and pyroclastic rocks, predominantly tuffs, breccias and ashes, often with buried soil horizons contained within multiple sequences. These sequences are approximately horizontal, although in places they have become inclined locally at up to 20° due to post-depositional tectonic

Fig. A3.37. Incipient toppling failure.

folding. The tuffs and breccias have often weathered to form silty clays. Clay-rich residual soils (including black cotton (vertisol) soils) have developed on the weathered basalts. Groundwater and rainwater percolating through the jointed basalts lead to the development of pore pressures within underlying clay horizons, creating conditions of slope instability wherever the inclined bedding intercepts the slope surface (Fig. A3.18).

Some of these failures have been described by Hearn & Massey (2009). As illustrated in Figure A3.43, landslides have developed on relatively gentle slopes in the weathered volcanic sequence close to ridge crests. The failure surfaces have tended to be located within weathered tuff beneath overlying basalt, or within black cotton soil. Maximum stable slope angles formed on the tuff are approximately 20° and failure mechanisms are planar and circular through these fine grained materials. Large deep-seated landslides have occurred in association with these sequences, and are especially evident on the margins of the Rift Valley where earthquake accelerations contribute to slope instability.

By contrast, in the Blue Nile gorge of central Ethiopia, major rockfalls and topples of between 10^4 and 10^5 m^3 in volume are dominant in the landscape. These failures have occurred in basalt overlying pyroclastic materials and in underlying sedimentary sequences (predominantly limestone overlying marl and mudstone). Weathering of pyroclastic material in the Blue Nile gorge is not as advanced as it is in the less steep terrain and wetter climate bordering parts of the Rift Valley, and the only clays that have been encountered are associated with the residual soils developed on the basalt and the clayey deposits within the marl and mudstone.

A3.7.2 Philippines

Figure A3.44 shows a section of the Halsema Highway (Section A3.6) where almost the entire carriageway (save

Fig. A3.38. Columnar jointing in basalt promotes small-scale planar, topple and wedge failures on steep slopes and in excavations.

Fig. A3.39. Talus slopes formed beneath retreating cliff faces.

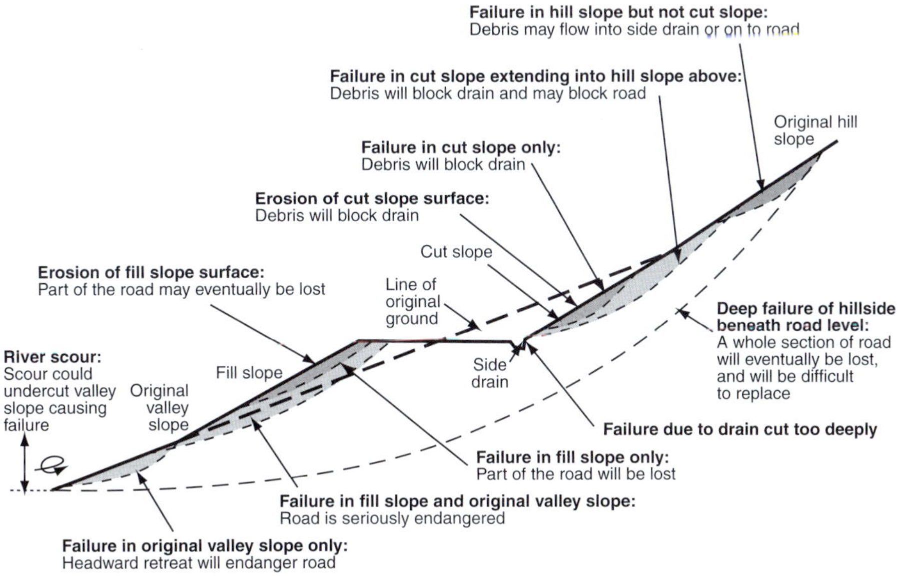

Fig. A3.40. Typical range of slope failures in relation to a road constructed across side-long ground.

Landslide Number 1:
Deep-seated,
slow moving mudslide,
with ongoing reactivation
due to removal of toe
support by river erosion

Landslide Number 2:
Shallow flow-type
landslide with
retrogressing head scarp.

Landslide number 1 poses
the greatest risk to the
road as the failure surface
passes beneath the road
foundation

Fig. A3.41. Mudslide and mudflow impacts on mountain roads (two photographs shown; one with and the other without annotation).

Fig. A3.42. Fill slope erosion due to lack of road runoff control.

Fig. A3.43. Typical low-angle slope failures developed on residual clay soils bordering the Rift Valley in Ethiopia.

Fig. A3.44. Extreme slope conditions along the Halsema Highway, Philippines.

Fig. A3.45. Slope failures and excavations make access difficult for traffic (1996).

Text box A3.5. Typhoon Ondoy causes further damage to the Halsema Highway

On 25 September 2009 Typhoon Ondoy resulted in major flooding in Manila and the neighbouring region. 410 mm of rain were recorded in 24 hours with more than 340 mm falling during a 6 hour period. Several landslides occurred along and in the vicinity of the Halsema Highway. Over a 20 km section, the road was blocked in 30 locations for an estimated total length of 800 m. The road edge was undermined at six locations while the entire road width failed at eight locations over a total length of *c.* 500 m. Total slope failure occurred in two of these locations causing complete loss of the original road cross-section. The majority of these failures occurred in areas where significant slope movement and road damage had been hitherto unrecorded and in areas that were largely unaffected by the 1990 earthquake and the typhoons that immediately followed (Hart *et al.* 2002). This is undoubtedly due to the marked variations in rainfall intensities that occur over very short distances during typhoons arising from the combined meteorological and orographic effects associated with each storm. The ability to predict the locations of these failures is severely limited by these natural variations.

for the inside edge of the road) has been removed by slope failure initiated by erosion beneath a culvert outlet. Figure A3.45 illustrates the difficulty and danger that these conditions pose to traffic. The underlying geology is extremely complex and comprises intrusive and extrusive volcanic rocks, limestone and conglomerate. Faults and shear zones are frequent, and the high levels of seismicity continue to cause disturbance and weakening of surface rocks while triggering landslides and ground movements. This inherent geological instability, combined with high-intensity typhoon rains, results in a rapid rate of valley side retreat through surface erosion, rockfall and slope failure. Continual loss of support to the road formation is the outcome for sections of road constructed on side-long ground (Hart *et al.* 2002). Text box A3.5 describes the effects of Typhoon Ondoy on parts of the Halsema Highway in 2009.

A3.7.3 Himalayas

In the Himalayan foothills sedimentary rocks such as siltstone, mudstone, sandstone and limestone commonly occur. The Tertiary Murree Formation in Pakistan, for example, contains a sequence of sandstone, siltstone, mudstone and shale, and is highly susceptible to landsliding. The road between Murree and Muzaffarabad is regularly cut and blocked by large deep-seated circular slides, mudslides and mudflows in these materials (Figs A3.46 & A3.47). Slope instability is exacerbated by down-cutting and lateral erosion in the Jhelum River and by the frequent seismicity associated with the adjacent Main Boundary Thrust. In Nepal, siltstone, sandstone and mudstone form part of the Siwalik sequence of rocks. These have been extensively folded and faulted due to their proximity to major thrust faults and are especially prone to landslides.

Fig. A3.46. Circular (rotational) failure in weathered Murree Formation mudstone, Pakistan, triggered by stream erosion.

Fig. A3.47. Mudslide in the Murree Formation, Pakistan.

Fig. A3.48. Rock avalanche in limestone (with flow tracks indicated) following the 2008 Sichuan earthquake, China.

The Main Frontal Thrust, Main Boundary Thrust and the Main Central Thrust run through the entire range of the Himalayas; in particular, movement along the Main Central Thrust has resulted in extensive fragmentation of moderate and high grade metamorphic rocks at outcrop. Major deep-seated slope failures are encountered along much of the length of this fault.

A3.7.4 China

In China, limestone forms much of the high and steep mountain slopes of Sechuan Province. The May 2008 earthquake triggered many large and deep-seated rockslides and rockfalls. Many of these originated at sharp convex breaks of slope, such as at the top of steep slopes and along ridge lines (Fig. A3.48) possibly as a result of topographic amplification of seismicity (Section A3.4.4).

References

Ambrosi, C. & Costa, G. B. 2011. Valley shape influence on deformation mechanisms of rock slopes. *In*: Jaboyedoff, M. (ed.) *Slope Tectonics*. Geological Society, London, Special Publications, **351**, 215–233.

ANON 1977. The logging of rock cores for engineering purposes. *Quarterly Journal of Engineering Geology*, **10**, 45–52.

ANON 1995. The description and classification of rocks for engineering purposes. Geological Society Engineering Group Working Party Report. *Quarterly Journal of Engineering Geology and Hydrogeology*, **28**, 207–242.

Bromhead, E. N. 1986. *The Stability of Slopes*. Surrey University Press, London.

BSI 1999. *Code of Practice for Site Investigations*. BSI 5930, British Standards Institution, HMSO, London. First published in 1981 (Amendment 2, 2010).

Chen, H. & Hawkins, A. B. 2009. Relationship between earthquake disturbance, tropical rainstorms and debris movement: an overview from Taiwan. *Bulletin of Engineering Geology and the Environment*, **68**, 161–186.

Cruden, D. M. & Varnes, D. J. 1996. Landslide types and processes. *In*: Turner, A. K. & Schuster, R. L. (eds) *Landslides Investigation and Mitigation*. Special Report 247 of the Transportation Research Board, National Research Council, National Academy Press, Washington DC, 36–75.

Dahal, R. J. & Hasegawa, S. 2008. Representative rainfall thresholds for landslides in the Nepal Himalaya. *Geomorphology*, **100**, 429–443.

Dahal, R. J., Hasegawa, S., Bhandary, N. P. & Yatabe, R. 2010. Low cost road for the development of Nepal and its engineering geological consequences. *In*: Williams, A. L., Pinches, G. M., Chin, C. Y., McMorran, T. J. & Massey, C. I. (eds) *Geologically Active. Proceedings of the 11th IAEG Congress, Auckland, New Zealand, 5–10 September, 2010*, Taylor and Francis Group, London, 4085–4094.

Dikau, R., Brunsden, D., Schott, L. & Ibsen, M. L. 1996. *Landslide Recognition*. John Wiley and Sons, Chichester.

Fannin, R. J. & Bowman, E. T. 2010. Debris flows – entrainment, deposition and travel distance. *In*: Williams, A. L., Pinches, G. M., Chin, C. Y., McMorran, T. J. & Massey, C. I. (eds) *Geologically Active. Proceedings of the 11th IAEG Congress, Auckland, New Zealand, 5–10 September, 2010*, Taylor and Francis Group, London, 1111–1118.

Fell, R., Glastonbury, J. & Hunter, G. 2007. Rapid landslides: the importance of understanding mechanisms and rupture surface mechanics. The Eighth Glossop Lecture. *Quarterly Journal of Engineering Geology and Hydrogeology*, **40**, 9–27.

Fookes, P. G. (ed.) 1997. *Tropical Residual Soils*. Geological Society of London, Engineering Group Working Party Revised Report, London.

Grace, J. M. 1999. Erosion control techniques on forest road cutslopes and fillslopes in North Alabama. *In*: *Transportation Research Record*. 1652, 2, TRB National Research Council, Washington DC, 227–234.

Griffiths, J. S. 2005. Landslides. *In*: Fookes, P. G., Lee, E. M. & Milligan, G. (eds) *Geomorphology for Engineers*. Whittles Publishing, Dunbeath, UK, 173–217.

Hart, J., Hearn, G. & Chant, C. 2002. Engineering on the precipice: mountain road rehabilitation in the Philippines. *Quarterly Journal of Engineering Geology and Hydrogeology*, **35**, 223–231.

Hearn, G. J. & Massey, C. I. 2009. Engineering geology in the management of roadside slope failures: contributions to best practice from Bhutan and Ethiopia. *Quarterly Journal of Engineering Geology and Hydrogeology*, **42**, 511–528.

Hearn, G. J., Hunt, T., Aubert, J. & Howell, J. H. 2008. Landslide impacts on the road network of Lao PDR and the feasibility of implementing a slope management programme. *International Conference on Management of Landslide Hazard in the Asia-Pacific Region*, Sendai, Japan.

Hencher, S. 2006. Weathering and erosion processes in rock – implications for geotechnical engineering. *In*: *Proceedings of the Symposium on Hong Kong Soils and Rocks*. March 2004. Institution of Mining, Metallurgy and Materials, London, 29–79.

Hencher, S. 2008. Eurocodes: the 'new' British and European standard guidance on rock description. *Ground Engineering*, **41**, 7, 17–21.

Hencher, S. & Lee, G. 2010. Landslide mechanisms in Hong Kong. *In*: Calcaterra, D. & Parise, M. (eds) *Weathering as a Predisposing Factor to Slope Movements*. Geological Society, London, Engineering Geological Special Publications, **23**, 77–103.

HO, K. K. S. & LAU, J. W. C. 2010. Learning from slope failures. *Quarterly Journal of Engineering Geology and Hydrogeology*, **43**, 33–68.

HOEK, E. & BRAY, J. 1981. *Rock Slope Engineering*. Revised 3rd edn. Institution of Mining and Metallurgy, London.

HOWELL, J. 2005. Study of slope stability in Trinidad and Tobago with particular reference to the impacts on roads. *In*: *Proceedings of the Symposium Landslide Hazards in the Orogenic Zones, from the Himalayas to Island Arcs in Asia*. Nepal Landslide Society, Kathmandu, 345–359.

HUNGR, O. 1995. A model for the runout analysis of rapid flow slides, debris flows and avalanches. *Canadian Geotechnical Journal*, **32**, 610–623.

HUNGR, O., COROMINAS, J. & EHERHARDT, E. 2005. Estimating landslide motion mechanism, travel distance and velocity. *In*: HUNGR, O., FELL, R., COUTURE, R. & EHERHARDT, E. (eds) *Landslide Risk Management*. Taylor and Francis Group, London, 99–128.

HUTCHINSON, J. N. 1968. Field meeting on the coastal landslides of Kent. *Proceedings of the Geologists Association*, **79**, 227–237.

IAEG 1981. Rock and soil description and classification for engineering geological mapping. *Bulletin of the International Association of Engineering Geology*, **24**, 253–274.

IRFAN, T. Y. 1996. Mineralogy, fabric properties and classification of weathered granites in Hong Kong. *Quarterly Journal of Engineering Geology*, **29**, **1**, 5–35.

JABOYEDOFF, M. (ed.) 2011. *Slope Tectonics*. Special Publications 361, The Geological Society, London.

JAISWAL, P. & VAN WESTEN, C. J. 2009. Estimating temporal probability for landslide initiation along transportation routes based on rainfall threshold. *Geomorphology*, **12**, 96–105.

JONES, D. K. C., BRUNSDEN, D. & GOUDIE, A. S. 1983. A preliminary geomorphological assessment of part of the Karakoram Highway. *Quarterly Journal of Engineering Geology*, **16**, 331–355.

KOOR, N. P., PARRY, S. & YIN, J. H. 2000. The shear strength of infilled and slickensided discontinuities of some Hong Kong saprolites. *In*: *Proceedings of the Conference on Engineering Geology*, Hong Kong. Institution of Mining and Metallurgy, Hong Kong Branch, 187–192.

LEGROS, F. 2002. The mobility of long-runout landslides. *Engineering Geology*, **63**, 301–331.

LIN, D. G., HSU, S. Y. & CHANG, K. T. 2009. Numerical simulations of flow motion and deposition characteristics of granular debris flows. *International Symposium on Geotechnical Engineering, Ground Improvement and Geosynthetics for Sustainable Mitigation and Adaptation to Climate Change, Including Global Warming*. 3–4 December 2009, Bangkok, Thailand.

LITTLE, A. L. 1969. The engineering classification of residual tropical soils. *Proceedings of the 7th International Conference on Soil Mechanics and Foundation Engineering, Special Session on Engineering Properties of Lateritic Soils*, Mexico, 1–10.

LUMB, P. 1975. Slope failures in Hong Kong. *Quarterly Journal of Engineering Geology*, **8**, 31–65.

MOYE, D. G. 1955. Engineering geology for the Snowy Mountain Scheme. *Journal of the Institution of Engineers*, Australia, **27**, 287–298.

NORBURY, D. 2010. *Soil and Rock Description in Engineering Practice*. Whittles Publishing, CRC Press, Taylor & Francis Group, Caithness, Scotland.

PETLEY, D. N., HEARN, G. J. & HART, A. B. 2005. Towards the development of a landslide risk assessment for rural roads in Nepal. *In*: GLADE, T., ANDERSON, M. & CROZIER, M. J. (eds) *Landslide Hazard and Risk*. John Wiley & Sons Ltd, Chichester, 597–619.

PLAFKER, G. & ERICKSEN, G. E. 1978. Nevados Huascaran avalanches, Peru. *In*: VOIGHT, B. (ed.) *Rockslides and Avalanches, 1: Natural Phenomena*. Elsevier, Amsterdam, 277–314.

REEVES, G. M., SIMS, I. & CRIPPS, J. C. 2006. *Clay materials used in construction*. Geological Society, London, Engineering Geology Special Publications, **21**.

RIVAS, T. 2003. Improved erosion control selection for steep slopes with the USDA Forest Service Erosion Control Selection Guide. *In*: *Transportation Research Record*. 1819, Paper no. LVR8-1094, 182–186.

RUXTON, B. P. & BERRY, L. 1957. Weathering of granite and associated erosional features in Hong Kong. *Geological Society of America Bulletin*, **68**, 1263–1291.

SARKAR, S., KANUNGO, D. P. & CHAUHAN, P. K. S. 2011. Varunavat landslide disaster in Uttarkashi, Garhwal Himalaya, India. *Quarterly Journal of Engineering Geology and Hydrogeology*, **44**, 17–22.

SORALUMP, S. 2009. Landslide mitigation in Thailand – from research to practice. *International Symposium on Geotechnical Engineering, Ground Improvement and Geosynthetics for Sustainable Mitigation and Adaption to Climate Change, Including Global Warming*, 3–4 December 2009, Bangkok, Thailand, 399–409.

TOLL, D. G., LOURENCO, S. D. N. *ET AL*. 2011. Soil suction monitoring for landslides and slopes. *Quarterly Journal of Engineering Geology and Hydrogeology*, **44**, 23–33.

VARNES, D. J. 1978. Slope movement types and processes. *In*: SCHUSTER, R. L. & KRIZEK, R. J. (eds) *Landslides – Analysis and Control*. National Academy of Sciences, Washington DC, 11–33.

WALTHAM, T. 2009. *Foundations of Engineering Geology*. 3rd edn. Spon Press, London.

WESLEY, L. D. & IRFAN, T. Y. 1997. The classification of residual soils. *In*: BLIGHT, G. E. (ed.) *Mechanics of Residual Soils*. International Society of Soil Mechanics and Foundation Engineering, Technical Committee 25 on the Properties of Tropical and Residual Soils. A. A. Balkema, Rotterdam.

WILSON, G. 1982. *Introduction to Small-Scale Geological Structures*. George Allen and Unwin Publishers Ltd, London.

WYLLIE, D. C. & MAH, C. W. (eds) 2004. *Rock Slope Engineering*. 4th edn, Taylor and Francis, Suffolk.

A4 Landslide risk management for mountain roads

G. J. Hearn

URS Scott Wilson Ltd, Scott House, Alençon Link, Basingstoke, Hampshire RG21 7PP, UK
(e-mail: gareth.hearn@scottwilson.com; garethhearn@talktalk.net)

A4.1 Decision-making

When the design, construction and maintenance of mountain roads are required to accommodate landslides and difficult ground conditions, decisions need to be made based on the assessment of risk. Risk management requires a balance to be struck between acceptable risk of blockage, damage or loss and affordable cost of risk reduction. However, before these decisions can be made, an assessment of landslide susceptibility and hazard is usually required.

A4.2 Landslide susceptibility, hazard and risk

These terms are often used synonymously in the published literature, and consequently there is sometimes some confusion in how they apply. As far as this book is concerned, the hazard and risk definitions originally proposed by Varnes (1984) and more recently elaborated upon in publications such as those of the Australian Geomechanics Society (2000, 2007), Hearn & Griffiths (2001), Lee & Jones (2004) and Fell *et al.* (2008) are followed. Landslide susceptibility, hazard and risk mapping techniques are discussed in Section B2.6.

Landslide susceptibility refers to the potential for a given slope to fail compared to others, and the term is usually used in the context of the opportunity for first-time failures to occur. If a factor of safety (Section C3.2) could be calculated, then it would be this value that would determine the *absolute* susceptibility of each slope. This information is usually unavailable, however, and *relative* susceptibility is normally assessed by reference to conditioning and triggering factors (Table A3.2) either on a site-by-site judgemental basis or by using a formal mapping approach (Section B2.6).

Landslide hazard defines the potential posed by an existing or future landslide to cause damage or loss (economic and societal). Hazard combines components of failure volume and speed of movement (sometimes referred to collectively as landslide intensity, see Hungr 1997; Australian Geomechanics Society 2007) and frequency of movement, or probability over a given area in a given time period. Due to lack of information and computational uncertainty, speed of movement is usually excluded from the assessment and this represents an important limitation. Commonly, therefore, hazard is considered as the product of magnitude (volume) and probability of movement over a given time, such that

$$\text{Hazard (H)} = \text{Magnitude (M)} \times \text{Probability (P)}.$$

Clearly, the opportunity to assess the magnitude and probability of a future first-time failure is considerably less than it is for existing landslides that can be defined, investigated and monitored.

Landslide risk defines the actual or potential damage or loss that may occur as a result of a landslide movement taking place. Risk combines hazard (H) with the value of the assets (engineering, environmental and societal) at risk and their vulnerability (degree of loss) to the landslide movement should it take place. Risk is therefore commonly considered as the product of hazard and value and vulnerability, such that

$$\text{Risk (R)} = \text{Hazard (H)} \times \text{Value (Va)} \times \text{Vulnerability (Vu)}.$$

The *risk* posed by future first-time landslides can rarely be fully evaluated because there are:

- multiple and usually indeterminate parameters that ultimately dictate hazard, including areal extent and volume, rate and extent of movement, frequency and timing of movement and runout or displacement distance;
- multiple assets at risk, including engineering structures, traffic, land use and agricultural resources, population and social infrastructure; and
- multiple vulnerabilities to hazard, including damage, partial loss or complete loss, either repairable or irreparable (replacement required).

Where past landslide occurrence and impact data exist, assessments can be made of probabilities and losses on an annual and per kilometre basis (e.g. Bonachea *et al.* 2009; Jaiswal *et al.* 2010), and this can be used to forecast future risk outcomes. If it can be demonstrated that significant

From: HEARN, G. J. (ed.) *Slope Engineering for Mountain Roads*. Geological Society, London, Engineering Geology Special Publications, **24**, 59–66.
DOI: 10.1144/EGSP24.4 0267-9914/11/$15.00

landsliding is triggered by an earthquake or a rainstorm of a certain size (e.g. Ahrendt & Zuquette 2003; Dahal & Hasegawa 2008; Jaiswal & van Westen 2009; Wu & Chen 2009), then the probability of a landslide occurring over a given time period can be approximated through associated probability. However, landslide-triggering rainfall intensities vary significantly, even within the humid tropics and subtropics (Hencher & Lee 2010), and data are usually unavailable to make these linkages with any degree of certainty.

The assessment of probability is therefore qualitative in most cases, or semi-quantitative at best, and based largely on judgement (see illustrations in Lee & Jones 2004 for example). The Australian Geomechanics Society (2000, 2007) provides a structured judgemental approach to the assessment of probability for use where event data are unavailable; the approach has been reproduced in Fell *et al.* (2008) and used, for example, by Mote *et al.* (2010) for rockfall management on Christmas Island, Indian Ocean. With respect to the vulnerability of roads to landslides and ground movements, there are usually insufficient data available to facilitate statistical assessment, and recourse to judgement is also required (see e.g. Michael-Leiba *et al.* 2005; Jaiswal *et al.* 2010).

Table A4.1 shows a landslide hazard and semi-quantitative risk matrix developed for a slope management feasibility study along mountain roads in Laos (Hearn *et al.* 2008). The magnitude of hazard (H) has been defined in terms of surveyed or estimated volume, while the probability of movement during an average 20-year low-cost road design life is based on judgement. Road assets or elements at potential risk are categorized into three classes based on relative value and implications for traffic disruption, should they be destroyed, damaged or blocked by landslide movement. Vulnerability is assessed in qualitative terms to define the degree of loss or blockage that is anticipated as a result of the landslide movements taking place, and is usually expressed on a scale of between 0 and 1.0 (Finlay 1996; Leone *et al.* 1996; Section B2.6). However, where assessment of probability, value and vulnerability are undertaken on a relative rather than an absolute basis, assigned values of 0, 1, 2 and 3 may be more appropriate.

In Table A4.1 relative risk (R) is computed as the product of M, P, Va and Vu. For example, a *small* landslide (Magnitude = 1) immediately below a road that is *expected* to occur (Probability = 2) could undermine the foundations to road retaining structures (Value = 3) leading to their *total loss* (Vulnerability = 3). In this example, Risk = 18. By contrast, a *large* landslide (Magnitude = 3) above the road that might *possibly* occur (Probability = 1) would cause *blockage* to the entire road (Value = 3 and Vulnerability = 1), yielding a Risk number of 9. In this comparison, even small slope failures on the slopes beneath a road that have potential to cause foundation failure to road retaining walls pose a greater level of risk to the operation of the road than larger failures onto the road from the slopes above that can be cleared relatively easily.

Figure A4.1 illustrates the use of this matrix. The example shown is of a reinstated cut slope in weathering grade II–III (Section A3.1) rock along a recently upgraded road in southern Laos. The original failure was considered to have occurred along rock bedding and partially blocked the road. The landslide debris was cleared and the slope cut back to a shallower angle, approximately similar to that of the underlying dip of strata. The risk number was judged to have been reduced from 18 to 12 as a result of this action. The fact that the slope remains at the same angle as the dip of strata means that bedding plane failure (failure along one or a number of bedding planes in sedimentary rocks and volcanic sediments) remains a possibility, hence the continuing development of tension cracks and a risk number of 12.

This approach does not provide absolute values of risk (few seldom do for the reasons given), but it does provide a means by which slopes and landslides can be recorded and compared for prioritizing stabilization and protection. It is intended for use by road maintenance engineers, for example, who are required to make rapid assessments based on visual observations rather than by using mapping

Table A4.1. *Risk assessment matrix*

Risk Components	*Assigned Relative Values*			
	0	*1*	*2*	*3*
Magnitude of hazard (M)		Small (shallow and involving up to $500m^3$)	Moderate	Large (deep and involving $5000m^3$ or more)
Probability of hazard occurring during 20 year period (P)	Not expected to happen	Possible	Expected to happen	Definite
Value of road elements at risk (Va)		Existing slope works and side drain	Existing slope works, side drain, and up to 50% of carriageway width	Entire carriageway and adjacent structures
Vulnerability of elements to the hazard, should it occur (Vu)	No effect	Deformation or blockage	Partial loss	Total loss
Risk = M x P x Va x Vu				

Road	Km	Grid Ref E/N	Geology	Hazard Type	Risk without mitigation M.P.Va.Vu	Measures Taken	Degree of Success	Risk with mitigation M.P.Va.Vu
Road 1E	16.1	105°03'06" 17°47'24"	Sandstone mudstone WG II-III	Above road: Rock slide (bedding plane failures) in cut slope. 100m of road affected	Above road: M = 2 P = 3 Va = 3 Vu = 1 R = 18	Cut slope angle reduced	Tension cracks continue to develop	Above road: M = 2 P = 2 Va = 3 Vu = 1 R = 12

Fig. A4.1. Illustration of risk matrix application.

and geotechnical analysis (Sections B2.6 & C3.2). Analytical methods for assessing risk are described, for example, in Dai *et al.* (2002), Wong (2002), Lee & Jones (2004), Fell *et al.* (2008) and Uzielli *et al.* (2009). These methods require datasets and parameters that are rarely available in low-cost road situations, however, and many assumptions have to be made. For example, Fell *et al.* (2008: p. 89) recommend that 'quantitative hazard and risk zoning cannot be performed where data on frequency of landslides either do not exist or are so uncertain as to not be relied on.' Recourse to qualitative assessment is therefore required.

The matrix presented in Table A4.1 requires a judgement as to each of the four main parameters that combine to yield relative risk. This judgement can only be made on the strength of experience, and preferably with professional observational and interpretational skills provided by an appreciation of engineering geology. For reasons of very low population density and low traffic volumes on the Laos hill road network, the risk associated with potential loss of life due to landslides and earthworks failures was not considered (Table A4.1). This approach is not appropriate for very heavily-trafficked roads where a more rigorous assessment of risk is required. These issues are discussed further in Part D with respect to components of risk and the various categories of slope maintenance adopted for risk management during road operation.

A4.3 Risk management

A4.3.1 New road construction

At the feasibility stage of new road construction, an assessment must be made as to the level of investment required to achieve a given standard of road operation. In mountainous terrain this should apply equally to the stability of slopes as it does to the geometric design and the performance of the pavement. However, slopes are often cut steeply to minimize earthworks volumes in the expectation that most will remain stable while some may fail. The reasoning behind this approach is that it will be cheaper in the long term to clear up those failures that do occur than to cut all slopes to more conservative angles due to uncertainties in ground conditions and ground behaviour.

The Laos study found that approximately 70% of recorded landslides along the hill road network were of low risk, the majority of these being small cut slope failures into the roadside drain and adjacent carriageway. On low-volume roads there is no economic justification for investing in conservative earthworks designs in order to prevent or minimize these slope failures. However, the study concluded that long-term economic benefit can be derived from investments in selected slope stability improvements during design and construction (Text box A4.1).

These improvements might include reduced cut slope angles and a greater use of retaining walls and erosion control measures in areas of recognized hazard, such as in weak and erodible soils and areas where a road is required to cross landslides and taluvium deposits. However, it is important that these improvements remain essentially low-cost; the economic return on the investment can be marginal and will usually only be positive where the investment prevents a situation from developing that results in significant and recurrent traffic delays. Engineering geological assessments (Section B) are required for earthworks design and decision-making concerning levels of investment in slope improvement and stabilization works (Sections C3 & C4).

A4.3.2 Improving existing roads

Where an existing road is to be improved or upgraded, earthworks will be necessary to widen and, where necessary, improve the alignment. Where landslides already pose a significant hazard, consideration can be given to local realignments to avoid them (Fig. A4.2). Finding a suitable

Text box A4.1. Feasibility study for slope stability management in Laos

A study in Laos (Hearn *et al.* 2008) compared the estimated engineering costs with the anticipated engineering benefits that would be accrued from adopting a more conservative approach to earthworks design and construction. On the one hand are the additional engineering costs related to the design of less steep cutting angles, a greater use of retaining walls, slope drainage and erosion control and by adopting longer spoil haulage lengths to safe disposal areas. On the other are the potential engineering benefits to be derived from a reduction in (a) damage to engineering structures and (b) landslide debris clear-up costs during operation.

In order to derive the input data for the cost–benefit comparisons required, a number of generalizations and assumptions had to be made. These related to the following:

- the average hill road construction cost per kilometre (figures taken from Laos and Nepal);
- the design cost as a percentage of construction cost;
- the costs of routine and emergency maintenance (using data from Laos);
- the costs of earthworks and additional retaining walls and drainage measures required to increase the stability of roadside cut slopes (using data from Laos, along with assumptions regarding increased factors of safety); and
- calculated engineering costs associated with slope failures and road closures.

The following total costs were derived for road construction and maintenance, with and without slope stability improvement, using 2008 prices.

Scenario 1: Construction and operational maintenance costs with improved slope stability

- investment costs of US $ 27 million;
- maintenance costs of US $ 80 000 per annum; and
- one major landslide per annum, leading to a blockage of three hours.

Scenario 2: Construction and operational maintenance costs without improved slope stability

- investment costs of US $ 25 million;
- maintenance costs of US $ 110 000 per annum;
- two major landslides per annum, each leading to a blockage of six hours.

Economic modelling led to the conclusion that the introduction of the improved design and construction methods described would be marginally beneficial, leading to a 2% lower overall cost in net present value terms over the design life of a low-cost road. This analysis took only engineering costs and benefits into account. The commercial costs associated with traffic delays and vehicle operating costs on low-volume roads were calculated to be marginally greater per annum under Scenario 2 than Scenario 1, that is, the costs did not substantially justify the additional investment in terms of traffic delays alone. However, on more highly trafficked roads, the costs of traffic delays are significantly greater and the economic benefits to be gained from a more conservative design involving reduced cutting angles and slope improvement works during construction are more apparent.

Fig. A4.2. (a) Failed section of road and viaduct requiring total road realignment. (b) Failed section of road requiring local realignment.

cost-effective alternative in mountainous terrain can be difficult, however, and construction along a new alignment can encounter or create significant further instability. Roads are usually improved within their existing corridor, or right of way, and a decision will need to be made as to whether widening will take place primarily by excavation into the hillside or by filling onto the valley side (outside edge) of the road (Text box A4.2).

Excavation into the hillside may reactivate landslides and trigger new slope failures; widening onto fill will invariably require additional retaining wall construction with considerations of bearing capacity and foundation stability. There may also be issues with the stability of previous uncompacted construction spoil that has since become vegetated, giving the appearance of being *in situ* ground.

On balance, if suitable foundations and adequate compaction can be achieved it is preferable to widen onto fill, but each section of road will require its own assessment. If there is any uncertainty over bearing capacity and foundation stability for walls or the stability of natural slopes and fill slopes below the road, then it is preferable to widen into cut. A balance of cut and fill, either in cross-section or over relatively short alignment lengths, is the preferred solution if the cut material is suitable as fill (Section C2). On low-cost road improvement schemes, the ease of excavation and the costs and difficulties associated with fill and retaining wall construction usually mean that widening takes place as cut to spoil, frequently to the detriment of slope stability. Engineering geological assessments and ground investigations will be required (Section B) before such important decisions are made.

Upgrading from a gravel surface to a sealed (bituminous or concrete) pavement is a frequent component of road improvement schemes. Where slow-moving landslides affect a road formation, ongoing movement will become more noticeable when a road surface becomes sealed (Text box A4.3).

Text box A4.2. Flood damage to the Naubise to Mugling road, Nepal, following road improvement

In the case of the Naubise to Mugling road improvement scheme in Nepal, the decision was made, generally, to widen on to embankment or retaining wall rather than to cut into the hillside. The road is located alongside the Trisuli River over much of its length and the slopes above the road are high and steep, with the foliation of the metamorphic rock dipping steeply out of the slope, adverse to stability. Unfortunately, within a year of completing the works, major flooding during intense and prolonged rainfall took place in the Trisuli River and many road fill retaining walls were undermined by scour (see photograph below) where they had not been founded on rock. Attention during construction had been focused on achieving suitable bearing capacities for wall foundations rather than protecting against river scour at the extraordinarily high flood levels (up to 10 m above normal monsoon level) that occurred. However, the total length of flood-damaged road amounted to less than 5% of the entire alignment length. Had the widening scheme been based on hillside excavation instead, it is anticipated that a much longer length of road would have been affected by slope failure from above.

Text box A4.3. Landslide damage to the Hirna to Kulubi road, Ethiopia, following road improvement

Upgrading of the Hirna to Kulubi road in Ethiopia took place between 1996 and 2005. The construction contract was due to be completed in 2003, but heavy rains during that year resulted in over 26 slope failures that affected the road carriageway. Some of these slope failures had caused previous road subsidence, but their full impact was not realized until after the road pavement was sealed just prior to the 2003 rains. The cost of reinstating the road in the landslide locations amounted to 5% of the total final contract price for the 90 km road.

Whereas previously this situation would have been rectified by additional filling and re-gravelling, once cracking and subsidence has occurred to a sealed surface the maintenance options are fewer and the desire for slope stabilization becomes more pressing.

A4.3.3 Landslide stabilization on existing roads

When deciding on where and when to invest in slope stabilization works, many road authorities prefer to adopt the philosophy of 'wait and see'. Although this is principally due to a lack of resources with which to respond fully and immediately to existing or potential landslide situations, there can be good technical and economic justification in adopting this approach. A major rainstorm, flood or landslide event can significantly alter the geomorphology of a slope, valley side or even an entire valley, and it can take some time for slopes to fully readjust (Brunsden 2002). Attempts to stabilize large landslides through early engineering intervention in terrain that is highly active geomorphologically may prove futile until an equilibrium condition is reached naturally. Furthermore, many landslides and associated ground movements occur in response to an extreme rainfall event which might not be repeated again in that particular area or location for a considerable period of time. Continued slope movements of the same order of magnitude as the initial failure might, therefore, not occur, and major investment in engineering works might therefore be an over-reaction. Consequently, one of the more common lessons learnt in relation to the management of landslides on low-cost roads is the value of observation and monitoring compared to immediate engineering intervention.

In making these decisions, a judgement will need to be made as to which landslides are most likely to:

- self-stabilize over a given time period (perhaps 5 years is acceptable in the context of low-cost, low-volume road maintenance) with a reducing and acceptable level of hazard, thus avoiding the need to invest in stabilization works;
- continue to fail with a constant and acceptable level of hazard to the road and road users, thus avoiding the need to invest in stabilization works;
- continue to fail with a constant and unacceptable level of hazard to the road and road users, thus requiring engineering intervention; or
- continue to fail and develop into larger and more significant hazards if engineering works are not implemented prior to the next wet season.

These judgements can only be made effectively when ground conditions are fully evaluated (Section B). Clearly, those landslides that are in the final category listed above pose the greatest potential risk. An existing landslide might be expected to enlarge if any or all of the following conditions occur:

- the slope above the landslide is steep and long;
- the slope material is a soil and comprises previously failed and transported material;
- the slope material is rock and the bedding, foliation and other principal discontinuity planes are observed to dip out of the slope (adversely to stability);
- the slope has a high water table throughout the wet season;
- slope movements continue in response to moderate rainfall events;
- surface water and groundwater are observed to converge into the landslide mass; and
- the slope is being actively eroded at its toe by a stream or river.

Part D reviews and discusses methods of slope inspection and risk assessment for slope management purposes during road operation and maintenance.

References

AHRENDT, A. & ZUQUETTE, L. V. 2003. Triggering factors of landslides in Campos do Jordao city, Brazil. *Bulletin of Engineering Geology and the Environment*, **62**, 231–244.

AUSTRALIAN GEOMECHANICS SOCIETY 2000. Landslide risk management. Concepts and Guidelines. *Australian Geomechanics*, **35**, 51–100.

AUSTRALIAN GEOMECHANICS SOCIETY 2007. A National landslide risk management framework for Australia. *Journal and News of the Australian Geomechanics Society*, **42**, 1–12.

BONACHEA, J., REMONDO, J., GONZALEZ-DIAZ, A., DIAZ DE TARAN, J. R. & CENDRERO, A. 2009. Landslide risk modelling: an experience from northern Spain. *In*: Malet, J. P., Remaitre, A. & Bogaard, T. (eds) *Landslide Processes: From Geomorphological Mapping to Dynamic Modelling*. Strasbourg European Centre on Geomorphological Hazards (CERG), Strasbourg, 259–264.

Brunsden, D. 2002. Geomorphological roulette for engineers and planners: some insights into an old game. The Fifth Glossop Lecture. *Quarterly Journal of Engineering Geology and Hydrogeology*, **35**, 101–142.

Dahal, R. J. & Hasegawa, S. 2008. Representative rainfall thresholds for landslides in the Nepal Himalaya. *Geomorphology*, **100**, 429–443.

Dai, F. C., Lee, C. F. & Ngai, Y. Y. 2002. Landslide risk assessment and management: an overview. *Engineering Geology*, **64**, 65–87.

Fell, R., Corominas, J., Bonnard, C., Cascini, L., Leroi, E. & Savage, W. 2008. Guidelines for landslide susceptibility, hazard and risk zoning for land use planning. JTC-1. Joint Technical Committee on Landslides and Engineered Slopes. *Engineering Geology*, **102**, 85–98.

Finlay, P. J. 1996. *The risk assessment of slopes*. PhD thesis. School of Civil Engineering, University of New South Wales, Australia.

Hearn, G. J. & Griffiths, J. S. 2001. Landslide hazard mapping and risk assessment. *In*: Griffiths, J. S. (ed.) *Land Surface Evaluation in Engineering Practice*. Geological Society, London, Engineering Geology Special Publication, **18**, 43–52.

Hearn, G. J., Hunt, T., Aubert, J. & Howell, J. H. 2008. Landslide impacts on the road network of Lao PDR and the feasibility of implementing a slope management programme. *International Conference on Management of Landslide Hazard in the Asia-Pacific Region*, Sendai, Japan.

Hencher, S. & Lee, G. 2010. Landslide mechanisms in Hong Kong. *In*: Calcaterra, D. & Parise, M. (eds) *Weathering as a Predisposing Factor to Slope Movements*. Geological Society, London, Engineering Geology Special Publications, **23**, 77–103.

Hungr, O. 1997. Some methods of landslide hazard intensity mapping. *In*: Cruden, D. & Fell, R. (eds) *International Workshop on Landslide Risk Assessment*. Honolulu. Balkema, Rotterdam, 215–226.

Jaiswal, P. & van Westen, C. J. 2009. Estimating temporal probability for landslide initiation along transportation routes based on rainfall threshold. *Geomorphology*, **12**, 96–105.

Jaiswal, P., van Westen, C. J. & Jetten, V. 2010. Quantitative assessment of landslide risk along transportation lines in southern India. *In*: Williams, A. L., Pinches, G. M., Chin, C. Y., McMorran, T. J. & Massey, C. I. (eds) *Geologically Active. Proceedings of the 11th IAEG Congress, Auckland, New Zealand, 5–10 September, 2010*, Taylor and Francis Group, London, 1185–1193.

Lee, E. M. & Jones, D. K. C. 2004. *Landslide Risk Assessment*. Thomas Telford, London.

Leone, F., Aste, J. P. & Leroi, E. 1996. Vulnerability assessment of elements exposed to mass-moving: working towards a better risk perception. *In*: Senneset, K. (ed.) *Landslides*. Balkema, Rotterdam, 263–269.

Michael-Leiba, M., Baynes, F., Scott, G. & Granger, K. 2005. Quantitative landslide risk assessment of Cairns, Australia. *In*: Glade, T., Anderson, M. & Crozier, M. J. (eds) *Landslide Hazard and Risk*. J Wiley & Sons Ltd, Chichester, England, 621–642.

Mote, T. I., Bowden, A. J. & Nation, K. E. 2010. Optimised rockfall hazard monitoring and assessment on Christmas Island, Indian Ocean. *In*: Williams, A. L., Pinches, G. M., Chin, C. Y., McMorran, T. J. & Massey, C. I. (eds) *Geologically Active. Proceedings of the 11th IAEG Congress, Auckland, New Zealand, 5–10 September, 2010*, Taylor and Francis Group, London, 523–530.

Uzielli, M., Lacasse, S. & Nadim, F. 2009. Probabilistic risk estimation for geohazards: a simulation approach. *In*: Honjo, Y., Suzuki, M., Hara, T. & Zhang, F. (eds) *Geotechnical Risk and Safety. Proceedings of the 2nd International Symposium on Geotechnical Safety and Risk*, 11–12 June, 2009, Taylor & Francis Group, London, 355–362.

Varnes, D. J. 1984. *Landslide Hazard Zonation. A Review of Principles and Practice*. International Association of Engineering Geology, Commission on Landslides and Other Mass Movements on Slopes, UNESCO, Paris.

Wong, W. K. 2002. Risk assessment for slopes. *Quarterly Journal of Engineering Geology and Hydrogeology*, **35**, 213–222.

Wu, C.-H. & Chen, S.-C. 2009. Determining landslide susceptibility in Central Taiwan from rainfall and site factors using the analytical hierarchy process method. *Geomorphology*, **112**, 190–204.

Part B: Site Investigation

B1 Scope and Programming

G. J. Hearn

URS Scott Wilson Ltd, Scott House, Alençon Link, Basingstoke, Hampshire RG21 7PP, UK
(e-mail: gareth.hearn@scottwilson.com; garethhearn@talktalk.net)

B1.1 Range of techniques

'... if you do not know what you should be looking for in a site investigation, you are not likely to find much of value' (Glossop 1968, p. 113).

The term *site investigation* is conventionally used in civil engineering practice (e.g. Dumbleton & West 1974; Weltman & Head 1983; Hawkins 1986; Fookes 1997; BSI 1999; Simons *et al.* 2002) to describe a range of studies and investigations undertaken to assess the topography, geology, geomorphology and geotechnical ground conditions of a site or an area for the purposes of engineering design. In hilly and mountainous areas, landslide and slope stability assessments usually form important elements of these studies, and are often undertaken as part of a *terrain evaluation*. This terrain evaluation includes office-based desk studies and field-based assessments, and comprises techniques designed to investigate, classify and interpret:

- landscape and landforms;
- geological structure, rock types and soil types;
- geomorphological processes, ground conditions and geohazards (including landslides) prior to embarking on any subsurface ground investigation;
- groundwater conditions; and
- surface drainage patterns.

As described by Lawrance *et al.* (1993), a site investigation comprises terrain evaluation followed by subsurface (or *intrusive*) ground investigation principally by trial pitting, drilling and boring and laboratory testing. In all applications, it is important to review and interpret existing information and to carry out remote sensing and field mapping (Sections B2.2, B2.3, B3.3 & B3.4) to the maximum extent possible before significant investments are made in ground investigation. However, there may be justification for some preliminary ground investigation during the terrain evaluation in order to:

- help calibrate anticipated ground conditions for terrain classification and reference condition mapping (Section B2.5 and B3.2); and
- investigate features and locations that are critical to decision-making during the project feasibility study and route corridor selection (these might include bridge locations and major landslides, for example).

Part B describes and illustrates the techniques that are available for use as part of a site investigation for road construction, road improvement and road maintenance. Although the emphasis is on landslide assessment, these techniques yield significantly more information on geology, geomorphology and ground conditions than is required purely for slope stability assessment alone, and they can make critically important contributions to option studies, engineering design, the assessment of construction material sources and maintenance management.

Site investigation techniques (Simons *et al.* 2002) are grouped here into those that are applied essentially as desk studies and those that form part of field investigations. Field investigations can be subdivided further into field mapping techniques and subsurface ground investigation. Even this simple subdivision is, however, somewhat misleading. Ground verification exercises should be carried out in conjunction with desk study interpretations of remote sensing data, and the site investigation as a whole should be regarded as an iterative process combining desk studies and field investigations with each element confirming the earlier work and informing the subsequent phases.

Table B1.1 indicates the relative importance of desk studies and field investigations as sources of information at the various phases of road construction, improvement and maintenance described in Section A2. Table B1.2 takes this a stage further by considering in more detail the individual techniques that make up desk studies and field investigations in relation to four principal areas of engineering application:

- the design and construction of new roads;
- the design and implementation of road improvement works;
- road reinstatement following major landslides or other damage; and
- the maintenance of existing roads.

Table B1.3 summarizes the type of information commonly obtained by desk study and field investigation techniques, and identifies the sections of this book in which these techniques are discussed. It should be noted that, for the majority of low-cost road construction and improvement

From: Hearn, G. J. (ed.) *Slope Engineering for Mountain Roads*. Geological Society, London, Engineering Geology Special Publications, **24**, 67–70.
DOI: 10.1144/EGSP24.5 0267-9914/11/$15.00

Table B1.1. *Relative importance of desk studies and field investigations according to project stage*

Project types	***Project stage***	***Techniques***		
		Desk Studies	***Field investigations***	
			Field mapping	***Ground investigations***
Road construction	Feasibility Study	High	High	Low
	Preliminary Design	High	High	Moderate
	Detailed Design	Moderate	High	High
	Implementation	Low	Mod/High	Low
Road improvement	Feasibility Study	High	High	Low
	Preliminary Design	Moderate	High	Moderate
	Detailed Design	Moderate	Moderate	High
	Implementation	Low	Low	Moderate
Road maintenance		Low	Moderate	Moderate

Table B1.2. *Use of site investigation techniques for low-cost road projects*

Site investigation	***Stage****	***Technique***	***Engineering application and level of importance***			
			New road design and construction	***Road improvement works***	***Landslide damage reinstatement***	***Operation and maintenance***
Terrain evaluation	Desk studies	Desk study data compilation and assessment	High	Moderate	Moderate	Low
		Aerial photograph and LiDAR interpretation	High	Mod – High	Moderate	Low
		Satellite image interpretation	High	Mod – High	Moderate	Low
		Landslide hazard mapping	Moderate	Low	Low	Low
		Terrain modelling	Moderate	Low	Low	Low
		Terrain classification	High	Moderate	Low	Low
	Field mapping	Reference condition mapping	High	Moderate	Low	Low
		Engineering geological and geomorphological mapping	High	High	High	Low
Ground investigation		Ground investigation	High	Moderate	High	Low
		Slope monitoring	Moderate	Moderate	High	High

* Many techniques are implemented through combined desk study, ground truthing and field mapping exercises, and so distinction between the two stages is rarely clear-cut

Table B1.3. *Typical information provided by site investigation techniques at each project stage*

Site investigation	*Technique*	*Project phase applications*				*Relevant text section*
		Feasibility study	***Preliminary and detailed design***	***Construction***	***Maintenance***	
Desk studies	***Review of published data, construction and maintenance records***	May provide useful data on geology, hydrology and case histories	May provide data useful for design of earthworks and drainage	Not normally used	Maintenance records* reviewed as part of ongoing maintenance management	B2.1
	Interpretation of published topographical maps	Identification and assessment of corridor options, river crossings, river catchment mapping	River catchment mapping for sizing of bridges and culverts. Project-specific topographical mapping normally used for alignment and cross-section design	Project-specific topographical mapping normally used for construction and maintenance detailing		B2.1
	Interpretation of published geological maps	Identification of areas of strong and weak ground and adverse structure	Provides basis for preliminary rock cut design and sources of materials	Not normally used; emphasis placed on use of project-specific engineering geological mapping and ground investigation results		B2.1
	Interpretation of LiDAR and aerial photography	As above, but also enables interpretation of geomorphology: landforms; materials; geohazards. Allows comparison of corridors	Detailed topography and geomorphology, assists design of detailed alignment and cross-section	May be required in areas of unforeseen ground conditions and/or slope instability exposed/caused by construction	Newly acquired aerial photography may be required following extreme events (storm or earthquake)	B2.2
	Interpretation of satellite imagery	Varies with resolution, helps interpret topography, landslide erosion and flood prone areas for corridor assessment	Updated imagery can help monitor changing slope and drainage patterns, plus as per feasibility study	Updated imagery can monitor changing slope and drainage patterns and construction effects, though rarely used	Updated imagery can help monitor changing slope and drainage patterns and construction/maintenance effects	B2.3
	Terrain modelling	Can assist in the interpretation of geomorphology and geology and provide a means of helping predict ground conditions for alignment selection	Helps to interpret and classify results of field observations and ground investigations	Helps interpret ground conditions as materials are exposed in excavations. Can enable ground conditions to be anticipated in advance of excavation and explain unforeseen ground conditions	Not normally derived during this phase, but can assist maintenance engineers to understand the terrain and ground conditions in which they are working	B2.4
	Terrain classification (TC)	Used as a tool to subdivide the study area into units of similar geology and geomorphology for route corridor selection and prediction of ground conditions	Classification of terrain aids definition of ground conditions and possible sources of materials. Assists the development of reference conditions	Classifications confirmed and modified following observed ground conditions, though not used if reference condition mapping is employed instead (see below)	Can be used as part of maintenance management subdivisions, though not used if reference condition mapping is employed instead	B2.5
	Hazard maps	Used for the assessment of corridor options	Updated for designed alignment	Updated for designed/constructed alignment	Updated for constructed alignment	B2.6/B3.3
Field investigations	***Reconnaissance surveys***	Allow desk studies to be checked and calibrated, provide an aid in decision-making over alignment selection	Usually undertaken during feasibility study, but may form part of preliminary design	Not used	Not used	B3.1
	Reference condition mapping	Not usually employed during feasibility study	Uses TC and EGM to divide alignment into units of similar anticipated ground conditions	Classifications confirmed and modified following observed ground conditions	Classifications can be used as part of maintenance management subdivisions	B2.5/B3.2
	Engineering geological/ geomorphological mapping (EGM)	Carried out at rapid reconnaissance-scale level	Detailed mapping of landslides in selected corridor, identification of susceptible slopes, derivation of engineering geological data for design	Logging of excavations and in areas of unforeseen ground conditions and/or slope instability triggered by construction	Provides data for reinstatement if new landslides or earthworks and wall failures occur	B3.4
	Ground investigation	Not usually part of feasibility study, except at critical locations to confirm feasibility or for general ground conditions if there are no geological maps or rock outcrops to observe	Used to determine depth of landslides and ground movements, to investigate soils and rocks, assess foundations for walls and other structures and permit sampling, testing and movement/groundwater monitoring	Required for unforeseen ground conditions to inform design changes and/or slope instability caused by construction and where access was previously not possible	Provides data for reinstatement if new landslides or wall failures occur	B4
	Slope monitoring	Not normally part of feasibility study, but could be established early on in critical areas, eg active or suspected landslides	Assists in assessment of stability for design and longer term observations	Used to assess slope stability during construction	Used to assess performance of works, residual or new slope problems	B5

*A system of maintenance records must be set up before completion of the works as part of a hazard register.

projects, it is usual for a highway engineering team to select an alignment or develop a design that is then studied in detail by an engineering geologist or similar specialist. This specialist may then make recommendations for alternatives or adjustments to the design based on desk studies and field observations. The review of appropriate published geology, the interpretation of aerial photographs (where available) and the use of field mapping remain among the most reliable techniques used in this process.

B1.2 Programming of techniques

B1.2.1 New road construction

The techniques listed in Table B1.3 are variously applicable to all project phases, but they offer the greatest application to new road construction projects as an aid to route corridor selection and the development of the engineering design. The order in which the techniques are listed in the table, and described in Sections B2–B5, is the approximate order in which they should be applied.

B1.2.2 Road improvement

In the case of road improvement projects, principally road widening, resurfacing and alignment improvement schemes, considerable information concerning slope stability affecting the original road should already exist. Nevertheless, remote sensing (and in particular the interpretation of aerial photography and LiDAR if available) combined with field mapping, ground investigation and geological/geotechnical analysis (Sections C3 & C4) should form necessary components in the development of the design.

B1.2.3 Landslide damage reinstatement

In the event of widespread landslide damage following an earthquake, heavy rain or flooding, the engineering response should focus on the use of field mapping and ground investigations. Recently acquired or specially commissioned satellite imagery, aerial photography or LiDAR will considerably enhance the landslide interpretation and the design of reinstatement and remedial works, especially if the damage extends over large areas.

B1.2.4 Road operation and maintenance

During road operation and maintenance, the focus of attention will be directed towards existing cut and fill slopes and the management of drainage. Systematic routine observation, slope monitoring and condition surveys will form the basis of the records necessary for ongoing assessment of slope stability. Field mapping, cross-section survey and ground investigation or monitoring at high-risk sites may be required for the design of reinstatement and remedial works for slopes and sections of road that have failed (Part D).

References

BSI 1999. *Code of Practice for Site Investigations*. BSI 5930, British Standards Institution, HMSO, London. First published in 1981 (Amendment 2, 2010).

Dumbleton, M. J. & West, G. 1974. *Guidance on Planning, Directing and Reporting Site Investigations in Britain*. Transport and Road Research Laboratory, LR625, Crowthorne, UK.

Fookes, P. G. 1997. Geology for engineers: the geological model, prediction and performance. The First Glossop Lecture. *Quarterly Journal of Engineering Geology*, **30**, 293–424.

Glossop, R. 1968. Eighth Rankine Lecture: the rise of geotechnology and its influence on engineering practice. *Géotechnique*, **18**, June. The Institution of Civil Engineers, London, 107–150.

Hawkins, A. B. (ed.) 1986. *Site Investigation Practice: Assessing BS 5930*. Geological Society, London, Engineering Geology Special Publication, **2**.

Lawrance, C. J., Byard, R. J. & Beavan, P. J. 1993. *Terrain Evaluation Manual*. State of the Art Review 7. Transport Research Laboratory, HMSO, London.

Simons, N., Menzies, B. & Matthews, M. 2002. *A Short Course in Geotechnical Site Investigation*. Thomas Telford, London.

Weltman, A. J. & Head, J. M. 1983. *Site Investigation Manual*. PSA Technical Guide No **35**, CIRIA, London, Special Publication No 25.

B2 Desk studies

G. J. Hearn

URS Scott Wilson Ltd, Scott House, Alençon Link, Basingstoke, Hampshire RG21 7PP, UK
(e-mail: gareth.hearn@scottwilson.com; garethhearn@talktalk.net)

B2.1 Traditional data sources

Desk studies are most critical at the initial feasibility and planning phase of new road construction projects (Dumbleton & West 1974). Decisions made at this early stage on the selection of alignment and the approach to design and construction are critical to scheme costs and to the future stability and operation of a road. While the advent of satellite imagery and geographical information systems (GIS) technology in particular (Section B2.7) has meant that desk studies have become potentially far more wide-ranging than they were even 10 years ago, the traditional desk study still remains valid and important, and often provides the bulk of information.

The traditional desk study essentially combines data sources that are conventionally available in paper format, namely topographical maps, geological maps and aerial photographs (although digital maps and orthorectified photographs are now much more common place). Table B2.1 lists typical information that can be obtained from these three principal data sources, though their availability varies significantly (Hearn 2004). Topographical and geological maps are normally available through government agencies and usually small-scale mapping can be downloaded from the internet prior to embarking on field investigations, either under license or for a fee. Unfortunately, published geological maps in many countries are small scale and show Formation-level (stratigraphic age) information only; the distribution of rock types, information that is most relevant to engineering, is often not shown. Even where larger scale geological mapping is available it usually pays 'very little attention to surface formations, soil cover and soft deposits, which are of the greatest importance for roads and other engineering works' (Rodriguez Ortiz & Prieto 1979, p. 139). Aerial photographs can be restricted in some countries for security reasons, and age, coverage, scale, quality and accessibility vary from country to country and region to region.

Rainfall and river gauging records are other sources of desk study data that are required to assess the hydrological regime of an area during feasibility study. They form the basis for the sizing of drainage structures during design. These aspects are described in more detail in TRL (1997).

B2.2 Airborne imagery

B2.2.1 Aerial photograph interpretation

Methods of terrain and landslide interpretation from stereoscopic ('stereo') aerial photography are described, for example, in Lawrance *et al.* (1993) and TRL (1997). Overlapping photographs, usually taken vertically from the air, provide a 3D image of the terrain within the overlap when viewed through a stereoscope. Figure B2.1 illustrates the typical landslide features commonly identifiable in stereo aerial photographs. Table B2.2 describes in more detail how landslide activity and landslide mechanism can be assessed from stereo aerial photographs. The level of interpretation detail that can be obtained depends on photographic scale, shade effects, atmospheric clarity, camera quality and the experience of the interpreter. Even 1:50 000 scale photographs can allow objects of less than a metre on the ground to be seen through a magnifying stereoscope. However, it is recommended that 1:40 000 is regarded as the smallest scale of photography for landslide mapping, with 1:10 000–1:20 000 being the preferred range. The interpretation of stereo aerial photographs in mountainous terrain can be made difficult due to scale, topographical distortion away from the centre of the image and the effects of vertical exaggeration of relief. Text box B2.1 provides a general discussion of the technique.

Despite the importance of aerial photography as a source of interpretation data, Hart *et al.* (2009) conclude that as many as 50% of landslides contained in a landslide inventory for an area of Spain could only be recognized from field mapping, and not aerial photography. The relatively poor success rate in landslide mapping from aerial photography was due to the small scale of aerial photographs used (1:50 000), variable photograph quality, shadow effects and the number of landslides that had occurred since the aerial photography was taken. In addition, many landslides

From: HEARN, G. J. (ed.) *Slope Engineering for Mountain Roads*. Geological Society, London, Engineering Geology Special Publications, **24**, 71–101.
DOI: 10.1144/EGSP24.6 0267-9914/11/$15.00

Table B2.1. *Data typically derived from traditional desk study sources*

For Road Alignment and General Engineering Purposes			*For Landslide Identification and Assessment Purposes*		
Topographical Mapping	***Geological Mapping****	***Stereo Aerial Photographs***	***Topographical Mapping***	***Geological Mapping***	***Stereo Aerial Photographs***
Review of route corridor options in terms of topography. Identification of steep terrain. Locations of rivers and potential river crossing points. Locations of towns and villages and existing infrastructure.	Locations of major geological features (faults, shear zones etc). Locations of weak rocks and unstable rock structures. Potential for construction material sources to be identified (for example Fookes & Marsh (1981)).	Review of route corridor options in terms of overall topography. Identification of steep terrain. Location of rivers and potential river crossing points. Location of towns and villages and land use and existing infrastructure.	Few topographical maps show landslides, though some may show major erosion areas. Contour patterns may indicate landslide morphology.	Few published geological maps show landslide areas. Potential instability can sometimes be inferred from rock structure and rock types. Most maps indicate bedding, foliation and some joint orientations, useful for preliminary slope stability assessment.	Identification of landslides and taluvium deposits. Identification of areas of slope erosion and river scour. Tones and hues in the photography can allow wet areas to be identified, potentially relevant to landslide studies. Structural geology lineaments, bedding and other major discontinuity sets may be interpreted and linked to landslide potential. Repeated aerial photography can provide information on rates of change.

* Showing rock types and structure.

mapped as individual features from the photographs turned out to be two or three coalescing landslides from field observation (Hart pers. comm. 2010). This is probably an uncommon outcome, with the ability of aerial photograph interpretation to detect landslides usually being much greater than this. However, it does emphasize the need to combine desk study-based aerial photograph interpretation with ground verification and field mapping in order to yield maximum results (Table B3.2). Dense vegetation cover, typical of the humid tropics and subtropics, may hide the more subtle landslide features that can only be detected by ground observation and some LiDAR (Section B2.2.2).

Aerial photograph interpretation of landslide-prone terrain for route alignment and road design has important applications at both small and large mapping scales. Figure B2.2 shows a small-scale (original mapping at 1:63 360) engineering geomorphological map developed from aerial photograph interpretation and rapid field reconnaissance survey for route corridor assessment in west Nepal. The main topographical, landslide and drainage hazards identified were used to determine the preferred route that was later confirmed by a detailed feasibility study (Hearn 1993).

Figures B2.3 and B2.4 show an example of larger scale (originally 1:30 000) terrain and landslide mapping for an area of east Nepal using aerial photography. A pocket stereoscope can be used to obtain a 3D image of the aerial photographs spliced together in Figure B2.3. This figure contains two stereo images derived from four contributing overlapping aerial photographs. The photographs have been arranged in such a way that, from left to right, the first and third vertical strips are stereo images (when viewed through a stereoscope) and abut one another. This can be compared to the interpretation shown in Figure B2.4 at the same scale. The illustration demonstrates how aerial photographs can yield valuable information on topography, drainage and slope stability for purposes of route corridor selection and alignment design. For example, areas of cliff and steeply sloping ground may prove difficult for road construction and will most probably be formed entirely in rock. Structurally controlled benches are likely to be stable and composed of *in situ* weathered soil overlying rock. Gently sloping cultivated land will pose few problems for detailed alignment design, but may present difficulties in terms of drainage management and could be old taluvium or colluvium with the potential for reactivated movement if disturbed. The *observed* landslide areas shown on Figure B2.4 are considered to be almost certainly failed slopes based on morphology, drainage and vegetation patterns. The *possible* landslide areas possess a morphology that is indicative of failed ground, but the features are too

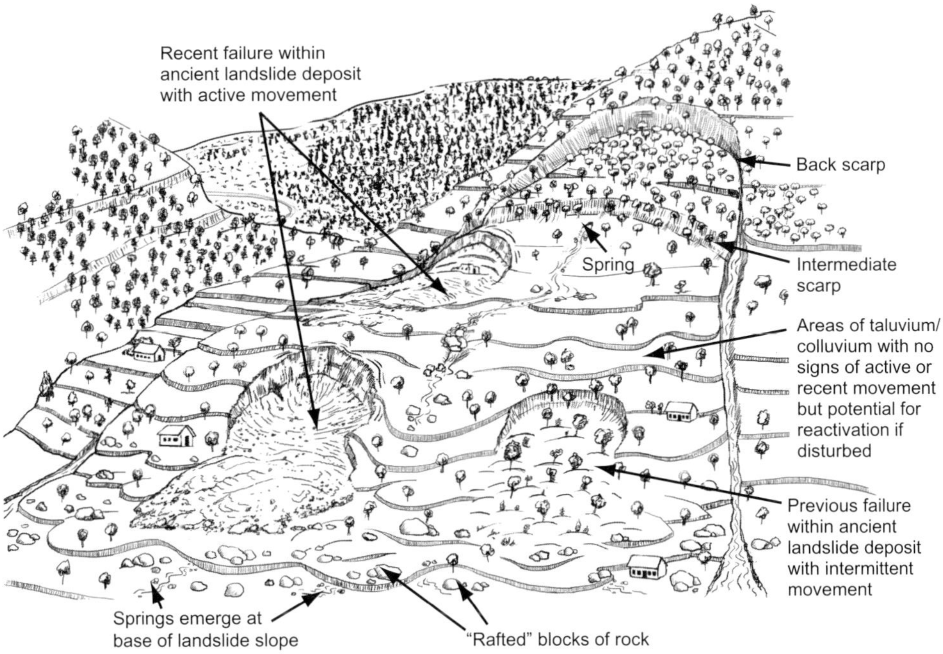

Fig. B2.1. Features commonly observed in stereo aerial photography of landslide areas.

subtle for these areas to be confirmed as areas of ground movement without field verification.

Usually, in the context of low-cost road projects, it may be necessary to rely on existing aerial photography rather than the acquisition of project-specific imagery. This is particularly the case for the maintenance and improvement of existing roads, where the majority of landslides can be observed quite readily on the ground and can be interpreted using existing photography. For the construction of new roads through complex terrain, and where alignment selection is required over large areas, the commissioning of new stereo photography may be necessary. This may be required for ground modelling and alignment design purposes anyway, and its use in terrain and landslide interpretation can be a very useful spin-off from this.

The current trend in the interpretation of aerial photography is moving away from the traditional stereoscope approach towards on-screen visualization and digitizing. This is taking place through the acquisition of airborne 'scanning' systems that collect various datasets which allow both digital aerial imagery and digital elevation model (DEM) data to be derived. This data can then be used (with the relevant hardware and software) to create a 3D image on a computer screen which enables features, such as landslides, to be mapped directly into a GIS. The main advantage of this is that it avoids the errors and inaccuracies associated with overlays and the transposition of information onto a paper map for later digitizing. Digital aerial photographs can also be adjusted by georectification (orthophotographs) using ground control points to remove scale distortions due to edge effects in a way that cannot be achieved through manual interpretation and mapping.

B2.2.2 LiDAR and other airborne digital imagery

Airborne laser scanning (LiDAR) equipment is also being increasingly used to provide submetre accuracy elevation data and to help map landslide morphology (e.g. Booth *et al.* 2009; Chigira *et al.* 2010; Miner *et al.* 2010; Ellis *et al.* 2011) with the added advantage that it can be commissioned to achieve maximum penetration of the vegetation cover (Booth *et al.* 2009).

Figure B2.5a compares landslide mapping for an area in Papua New Guinea using hillshade LiDAR and large-scale (1:5000) stereo aerial photography. There is general agreement between the two independent interpretations. Although LiDAR may be able to record the topography in far greater detail due to tree canopy penetration it does not allow 3D visual interpretation in the way that stereo air photographs do. Furthermore the direction of illumination chosen to generate the hillshade effect can have a significant influence on the interpretation of topography. Figure B2.5b shows how the topography is variously depicted when illuminated from the direction of the eight cardinal points of the compass rose. It is recommended to experiment with illumination

Table B2.2. *Landslide and related features commonly identifiable on stereo aerial photographs*

Feature	*Appearance in aerial photographs*
Landslides by mechanism	
Progressive Soil creep	Immature / uncharacteristic vegetation with disturbed / hummocky ground surface, small ridges / terracettes perpendicular to movement direction, discontinuous / uneven irrigation and cultivation
Debris slide	Exposed soils in well-defined back scarp/source area, clearly defined deposit containing boulders
Mudslides and mudflows (earth slides and earth flows)*	Long, narrow, planar flow track/deposit with lobate toe, clear lateral boundaries, disturbed vegetation, weathered rock or freshly exposed soils in back scarp, springs, ponds and wet ground indicative of high groundwater levels
Debris flow	Clearly defined flow track containing poorly-sorted, often chaotic deposits of debris, depositional fan at toe, flow lines composed of debris forming margins of flow track
Progressive rock creep	Difficult to identify on aerial photographs, often found in fractured rock masses, occupying high, steep slopes, ridges and trenches running across the slope may be visible and indicative, possible rock spalls on margins
Rockfall	Rockfall scarp, sometimes with arcuate form, freshly exposed rock in source area, progressive rockfall may form concave talus slope, boulders at toe or on less steep slopes below
Catastrophic rockslide/ avalanche	Usually large scar with failed mass of chaotic boulders and rafts of rock, slope angle of failed material usually lower than adjacent slopes, trail of boulders often extends considerable distance downslope/downstream
Slow rock slide	Hummocky and furrowed slopes in head of failure, boundaries of failing mass may be linear where joint-controlled, slope rupture and small vertical displacements along margins
Rotational landslide in soil	Arcuate back scarp in plan, concavo-convex slope profile in section from back scarp to toe, well-defined lateral shears, back-tilted blocks with reverse slope below scarps, ponds at junction of scarps and back-tilted blocks and in areas of spring seepage, in some cases, multiple-slipped blocks are seen forming a 'staircase'

Feature	*Appearance in aerial photographs*
Landslides by activity	
Active	Fresh failure scarp with high reflectance and low vegetation levels, slipped mass with immature / disturbed vegetation and areas of bare ground, landform disturbed and uncharacteristic of surrounding topography, possible rock spalls on margins, springs, ponds and wet ground in slipped mass, disruption of drainage patterns, areas of slope in tension (cracks) and in compression (hummocks and bulges)
Dormant or intermittent	As above, but vegetation more mature with scarps beginning to revegetate Possibly cultivated
Relict	As for above, but more subdued topographic distinctions, vegetation may be well-established, land may be cultivated and inhabited
Other important features	
Thin Soil cover	High percentage of rock outcrop, marked structural control on topography, patchy vegetation
Deep soil cover	Concavo-convex slope profile with lobate and gently rounded lower slopes Dendritic drainage pattern
Residual Soil	Red / red-brown appearance in colour aerial photographs Often occupies rounded ridge and spur summits and/or flat/gently sloping ground Benches often intensely cultivated, soil prone to erosion
Rockfall/ rockslide taluvium	Deposits of boulders below rock cliffs, unsorted with chaotic arrangement, steep slopes common, low levels of vegetation, often of shrub-type
Undifferentiated taluvium/ colluvium	Long slopes, often with constant slope angle, shallow landslide scarps may be visible where deposits are unstable, boulders at toe due to longer travel distances, immature drainage patterns with water seepage on lower slopes
Rock outcrop	Steep slopes, high reflectance from bare surfaces and low levels of vegetation, repeated pattern of structural control on topography and drainage
Strong rock	Steep and rugged topography, V-shaped gullies and valleys, knife-edge ridges
Weak rock	Gentle slopes with rounded spurs and ridges, and concave slopes No visible outcrop
Springs	Wet ground with dense vegetation in many cases, often located at concave breaks of slope and/or geological boundaries, often located at the head of a stream

* Cruden & Varnes (1996).

Text box B2.1. Aerial photographs

Aerial photographs are typically acquired using specially designed cameras mounted on an adapted light aircraft that is flown along a controlled flight path during good weather, so that the area is free of cloud. The quality of the film in the camera and the flying height of the aircraft determine the resolution of the photographs. Typically, the aircraft flies at a height of 500–2000 m, providing a photographic scale of 1:12500–1:50 000.

Aerial photographs have many advantages over other types of imagery. These include:

- usual availability in archive form, with many areas having several sets taken over a period of time;
- in most countries it is relatively easy for national agencies to commission new photography (it may not always be easy for external agencies to do so, and in some cases impossible);
- good resolution of ground detail;
- usual availability of stereo coverage;
- general interpretation skills are quite easy to learn, although subtle landform and landslide interpretation requires more experience; and
- analysis is intuitive, that is, interpretation is based on visual recognition.

directions in order to ensure the overall interpretations are correct. This can be assisted by overlaying contour information onto the hillshade LiDAR image as a supplementary means of interpretation. The strength of the technique is clear from this illustration and it is worth noting that the stereo aerial photography was derived by digitally combining the LiDAR data with a set of orthophotographs taken during the LiDAR survey. The cost of the LiDAR survey with ground control amounted to US$ 0.5 million for a total area of 1000 km^2 (US$ 500/km^2) and the additional cost of the derived stereo photography was US$ 50/km^2.

Airborne remote sensing, deriving multispectral and thermal imaging data, has been used to map landslides by Whitworth *et al.* (2005) for example. Automatic classification, based on textural recognition of landslide features is claimed by these authors to provide a landslide identification accuracy of 83%. These techniques are likely to contribute significantly to landslide mapping in the future and, where practical and affordable, could provide a useful supplement to the conventional approach using stereo aerial photography.

B2.3 Satellite image interpretation

B2.3.1 Potential applications and choice of imagery

The use of satellite imagery for low-cost roads is often thought of as being too complex and expensive. With the ever-increasing number of satellites orbiting the Earth, however, the availability of relatively low-cost satellite imagery is increasing. This has also been matched by the availability of powerful low-cost computers and the development of user-friendly software systems with which to analyse, manipulate or simply view the imagery. Satellite imagery is available from a wide range of sensors operated by both governmental and commercial organizations (Table B2.3).

Satellite imagery has a number of benefits and limitations which need to be considered, for example, which satellite data to use and whether to use existing archive data or to commission new imagery (Table B2.4). Very often the limitations can be overcome with careful selection of the best data and the most appropriate processing options for the required application.

The choice of which software to use will be dependent on the format in which the satellite imagery is being procured, what processing and analysis might have to be undertaken and how it will ultimately be used. For example, if the satellite imagery comprises multispectral data (such as raw Landsat data), it will require processing before being used and the user will need a specialized software package. Software will also be required if the user wishes to carry out multispectral analysis. If, however, the satellite imagery is in a format that does not require any processing (such as a high-resolution image), it may be possible to carry out the necessary data analysis using GIS software (sometimes requiring additional 'extensions' or 'plug-ins'). The ability to import satellite imagery into a GIS means that it can be viewed, compared and analysed with other project data (Section B2.7). There are numerous texts available (e.g. Campbell 2007; Lillesand *et al.* 2008) on how to use and interpret satellite imagery. Further discussion is also provided in Morgenstern & Martin (2008).

When choosing which imagery to procure, the following points need to be considered:

- the required output from the image interpretation;
- the data required to deliver that output; and
- the imagery most likely to contain that data.

These considerations may yield a number of options and the final selection will be based on budget, the size of the area to be covered and whether imagery is available in

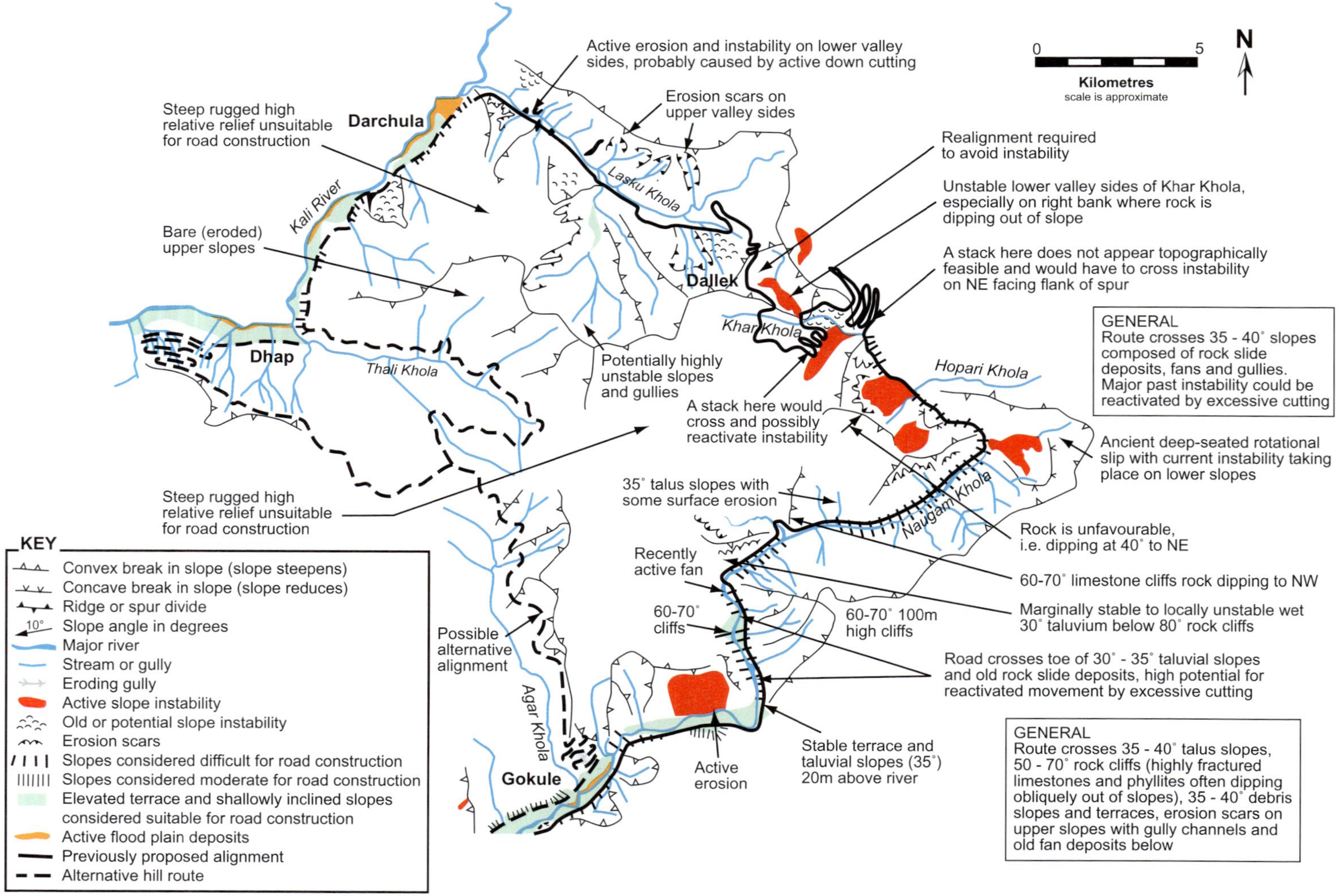

Fig. B2.2. Engineering geomorphological map for route corridor assessment in west Nepal, derived primarily from aerial photograph interpretation. Modified with kind permission of Springer Science & Business Media from Hearn (1993).

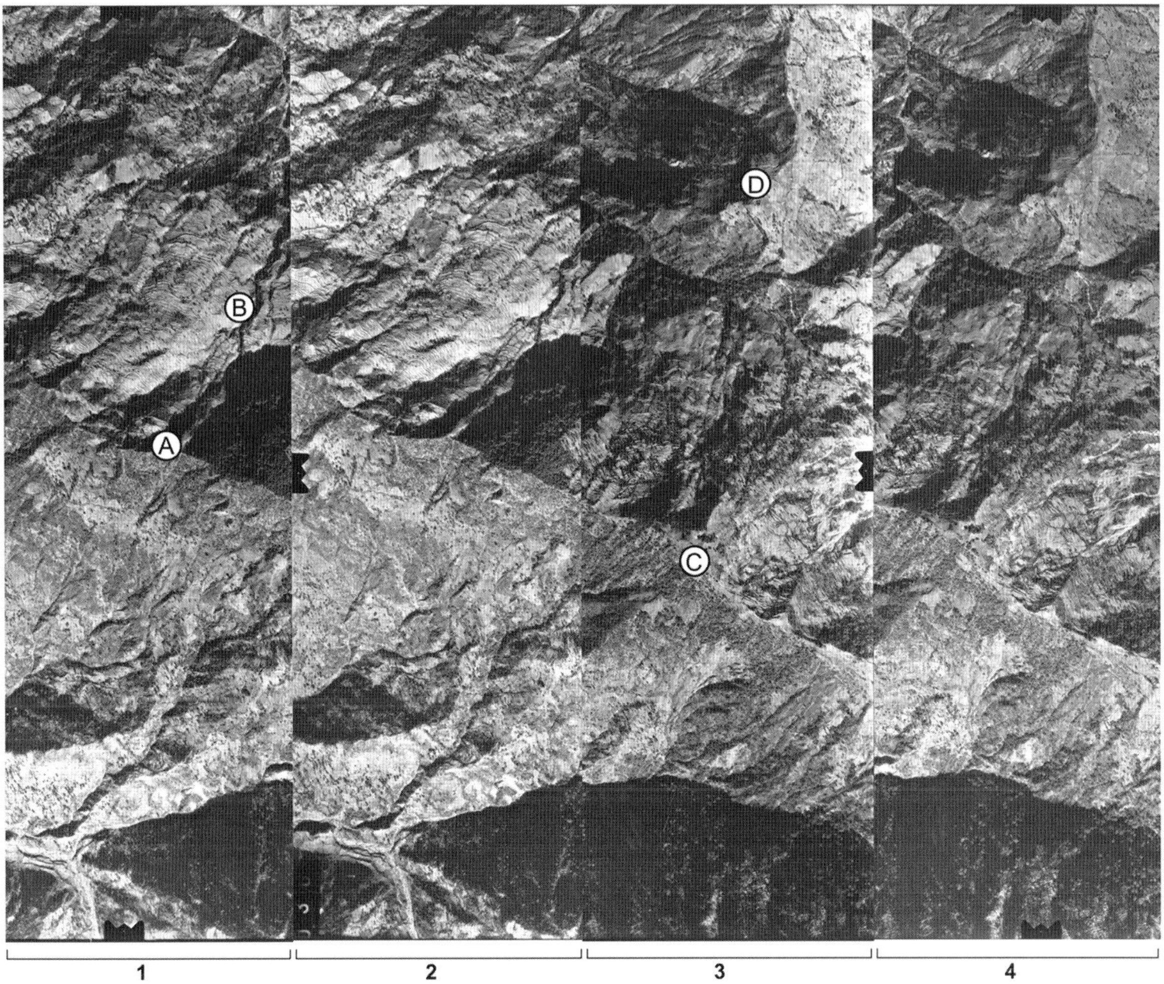

Fig. B2.3. Stereo aerial photographs of an area in east Nepal (A–D refer to Fig. B2.4).

archive or will need to be specifically procured. This latter point has important implications for project programming and budget, and it is usual to rely on archived data wherever possible. The computer hardware and software available may also have some influence on the final selection. A summary of typical data requirements and sources of satellite data for low-cost road applications is given in Table B2.5.

Although the costs of utilizing satellite imagery will vary according to size of project area, image type and the degree of processing required, a simple terrain interpretation from medium-resolution imagery could cost as little as $10 000 for a 100 km alignment, including image acquisition and interpretation time. This figure could double or triple if digital terrain modelling and complex image processing are required.

To illustrate some of the applications mentioned in Table B2.5, reference is made to rural access road studies in Nepal and Bhutan where satellite remote sensing was used to create data layers for landslide susceptibility mapping (Hart *et al.* 2003). Table B2.6 lists the imagery used to map structural lineaments (e.g. folds, faults and foliation), land use and vegetation and areas of taluvium, colluvium, erosion, standing water and wet ground.

The satellite image interpretation made use of the following techniques:

- visual identification and classification of ground surface features, assisted by:
  - ‘draping’ of the satellite imagery over DEMs, and
  - creation of false colour composite (FCC) images using the Landsat and SPOT data;

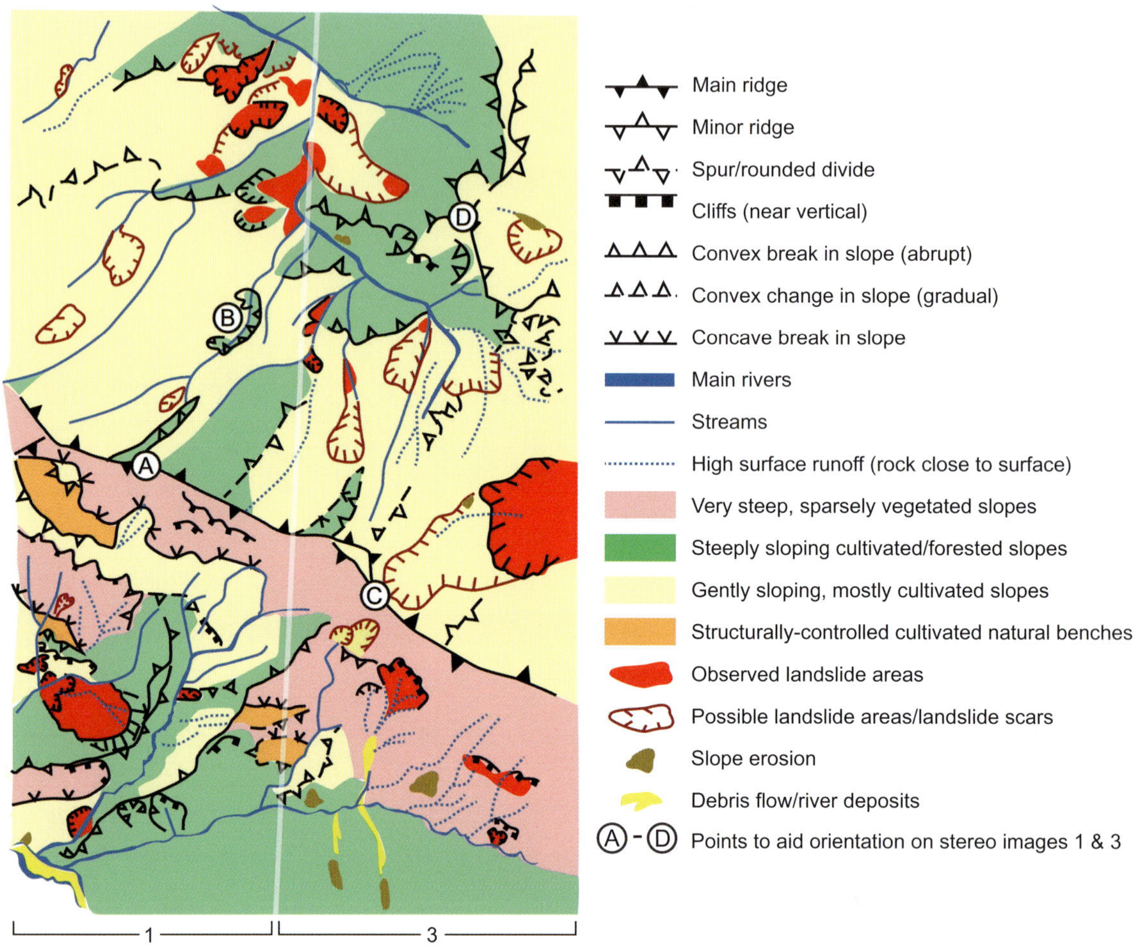

Fig. B2.4. Terrain and landslide interpretation of the air photography shown in Figure B2.3.

- comparing the different spectral bands of the Landsat data with each other to create 'band ratios', used to identify different slope materials within the study area.

Some of the output mapping is illustrated in Figure B2.6. The study concluded that:

- large areas could be mapped relatively quickly and cost-effectively using Landsat, although the spatial resolution of the imagery was a limiting factor;
- IKONOS and IRS-1D imagery were very useful for identifying geomorphological features;
- supervised (automatic) classification techniques were able to map wet ground or standing water although the mapping of soils, and especially colluvium, required specialist interpretation and extensive field verification;
- certain band ratios and FCC combinations worked better for some geological and vegetation cover conditions than others (the FCC combinations found to be the most useful were Landsat RGB 542 and SPOT RGB 321 and RGB 431);
- only the IKONOS imagery allowed landslides to be mapped with any reliability;
- in the Nepal studies, stereo aerial photography proved more reliable for landslide mapping than any of the satellite imagery due to the availability of good-quality and high-resolution photographs; and
- in Bhutan, the availability of aerial photography was more patchy and was generally of too small a scale and affected by shade to be of any significant value in landslide mapping.

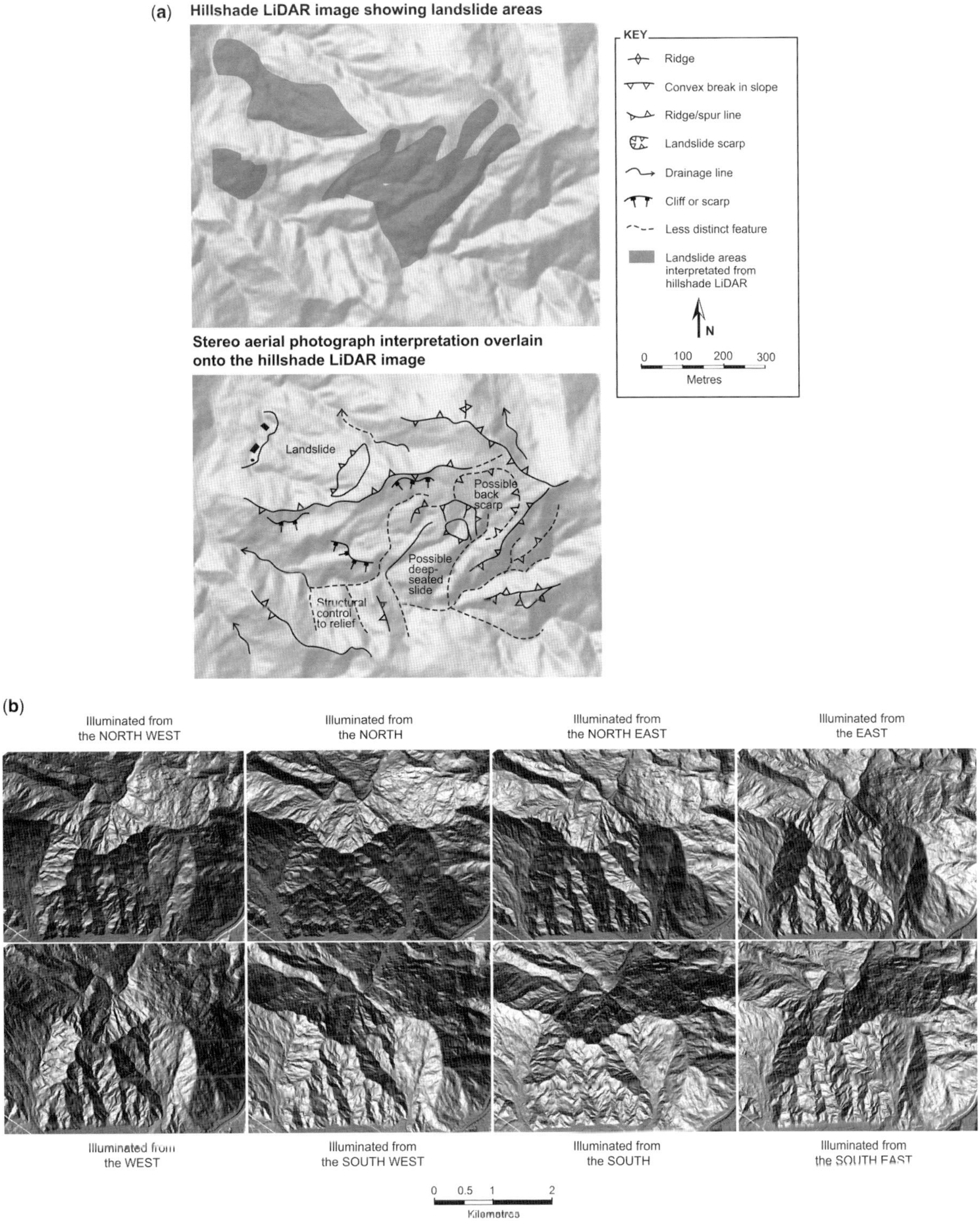

Fig. B2.5. Comparing slope interpretation from hillshade LiDAR and stereo air photography.

Table B2.3. *Range of government and commercial satellite data currently available (correct as of May 2011)*

Data Type	*Satellite data*	*Year of launch*	*Spectral bands*	*Spatial resolution*	*Typical scene size (km)*	*Re-visit time (days)*	*Stereo coverage*	*Typical indicative approximate cost (minimum order area may apply)*
Very High to High Resolution Optical Imagery	IKONOS	1999	5 bands: R, G, B, NIR, Pan	Pan: 1.0m MS: 4.0m	11 x 11	3 – 4	Yes	Arch: AOI dependent Prog: $20 per km^2
	Quickbird	2001	5 bands: R, G, B, NIR, Pan	Pan: 0.6m MS: 2.4m	16.5 x 16.5	3 – 4	No	Arch: $17 per km^2 Prog: $27 per km^2
	Orbview 3	2003	5 bands: R, G, B, NIR, Pan	Pan: 1.0m MS: 4.0m	8 x 8	3	No	Arch only: $5 per km^2
	WorldView-1	2007	1 band: Pan	Pan: 0.5m	16 x 16	1.7	Yes	Arch: $14 per km^2 Prog: $20 per km^2
	WorldView-2	2009	9 bands Pan, Coastal Blue, Blue, Yellow, Green Red, Red Edge, NIR1, NIR2	Pan: 0.5m MS: 2.0m	16.4 x 16.4	1.1 days at 1 metre GSD or less	Yes	Arch: $14 to $32 per km^2 Prog $20 to $38 per km^2
	GeoEye	2008	5 bands: R, G, B, NIR, Pan	Pan: 0.5m MS: 1.65m	15 x 15	3	Yes	Arch: $12.50per km^2 Prog: $25.00 per km^2
	Pleiades-1 and Pleiades-2	Pleiades 1 – end 2011 Pleiades 2 – Mid 2012	5 bands: R, G, B, NIR, Pan	Pan: 0.5m MS: 2.0m	20 x 20	Daily	Yes	Not available
	KOMPSAT-2	2006	5 bands: R, G, B, NIR, Pan	Pan: 1.0m MS: 4.0m	15 x 15	3	No	Arch: $7.50 - $16 per km^2 Prog: $15-$25 per km^2
	FORMOSAT-2	2004	5 bands: R, G, B, NIR, Pan	Pan: 2.0m MS: 8.0m	24 x 24	Daily	No	Arch: €4-5km^2 Prog: €6-8/km^2
	IRS CARTOSAT P5	2005	1 band: Pan	Pan: 2.5m	30 x 30	5	Yes	Mono = €2/km^2 Stereo = €4/km^2
	IRS CARTOSAT 2	2007	1 band: Pan	Pan: 0.8m (sold at 1m)	9.6 x 9.6	4	No	€1260/scene
	ALOS PRISM	2006	1 band: Pan	Pan: 2.5m	70 x 70	46	Yes	Arch: €0.1/km^2 Prog: €0.2/km^2
	Resurs-DK-1	2006	4 bands: Pan, R, G, NIR	PAN = 0.8 – 1.0m / Spectral = 0.8 – 3.0m		5 – 7	No	Not available
High to Medium Resolution Optical Imagery	ALOS AVNIR	2006	4 bands: R, G, B, NIR	MS: 10.0m	70 x 70	46	No	Arch: €0.1/km^2 Prog: €0.2/km^2
	SPOT V	2002	5 bands: R, G, NIR, SWIR, Pan	2.5m, 5m, 10m, 20m	60 x 60	2 – 3	Yes	Arch: €1900-€8100 per scene Prog: €2700-€8900 per scene
	RapidEye	2008	5 bands: R, G, B, NIR, 'Red Edge'	5.0	77km swathe width	Daily	Not available	Archive: €950 Prog: €4750
	IRS-1D	1997	Pan + 4 bands: 2 visible, 2 IR	23 m (6 m pan)	70 x 70	24	No	Pan €0.5/km^2 MS €0.5/km^2
	ASTER	1999	14 bands: R, G, NIR , SWIR x6, TIR x5	15m, 60m, 90m	60 x 60	4 – 16	Yes	Archive 3 JYen

Medium to Coarse Resolution Optical Imagery	Landsat 7 ETM+	1999	8 bands: R, G, B, NIR, MIR (x2),TIR, Pan	15m (Pan), 30m Multi-spectral), 60m (Thermal)	165 x 165	16	No	Data held at USGS is free of charge
	RESOURCESAT-1 (IRS-P6)	2003	LISS-4 = VNIR LISS-3 = 3 VNIR, SWIR AWiFS = 3 VNIR, SWIR	LISS-4 = 5m LISS-3 = 20m AWiFS = 60m	LISS-4 = 70 x 70 LISS-3 = 140 x 140 AWiFS = 370x370	5	No	LISS-4 = €0.5/km² LISS-3 = €0.1 – 0.15/km² AWiFS = €0.01/km²
	CORONA (Declassified Satellite Imagery)	1959-1972	B&W Panchromatic Photographs	Between approx. 1.8m and 7.5m	Variable	None	No	Arch: $30 per scene
Radar Satellite Data	ERS-1	1991	C-band	30m	100 x 100	35	No	Arch: €0.04/km²
	ERS-2	1995	C-band	30m	100 x 100	35	No	Arch: €0.04/km² Prog: €0.09/km²
	ENVISAT	2002	C-band	30-150	100 x 100 to 400 x 400	35	No	Arch: €400 per scene Prog: €900 per scene
	Radarsat-1	1995	C-band	8-100	50 x 50 to 500 x 500	24	No	$Can 3600 to $Can 4500
	Radarsat-2	2008	SAR (C-Band)	3-100 m	20 x 20 to 500 x 500	24	No	Arch: $Can 3600 to $Can 8400 Prog: $Can 3720 to $Can 8520
	IRS Risat-1	Pending	SAR (C band)	2m and 50m	Not available	Not available	Not available	Not available
	IRS Risat-2	2009	SAR (C band)	2m and 50m	Not available	Not available	Not available	Not available
	ALOS PALSAR	2006	L-band	10-100m	30 x 30 to 350 x 350	46	No	Arch: €500 per scene Prog: €900 per scene
	TERRASAR-X	2007	X-band	1-18m	10 x 5 to 100 x 1500	11 (2.5)	No	€2750 - €6720 per scene
	COSMO-SkyMed	2007	X-band	1-100m	10 x 10 to 200 x 2000	1	No	Arch: €825 - €4725 per scene Prog: €1650€ - €9450 per scene
Hyper-spectral Satellite Data	IMS-1	2008	Mx = 4 bands HySI = 64	Mx = 37m HySI = 506m	Mx = 151 x 151 HySI = 129.5 x 129.5	-	No	Not available
	EO1-Hyperion	2000	220 continuous VNIR and SWIR	30m	7.75 km²	16	No	Data available to download free of charge
	ENMAP	Expected launch 2012	213 spectral bands (VNIR and SWIR)	30m	30 km²	4	No	Not available

Arch = Archive; Prog = Programmed; AOI = Area of Interest; Pan = Panchromatic (black and white); MS = Multi-Spectral; LISS = Linear Imaging Self Scanner; Mx = Multi-spectral bands on the IMS-1 satellite; NIR = Near Infrared; Mono = no overlapping coverage; Stereo = overlapping coverage; MIR = Middle Infrared; TIR = Thermal Infrared; GSD = Ground Sampling Distance; VNIR = Visible and Near Infrared Region; SWIR = Short Wave Infrared; AWiFS = Advanced Wide Field Sensor; ASAR = Advanced Synthetic Aperture Radar; L & X bands = Microwave spectrum bands in SAR; MERIS = Medium Resolution Imaging Spectrometer Instrument; HySI = Hyperspectral Imaging Camera; $ = USD.

Table B2.4. *Key benefits and limitations of using satellite imagery for low-cost road projects*

Key benefits	*Key limitations*
• Depending on which satellite data is used, a single image can cover a relatively large area. For example, a Landsat image covers an area of 185 × 185 km. • Can provide a perspective or view of the landscape that cannot normally be achieved by other means. This could help to provide information about an area that may not be obvious from ground level, such as how a particular feature within the landscape relates to other parts of that landscape. • The availability of information that is beyond the visible part of the electromagnetic spectrum, such as infrared imagery. Such data can be used to enhance image interpretation. • The increasing availability of products derived from the satellite data, such as Digital Elevation Models (DEMs). • Frequent repeat collection of images: many satellites have the capability to collect an image for any given area at least once a month. • Low levels of distortion away from the centre of the image, particularly when compared with aerial photographs. • The potential for digital image analysis, such as automatic classification. • The increasing availability of high-resolution imagery and stereo capability using DEM or stereo images.	• Although there is almost complete global coverage of archived satellite imagery, there are gaps in some areas (particularly the tropics and high latitudes) where there is either limited data coverage or the data in the archive are not to the highest environmental quality standards (cloud cover, atmospheric haze, poor solar illumination, etc). • In some areas there is only low spatial resolution available, meaning that only large objects can be seen and identified. • The relatively high costs of high-resolution, large-scale imagery though these costs are becoming reduced significantly. • In most cases there is no ability to view the images in stereo (without also obtaining a DEM and draping the imagery over this). • The images can be difficult to interpret and sometimes require high levels of technology for processing. • Seasonal variations in lighting conditions or vegetation cover can influence image interpretation. • The interpretation of satellite imagery for landslide mapping (as with aerial photographs) does require a reasonably high level of skill, experience and familiarity of the area (or environmental conditions) being studied.

Draping non-stereo IKONOS high-resolution satellite imagery over an equivalent high resolution DEM, for example, can provide the basis for 3D visualization and interpretation (e.g. Wasowski *et al.* 2010).

B2.3.2 Satellite-derived topographical data

Increasingly, satellite data are being used to generate digital terrain or elevation data. These data can be used as topographical mapping for route corridor selection as well as for other applications, such as the derivation of slope angle, slope aspect, drainage and catchment area maps for geological and hydrological applications. The digital elevation data can also be used as a base on which to drape other satellite imagery or aerial photography in order to enhance 3D visualization. For detailed engineering design, high-resolution digital elevation data are currently available from certain satellites such as stereo IKONOS, but archive data is quite limited and the new information is expensive to acquire. Table B2.7 lists the main sources and accuracy of satellite-derived and airborne digital mapping.

There is an increasing amount of satellite imagery that is available through web-based applications such as Google Earth and Microsoft Virtual Earth (copyright protection may apply). These provide a very quick and easy method for viewing and navigating around a project area. Such applications have revolutionized how satellite imagery is used (e.g. Griffiths *et al.* 2010).

B2.4 Terrain models

'In areas where very little is known about the underlying geology, where published geological mapping is small scale and generalized and where engineering geological information is required on ground conditions that are anticipated to be complex and varied, geological modelling can provide the means of deriving important interpretations for engineering decision making' (Fookes 1997). Taken in its widest context, geological modelling interprets, classifies and portrays:

- the underlying geology and geological structure;
- the distribution of surface rocks and soils in relation to topography; and
- geomorphological process and geohazards.

Geological models summarize desk study information on topography and geology and are verified and strengthened through field validation, ground investigation (Section B4) and the logging of soil and rock exposures during construction. Although they are described by Selby (1993) and Fookes (1997) as geological models and Griffiths & Stokes (2008) as geomodels, they are referred to here under the general heading of *terrain models*. They include, for example, landscape evolution models, whereby the landscape is interpreted and described in terms of its geological history and contemporary geomorphology.

Table B2.5. *Potential applications of satellite imagery for low-cost road projects*

Application type	*How satellite imagery can be used*	*Suitable scales*	*Most suitable satellite data**
Mapping of route corridors or mapping for route alignment selection	▪ Mapping of geology, soils and land use ▪ Mapping of topography (if not available from elsewhere) ▪ Mapping of geohazards such as landslides, areas of erosion and flood prone areas ▪ Broad terrain classification ▪ Mapping of existing infrastructure	1:50,000 to 1:25,000	Combination of multi-spectral and optical data – i.e., Landsat, SPOT V, ASTER or any of the high resolution sensors
Landslide susceptibility and hazard mapping for existing or proposed road corridors	▪ Landslide identification ▪ Mapping landslide conditioning and triggering factors (Section A3.3) ▪ Mapping of land use and infrastructure	1:25,000 to 1:5,000	Combination of multi-spectral and optical data – i.e., Landsat, SPOT V, ASTER or any of the high resolution sensors
Mapping catchment areas and drainage patterns	▪ Derivation of topography and drainage network from DEM data	1:25,000 to 1:5,000	Combination of multi-spectral and DEM data – i.e., SPOT V or ASTER GDEM data
Management of existing road corridors	▪ Mapping changes in land use ▪ Mapping changes in drainage conditions ▪ Monitoring of geohazards such as landslides, areas of erosion and flood prone areas	1:25,000 to 1:5,000	Any of the high resolution sensors repeated at regular intervals
Identification of potential sources for suitable construction materials	▪ Mapping of geology and soils in suitable detail	1:25,000 to 1:5,000	Combination of multi-spectral and optical data – i.e., ASTER, SPOT V or any of the high resolution sensors
Management of existing roads	▪ Mapping of existing assets ▪ Monitoring of problematic areas (such as unstable slopes, landslides or areas of erosion)	1:10,000 to 1:5,000	Any of the high resolution sensors repeated at regular intervals
Road design and construction	▪ Detailed topographical mapping (if not available elsewhere) ▪ Mapping of geology, soils and land use in detail ▪ Mapping of geohazards such as landslides, areas of erosion and flood prone areas ▪ Detailed terrain classification ▪ Mapping of any existing infrastructure	1:5,000	Combination of multi-spectral and optical data – i.e., SPOT V, ASTER, or any of the high resolution sensors plus ground control points for topographical mapping
Landslide monitoring	▪ Detection of landslide movements	1:5,000	Combination of optical and SAR data – i.e., any of the high resolution sensors, as well as InSAR technology repeated at regular intervals

*ASTER, Advanced Spaceborne Thermal Emission and Reflection Radiometer; GDEM, Global Digital Elevation Model; SAR, Synthetic Aperture Radar; InSAR, Interferometric Synthetic Aperture Radar.

They offer their greatest potential during the early planning stages of new construction projects. It is at this time when an understanding of the landscape and its underlying geology is most critical to route corridor and alignment selection and the design and programming of field investigations.

For road alignments through complex topography, these models are best structured to enable them to be applied

Table B2.6. *Imagery used in the Nepal and Bhutan studies*

Imagery type	*Spectral bands*	*Resolution (m)*
Landsat ETM +	7 multispectral bands	30
	1 panchromatic* band	15
SPOT IV	4 multispectral bands	20
IKONOS	4 multispectral bands	4
	1 panchromatic band	1
Indian Remote Sensing (IRS) – 1D	4 multispectral bands	30
	1 Panchromatic band	5

*Black and white.

and extrapolated as part of a terrain classification exercise (Section B2.5). They can contribute the following to terrain evaluation for engineering purposes:

- an holistic interpretation of the landscape, drawing upon all available information that can be refined as the project proceeds;
- early identification of potential geohazards;
- the basis for terrain classification (Section B2.5) and reference condition mapping (Section B3.2);
- a structure upon which to plan field investigations (Section B4); and
- an effective method of conveying important technical considerations to the non-specialist.

Figure B2.7 illustrates a terrain model prepared for a road project across the Blue Nile gorge in Ethiopia. The model enables the terrain and its geomorphology to be visualized in relation to the underlying geology and formed the basis for the development of geotechnical reference condition mapping (Section B3.2). It was also a useful starting point for the interpretation of the terrain and ground conditions by other members of the project team, including engineering design and supervision staff. This model took approximately three man-days to prepare in draft using stereo aerial photography, published topographical and geological maps and site photographs.

Detailed models can be produced for specific areas, such as that shown in Figure B2.8. In this illustration, ground conditions were considered to be especially complex and slope stability hazards posed significant perceived risk to road construction and operation. The detailed model not only advanced the engineering geological interpretation of the area, but also helped engineering decision-makers to visualize ground conditions and the important factors to be considered in alignment selection. Note that Options 1, 2 and 3 are explained in Section C1.4. The model took approximately 3 man-days to prepare.

Figure B2.9 shows a terrain model used to derive terrain classification units (referred to in the model as *engineering terrain units*) for alignment corridor assessment and preliminary interpretation of ground conditions prior to field investigations. This model was prepared entirely from desk study interpretation of published topographical and geological mapping over a period of approximately 5 man-days. It enabled the geological structure of the study area to be summarized and interpreted for broad assessment of ground conditions and materials.

Where the area under consideration is a specific corridor, such as for a selected alignment or an existing road, a more alignment-specific approach may be required. This might include an interpretation and summary of terrain and ground conditions along the centre-line of the corridor based on available data, as illustrated in Figure B2.10. This figure has been derived entirely from desk study interpretation of published geological and topographical mapping for the purpose of assessing rock and soil profiles for preliminary estimations of earthworks quantities. It took approximately 5 man-days to prepare.

The use of terrain models in this way is an essential component of the *total geology approach* in the anticipation and interpretation of ground conditions for engineering purposes (e.g. Fookes *et al.* 2000; Baynes *et al.* 2005).

B2.5 Terrain classification

The purpose of a terrain classification is to identify and map zones within a project area that have similar topographical, material and geohazard characteristics (e.g. Dowling 1968; MEXE 1969; Anon 1972; Mitchell 1973; Hunt 1979; Rodriguez Ortiz & Prieto 1979; Lawrance *et al.* 1993; Waller & Phipps 1996; Phipps 2001). The terrain features usually embodied in a terrain classification are rock type and structure, slope angle and morphology, soil type and surface drainage (Lawrance *et al.* 1993). Terrain classification is usually based on the interpretation of remote sensing, topographical maps and geological maps of large areas with targeted field verification. Detailed terrain classifications can be prepared for specific alignments such as those developed by Rodriguez Ortiz &

Image 1 – SPOT FCC, supervised automatic classification to identify 'wet areas' – shown here by the red flecks (apart from the river which shows up as blue)

Image 2 – Landsat FCC, supervised automatic classification to identify areas of colluvium (areas in blue on sloping ground)

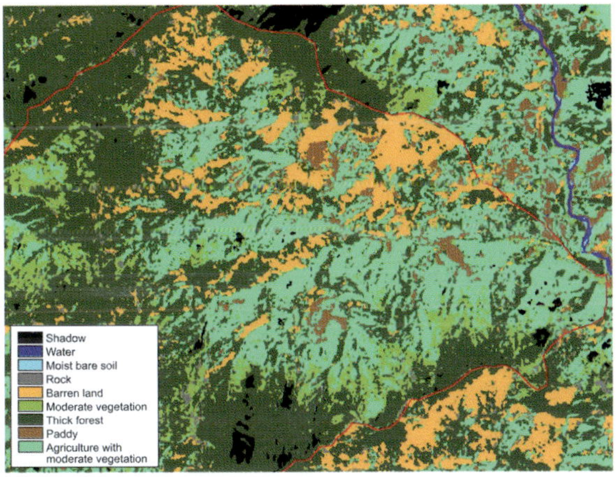

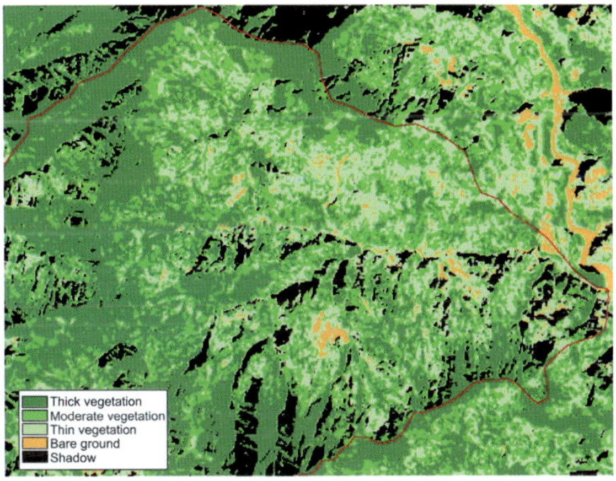

Images 3 and 4 – Use of Principle Component Analysis and supervised automatic classification techniques to identify different types of land use and vegetation density from Landsat.

Image 5 – Landsat FCC, visual mapping (i.e, using visible image) from the satellite imagery to map structural lineaments.

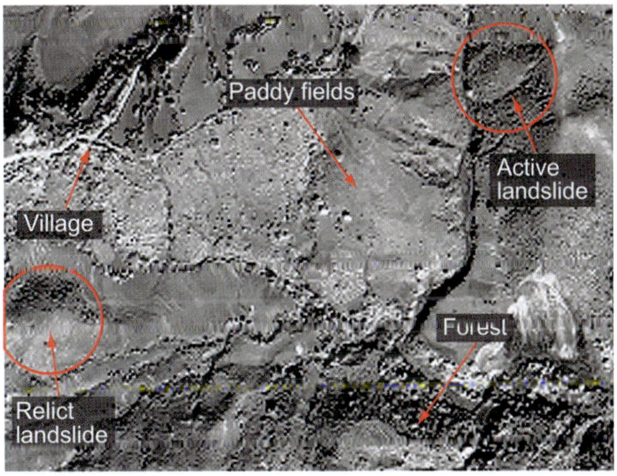

Image 6 – IKONOS, visual mapping from the satellite imagery to map geomorphology, land use, infrastructure and landslides.

Fig. B2.6. Illustration of some satellite image interpretation from Nepal (images 1 & 2 and 3 & 4 are of the same two areas).

Table B2.7. *Digital mapping data from the common sensors*

Sensor	*Horizontal resolution*	*Horizontal accuracy*	*Vertical accuracy*	*Scene size*	*Data collected*
ASTER GDEM	30 m	50 m	15–30 m	60 km × 60 km	Since 1999
SRTM	3 Arc seconds (90 m)	50 m	15 m	1° lat × 1°	Acquired 2000
SPOT	20 m	15 m	10–15 m	60 km × 60 km	Since 2002
IKONOS	1–2 m	1–2 m*	1–2 m*	11 km × 11 km	Since 2000
WorldView-2	0.5 m	1–2 m*	1–2 m*	16.4 km × 16.4 km	2009
GeoEye-1	0.5 m	1–2 m*	1–2 m*	15 km × 15 km	2008
Airborne LiDAR	0.5 m	0.5 m	0.10–0.25 m	Dependent on area of interest	On demand

*Dependent on the supply of Differential GPS (Global Positioning System) points.

Prieto (1979) and Das *et al.* (2010); these include classification of geology, geomorphology, geotechnical parameters and engineering criteria for roadworks based primarily on field observations.

Conventionally, terrain classification subdivides an area into a hierarchy of land systems, facets and elements (Lawrance *et al.* 1993), and is usually undertaken by desk study to enable large areas to be covered rapidly. *Land systems* comprise associations of land facets; each usually extends over an area of at least 100 km^2 and is typically mapped at a scale of 1:100 000–1:1 000 000. *Land facets* are terrain units containing similar landforms, slope, parent material, soils and hydrological conditions, and they are usually mapped at scales of 1:10 000–1:100 000. A *land element* is the smallest unit in the classification and represents an individual landform, such as a meander bend in a river or a cliff face within a larger slope.

Figure B2.11 shows the use of terrain classification mapping using land systems and land facets for a road project in northern Africa (Hunt 1979). Although located within an arid environment, the same principles would apply to the humid tropics and subtropics.

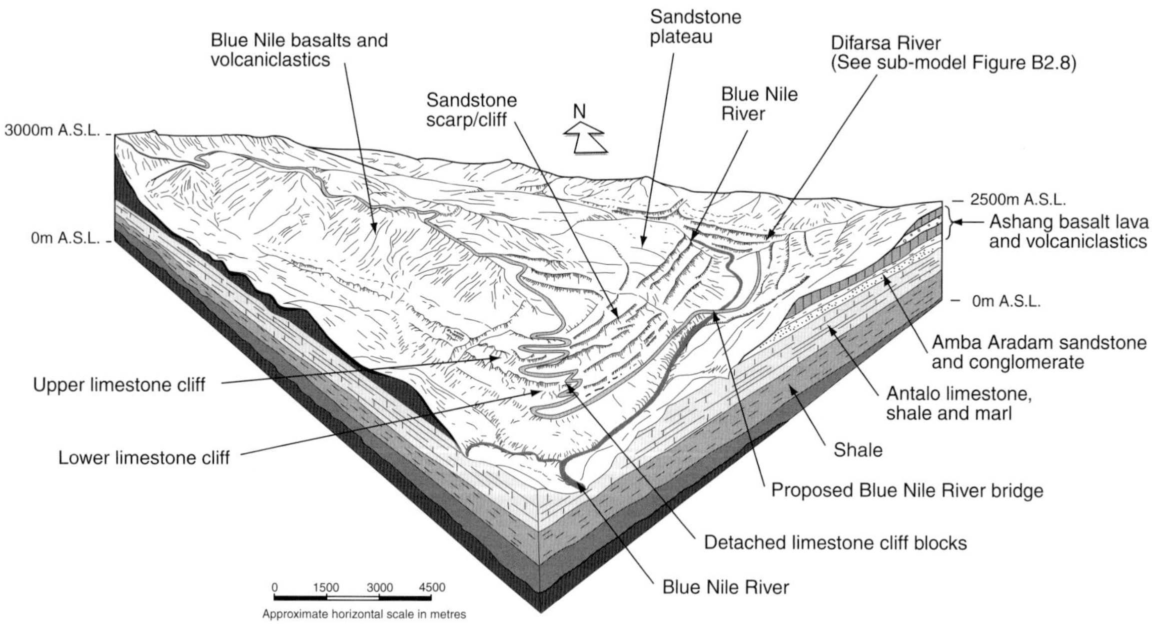

Fig. B2.7. Terrain model for the Blue Nile gorge, Ethiopia (drawn by G. Pettifer.)

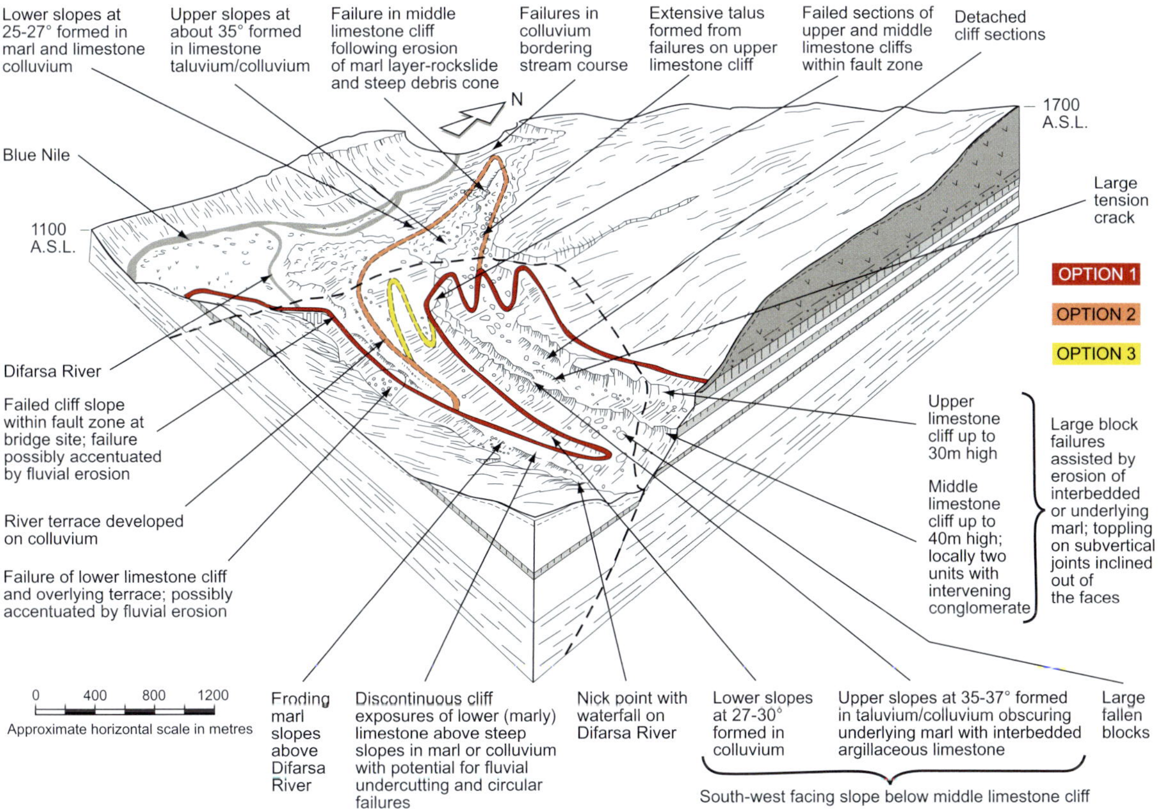

Fig. B2.8. Submodel for a problematic alignment section. (Model drawn by G. Pettifer.)

A short length of the proposed road corridor is shown passing through two land systems numbered 3 and 4. Land System number 3 is mainly underlain by near-horizontally bedded dolomite and dolomitic limestone, the landscape being predominantly flat to gently rolling or stepped plateaux with occasional deeply incised river valleys. Land System number 4 is mainly underlain by a sequence of marl, calcareous mudstone and shale with thin interbeds of limestone forming stepped plateaux. Each land system was subdivided into 17 facets (see Table B2.8), each facet being reasonably homogeneous and fairly distinct from the surrounding terrain. Field verification showed that soils and slope materials developed on each facet were relatively uniform. The terrain classification mapping was accompanied by tables to indicate the engineering attributes of the various facets, a typical extract of which is given in Table B2.9.

In this particular illustration, land elements were not derived as they represented features of the microrelief that were not considered relevant to the design.

While this systems based approach to terrain classification may be entirely relevant in mature and ordered landscapes, its value may be significantly reduced in complex and geomorphologically active mountain terrain. In this terrain, the emphasis should be on the identification of landforms, materials and processes that are of direct relevance to engineering. These include, for example, landslides, taluvium and colluvium, residual soils and slopes prone to river undercutting, flooding, debris flows and rapid sediment deposition. The terrain classification described by Anon (1972) and Rodriguez Ortiz & Prieto (1979) and the geomorphological mapping techniques described in Brunsden *et al.* (1975) and illustrated in Figure B2.2 are likely to prove more applicable in this respect.

Some countries possess terrain classification mapping for all or part of their territory (e.g. Verstappen 1983; Varnes 1984). However, many road projects will not have the benefit of this existing source of information and a project-specific terrain classification will therefore need to be established. The development of terrain models could form the basis of this classification; Figure B2.12 illustrates a terrain classification developed for a route selection study in Nepal. The figure is an extract of a larger mapping output (Hearn & Lawrance 2000) and has been simplified

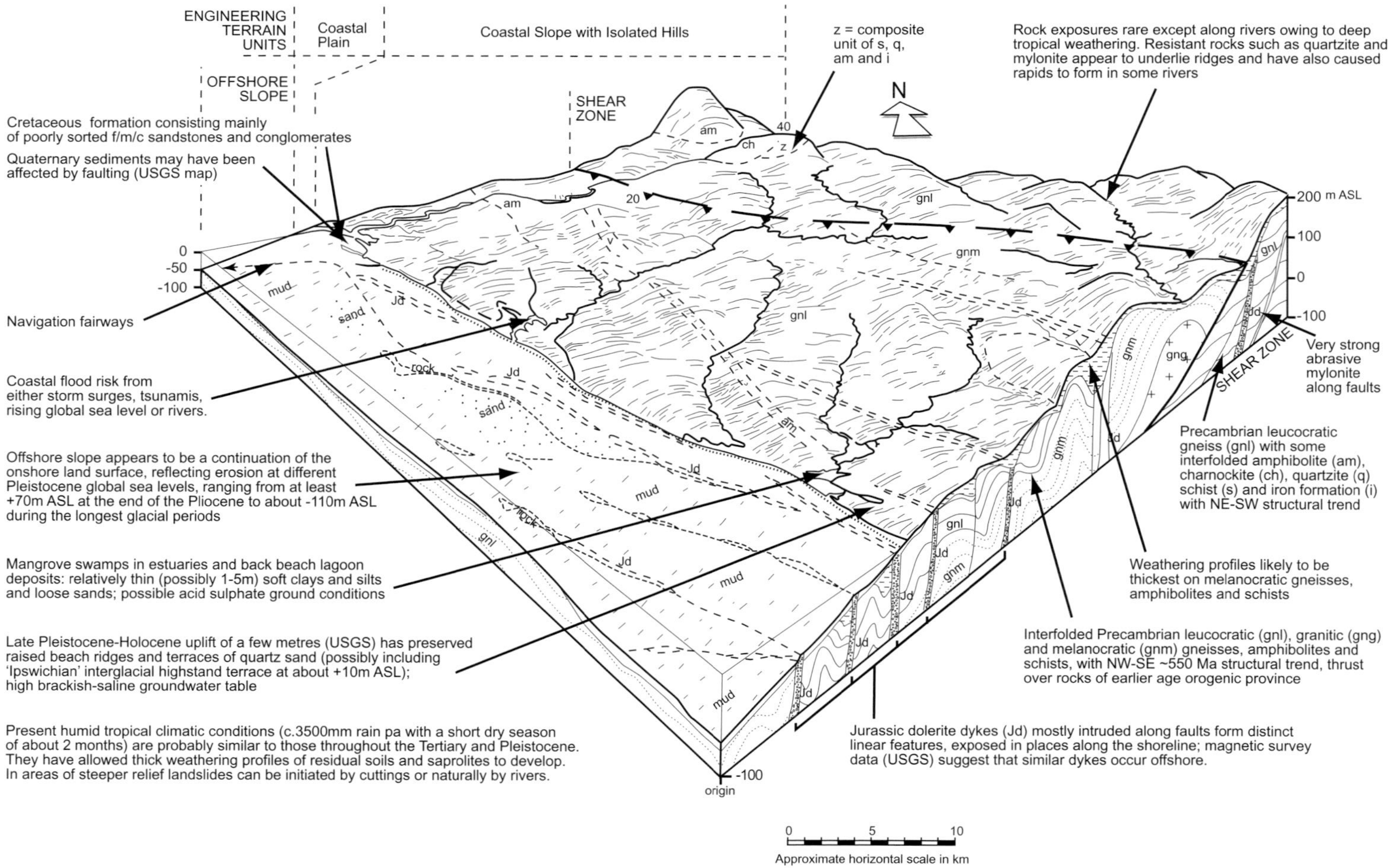

Fig. B2.9. Landscape evolution model for a route corridor feasibility study, West Africa (drawn by G. Pettifer.)

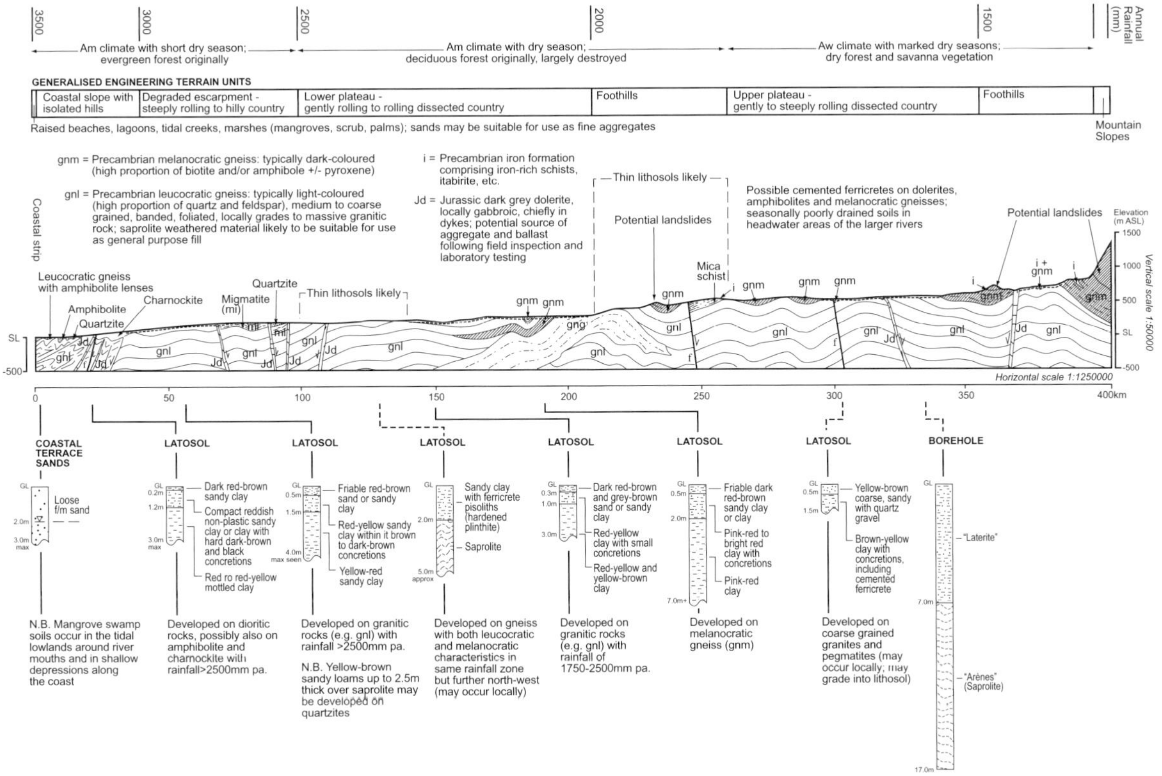

Fig. B2.10. Geological model developed for a specific alignment corridor in West Africa (drawn by G. Pettifer).

to be legible at the scale portrayed here. The classification is based principally on basic underlying geological structure, slope materials and geomorphological process, rather than the conventional terrain systems approach. It was used to estimate lengths of alignment at risk from landslides, fan crossings, flooding and erosion and to determine the proportion of rock, taluvium and residual soil likely to be encountered in road excavations for corridor comparison purposes. The mapping was derived essentially from published small-scale geological maps and the interpretation of aerial photography together with field verification.

Lawrance *et al.* (1993, p. 21) recommend that a terrain classification should '... start with an overview of the site and work towards a concentrated effort in the area in which the construction is to take place. Once set up a terrain classification can act as a referencing system for geotechnical data collected throughout the project period.' The development of reference conditions (CIRIA 1978) and their portrayal in *reference condition mapping* (e.g. Baynes *et al.* 2005) illustrates this process. As this is essentially a field-based application, reference condition mapping is described and illustrated in Section B3.2.

B2.6 Landslide susceptibility, hazard and risk maps

A decision will need to be made early on as to whether project-specific landslide mapping is to be developed and, if so, what mapping method to adopt. Methods of landslide susceptibility, hazard and risk mapping are described and illustrated in numerous publications such as those by Varnes (1984), Hansen (1984), Soeters & van Westen (1996), Aleotti & Chowdhury (1999), Guzzetti *et al.* (1999), Hearn & Griffiths (2001), Chacon *et al.* (2006) and Cotecchia *et al.* (2009). For the purpose of this discussion the following concepts are deemed to apply:

- landslide distribution maps depict the outline of existing landslides;
- landslide susceptibility maps provide an indication of where landslides are most likely to occur in the future;
- landslide hazard maps give an indication of the probability of landslides of a given size and extent or speed of movement occurring over a given time, such as the

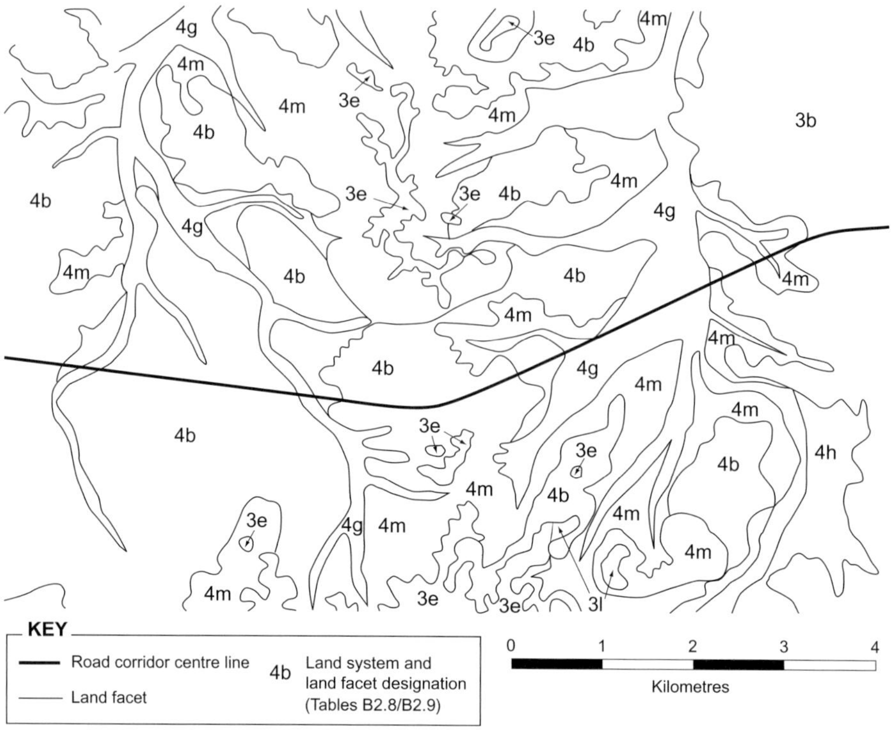

Fig. B2.11. Example of terrain classification mapping in an arid landscape (modified from Hunt 1979).

design life of a road, and therefore indicate the potential to cause damage; and

- landslide risk maps show the actual damage or loss that has occurred or is likely to occur as a result of landslides and ground movement taking place over a given period, and consider the value and vulnerability of road and other assets at risk.

According to the above and the discussion in Section A4.2, the majority of published hazard maps could be described more accurately as susceptibility maps; few published risk maps portray true risk and may not even qualify as hazard maps in the strictest sense of the term. Baynes (1997, p. 153) observed the need for 'some practical guidelines for the application of quantitative landslide risk assessment

Table B2.8. *Land facet designations (see Fig. B2.11)*

Land facet designation	*Type*	*Land facet designation*	*Type*
a	Mesa	k	Sand dunes
b	Stepped plateau	l	Coarse piedmont
c	Rolling plateau/plain	m	Footslope piedmont
d	Undulating plateau	n	Alluvial fan
e	Dissected plateau	p	High hills and ridges
f	Weathered dissected plateau	q	Low ridges, cols, saddles and foothills
g	Wadi plain	r	Localized drainage basin
h	Desert plain (stone desert)	s	Sabka
j	Desert plain (wind-eroded ridges and dissected rock)		

Table B2.9. *Extract from land facet table detailing engineering attributes (see Table B2.8 and Fig. B2.11)*

Land facet	*Form*	*Soils and materials*	*Engineering properties and comments*
4 m	Foot slope piedmont. Consists of gently sloping surfaces which link an upland to a lowland facet	Colluvial/taluvial soils of variable thickness overlying a stepped rock surface. The soil profile generally comprises coarser grained gap-graded cobbles, gravels and sands at the upper levels and finer grained materials at the lower levels. However, extensive intermediate levels can contain appreciable quantities of clayey material where the parent rocks are marl.	AASHTO (2004) classification group very variable depending on whether the material has been predominantly derived from the stronger limestone beds (generally A-1-b to A-2-4) or the more extensive weaker marls (generally A-4 to A-6).

methods ... to assist general practitioners in the future.' van Westen *et al.* (2005, p. 167) summarized the situation with risk mapping by concluding that '... the generation of quantitative risk zonation maps for regulatory and development planning by local authorities still seems a step too far ...'. Jaiswal *et al.* (2010, p. 1185) note that 'numerous publications are available that deal with the concept and possible methods to carry out risk analysis ... but the number of publications on the actual implementation of spatial landslide risk assessment in specific cases is still rather modest.' Quantitative risk assessments have been carried out in areas where the requisite data exist, such as by Ko Ko *et al.* (2003) in New South Wales for existing railway alignments and Bonachea *et al.* (2009) for infrastructure, buildings and land use in northern Spain. However, there appear to be few, if any, published maps that portray true landslide risk over large areas. Lee (2009, p. 445) concludes that a 'healthy scepticism is needed when using the results from a landslide risk assessment.'

As far as route selection is concerned, landslide distribution and susceptibility maps are of greatest value in helping to identify those corridors that are likely to be the most stable, both now and in the future. The former are usually derived from aerial photograph or LiDAR interpretation (Section B2.2) and field mapping (Section B3.4). The latter are most commonly developed by analysing the distribution of landslides in relation to landslide conditioning and triggering agents or factors (Section A3.3) and deriving a composite map showing those slopes that are considered to be more susceptible to future instability than others.

Figure B2.13 illustrates an extract of a simple landslide susceptibility mapping exercise developed for route corridor comparison in Nepal. The study area comprised three main drainage catchments covering a total area of 102 km^2 with 369 landslides mapped from aerial photographs. The variation in landslide density across the study area was assessed according to its relationship with mapped conditioning factors, including rock type, slope aspect, physiography, land use, slope angle and channel proximity. The statistical significance of the relationships was calculated using the Chi^2 test (e.g. Hammond & McCullagh 1978), whereby the observed landslide distribution (O) for each factor is compared with that which would be expected (E) from a random distribution. In the case of rock type, slope aspect and physiography, the highest value of O/E was four times or more greater than the lowest, demonstrating that the distribution of landslides was significantly correlated with the variation in these factors. The susceptibility rankings for each of these factors were overlain and a composite susceptibility map produced. The other factors (land use, slope angle and channel proximity) had lower O/E differentials and these factors failed the Chi^2 significance test. Slope angle had been measured using a parallax bar and stereo aerial photographs for each of a total of 2300 grid squares. This method did not always coincide with the morphological boundaries of the terrain and this was probably the main reason why slope angle was not significantly correlated with landslide distribution. However, because slope angle formed a fundamental element of the physiography classification, the overall susceptibility model was considered to be valid.

Although some ground verification was carried out, the study was undertaken as a desk-based exercise using published geological maps, topographical maps and stereo aerial photographs, and took two man-months to complete the entire 102 km^2 area (approximately 3 weeks for the extract shown in Fig. B2.13). It was undertaken several years after road construction as part of a research exercise (Hearn 1987). This was prior to the advent of GIS, a tool that would have increased the efficiency of the work still further (Section B2.7).

The constructed alignment had been selected by reconnaissance survey (Brunsden *et al.* 1975) and generally follows zones of moderate and highest stability defined by the susceptibility mapping. Two large and deep-seated slope failures took place during and soon after road construction (Fig. B2.14) and these were located either within or very close to the two least stable areas identified on the alignment by the susceptibility mapping. The full potential for slope failure at these locations had not been recognized during alignment design (Hearn 2002*a*).

The credibility of landslide susceptibility maps is called into question when the available information to derive

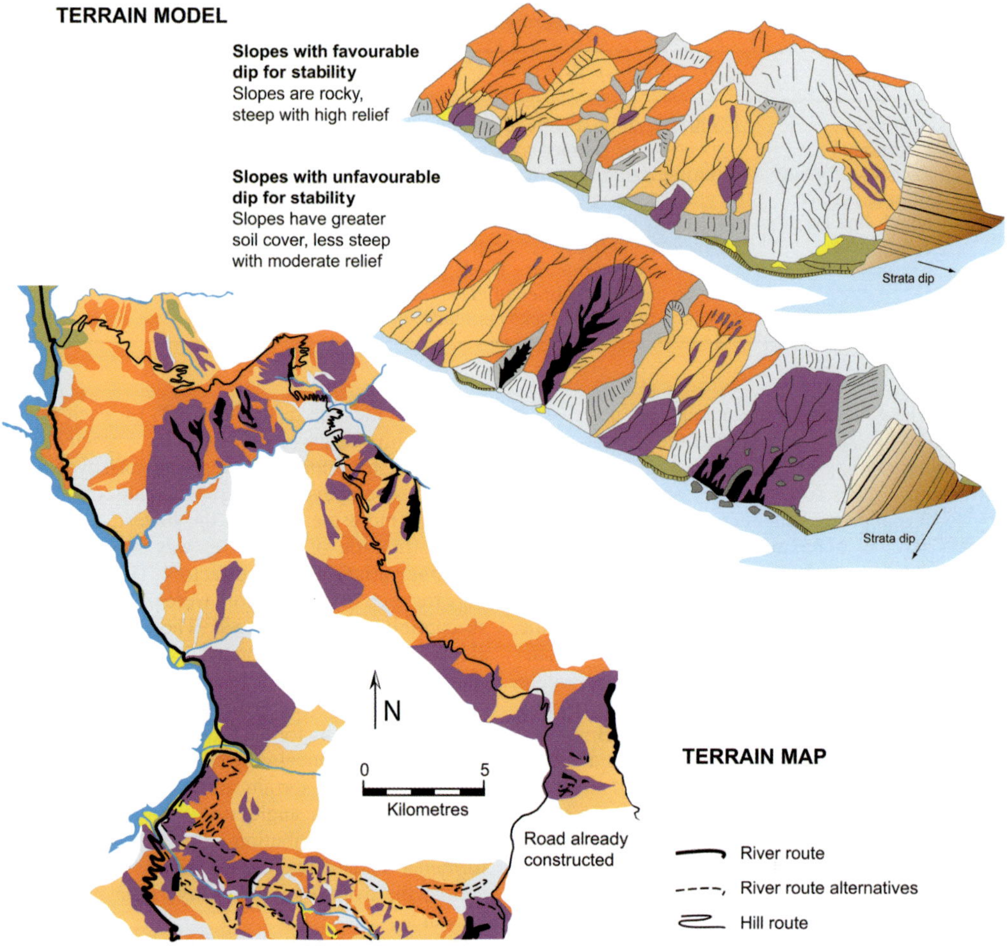

	LANDFORM TYPE	LANDFORM SUB-TYPE*
(orange)	Residual soil slopes	• Rounded ridge and spur summits, and gently sloping ground with residual soil cover • Mid slope erosional benches with residual soil cover • Structurally-controlled dip slopes with residual soil cover
(light grey)	Rock slopes: miscellaneous	• Miscellaneous rock cliffs • Rock cliffs forming landslide scarps • Irregular steep slopes (up to 45°)
(grey)	Rock slopes: structurally-controlled	• Bedding, foliation or joint-plane controlled slopes • Structural bench • Structurally-controllled ridge and ravine topography
(light orange)	Transported soil slopes	• Taluvium and colluvium (undifferentiated)
(black) / (purple)	Unstable slopes Previously failed slopes	• Mudslides, debris slides and rockslides on steep slopes adjacent to eroding rivers and streams • Mudslides and debris slides associated with local erosion and high groundwater • Large, deep-seated slope failures usually with adverse structural-control
(yellow)	Fan deposits	• Active fan surface, usually on flood plains • High level fan deposits, usually associated with high level terraces or deposited by catastrophic landslides
(olive)	Terrace gravels	• Flood plain terrace (periodically inundated). • Terrace flank • High level terrace

*In the original terrain classification and mapping these landform sub-types were shown separately but for the sake of clarity at this scale they are shown combined

Fig. B2.12. Terrain classification for route selection, Nepal. Modified from TRL 1997.

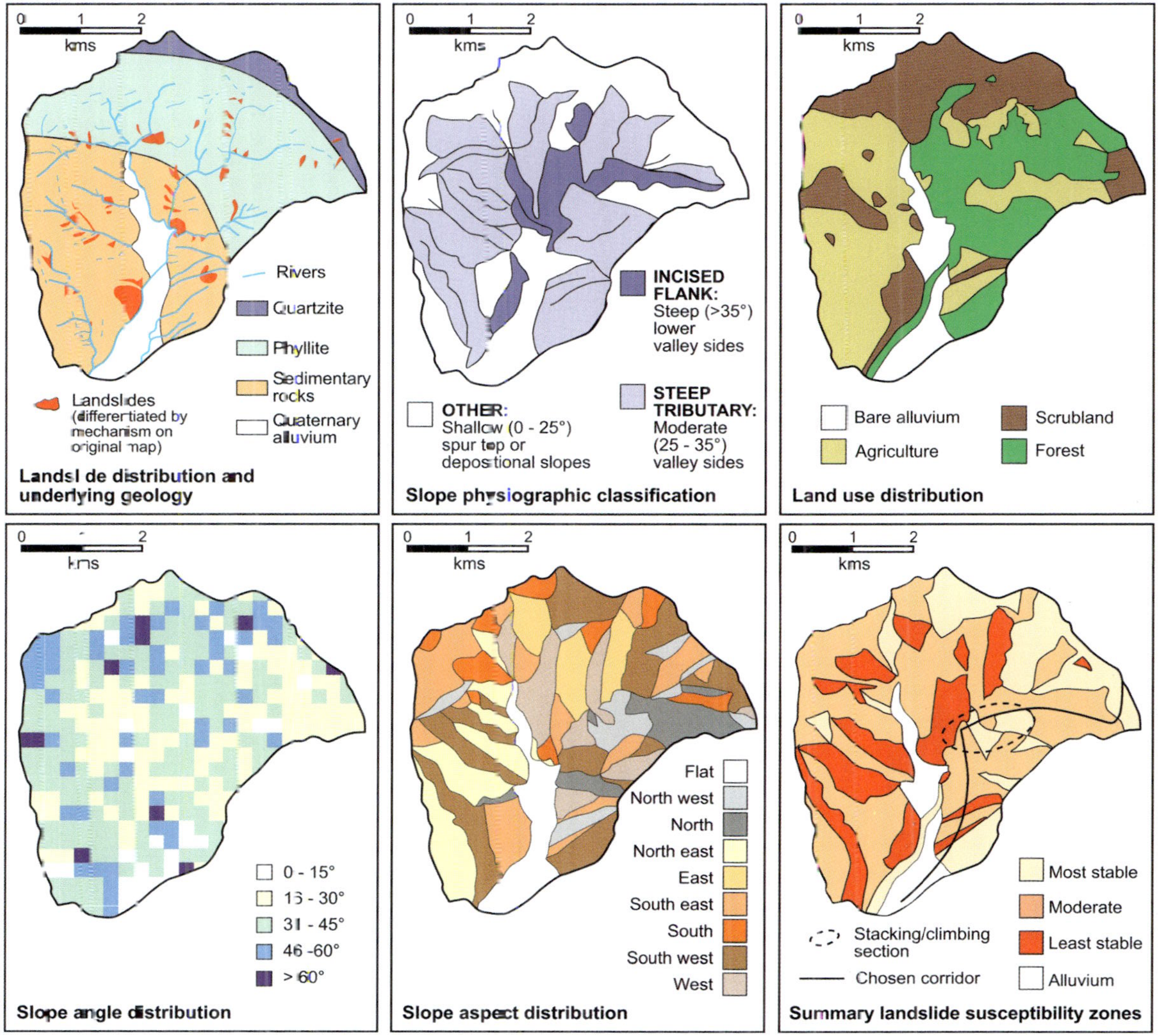

FACTOR	CATEGORY	O/E	SUSCEPTIBILITY RANK
ROCK TYPE	Sedimentaries	1.2	3
	Phyllite	0.8	2
	Quartzite	0.3	1
SLOPE ASPECT	North	0.7	1
	South	2.0	2.5
	East	2.2	3
	West	0.7	1
	North west	0.6	0
	North east	0.9	2
	South east	0.9	2
	South west	0.5	0
PHYSIO-GRAPHY	Incised flank	2.9	3
	Steep tributary	1.1	2
	Other	0.3	1
LAND USE	Scrub	1.0	NOT SIGNIFICANT
	Agriculture	1.1	NOT SIGNIFICANT
	Forest	0.9	NOT SIGNIFICANT
SLOPE ANGLE	0 - 15°	0.7	NOT SIGNIFICANT
	16 - 30°	1.0	NOT SIGNIFICANT
	31 - 45°	1.3	NOT SIGNIFICANT
	46 - 60°	0.8	NOT SIGNIFICANT
	> 60°	0.4	NOT SIGNIFICANT
CHANNEL PROXIMITY	Stream rank:		NOT SIGNIFICANT
	First order	0.9	NOT SIGNIFICANT
	Second order	1.1	NOT SIGNIFICANT
	Third order	1.5	NOT SIGNIFICANT
	Fourth order	0.9	NOT SIGNIFICANT
	Fifth order	1.0	NOT SIGNIFICANT

Build up of susceptibility rank for the illustrated catchment. Susceptibility ranks for each factor category are summed for every terrain unit and assigned to one of three susceptibility classes. 3 is the most unstable condition.

Fig. B2.13. Illustration of landslide susceptibility mapping for route corridor comparison, Nepal. Modified from Hearn 1987.

Fig. B2.14. Post-construction landslide that occurred in one of the least stable areas shown on Figure B2.13.

them is limited, giving rise to an incomplete analysis (see discussion in Huabin *et al.* 2005 and Petley 2010, for example). Methods of 'data mining' (Miner *et al.* 2010), whereby known landslide distributions are compared geographically with whatever information is available, offer some possibilities as long as any relationships that emerge can be considered to be genuine cause and effect (Huabin *et al.* 2005). While general guidelines on the preparation of landslide susceptibility maps are given in, for example, Fell *et al.* (2008), the detailed methods applied are, to an extent, controlled by the information that is available and any pre-conceived views on the factors that control slope stability. Soralump (2009), for example, describes how four different organizations in Thailand have applied five different approaches to susceptibility mapping in the country, illustrating the lack of a common procedure. Guzzetti *et al.* (2006) suggest a method for assessing the quality of a landslide susceptibility assessment based on the type of information available, estimates of uncertainty and the degree to which the predictive model reflects the landslide record. Comprehensive landslide susceptibility mapping studies are described, for example, by Nagarajan *et al.* (2000) and Acharya *et al.* (2006).

In steep and complex terrain, it is difficult to select an alignment that crosses the most stable terrain in its entirety, and usually some compromises need to be made (Section C1.2.5). In these circumstances, a comparison of the hazard posed by ground conditions along each option will be required. This will include, for example, the likelihood of activated or reactivated movement and the depth and extent of movement when it does occur. In mountain areas landslides can travel considerable distances downslope, thus significantly extending their potential impact zone (especially if they become channelized; Section A1.4). Methods for assessing landslide runout and debris flow travel distances are usually based on empirical relationships using the mechanism, volume, topographic conditions and observed displacements of past events (e.g. Scheidegger 1973; Hsü 1975; Corominas *et al.* 1988; Sousa & Voight 1991; Corominas 1993, 1996; Evans & Hungr 1993; Hearn 1995; Evans & King 1998; Hadley *et al.* 1998; Franks 1999; Dai *et al.* 2002; Hungr *et al.* 2005; Fell *et al.* 2007; Fannin & Bowman 2010) and analytical models taking rheological properties, flow resistance, entrainment and other factors into consideration (Hungr 1995; Hürlimann *et al.* 2008). The assessment of future landslide runout requires knowledge of the location and volume of the future landslide, the volume of material likely to be entrained as the landslide travels downstream (Fannin & Bowman 2010) and the location of the landslide with respect to downslope topography. In the common absence of such data, a judgement-based approach is likely to be required (Hadley *et al.* 1998).

Figure B2.15 illustrates a landslide susceptibility, hazard and risk mapping exercise undertaken as part of the same study in Nepal referred to in Section B2.3 (Hearn 2002*b*; Hart *et al.* 2003). Landslides were mapped from aerial photograph interpretation and their distribution was compared

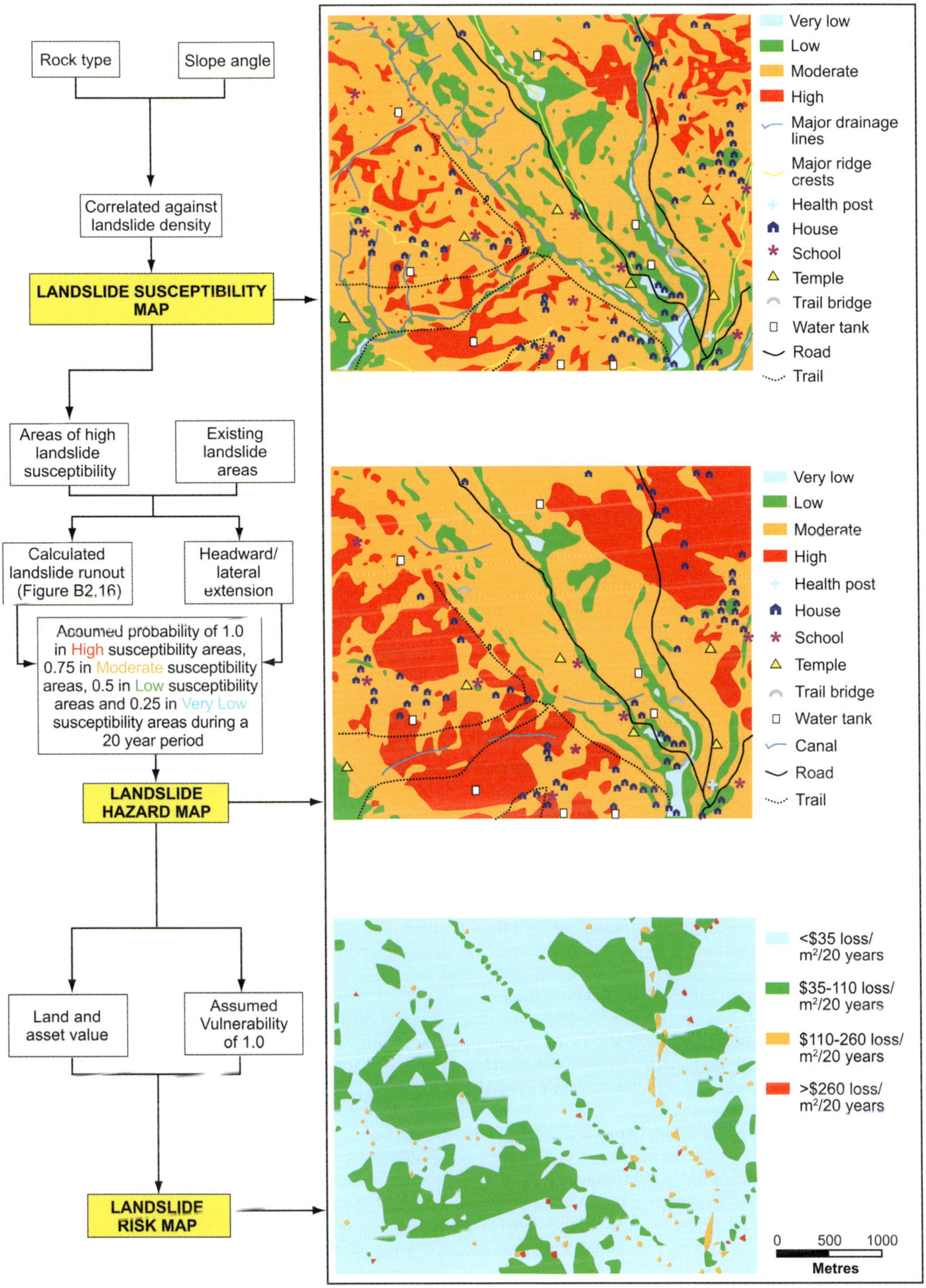

Fig. B2.15. Landslide susceptibility, hazard and risk mapping for route corridors, Nepal (see text discussion).

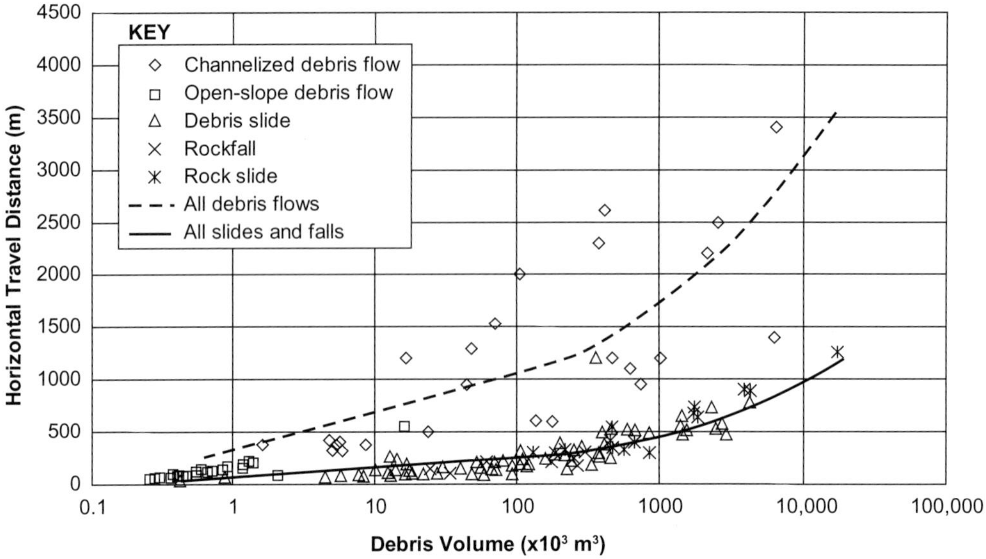

Fig. B2.16. Observed landslide runout distances according to failure mechanism and volume.

with fault pattern, rock type, soil type, slope angle, land use, rainfall distribution and drainage pattern. These factors were derived from published sources, remote sensing and field observations. Rock type and slope angle were found to be the factors that were most closely correlated with landslide locations. A susceptibility map was produced that indicated where, based on these two factors, future landslides might be expected to occur (Fig. B2.15). It is interesting to note on Figure B2.15 that virtually all of the housing areas, schools, trails and roads are located outside of those areas mapped as high susceptibility to landslides. This may be purely coincidental or, more likely, it may reflect community knowledge of which slopes have the greatest potential to become unstable.

The landslide database contained details of scarp location, landslide displacement and runout distance and estimated volume and topographic slope for each mapped landslide. Empirical analyses were undertaken to derive runout curves according to failure mechanism and volume (Fig. B2.16). The relationship between failure volume and travel distance is approximately similar for debris slides, rockfalls and rockslides, whereas debris flows show a much greater scatter in the relationship with some very long travel distances being recorded. These curves were then used to compute anticipated future runout distances from the high susceptibility areas using averaged landslide volumes. This was combined with the average rates of upslope and lateral regression (headward and lateral extension of the landslide scarp) to derive a map that showed the total areas anticipated to be affected by future landslide activity. Figure B2.15 shows that, while some housing areas are within the high hazard zones, the majority are outside; again, this could reflect local community knowledge as to where the safest parts of the landscape are likely to be.

In order to complete the assessment of hazard, an attempt was made to determine landslide frequency and hence probability. The landslide database contained information on the approximate date or date range for some of the mapped landslides, as determined from aerial photography and community knowledge. The temporal distribution of landslides was compared with the seismic and rainfall records, but both records were insufficient to provide any conclusive indication of the frequency of events and hence the annual probability of landsliding. Consequently, a probability of 1.0 was assumed for slope failure in all high landslide susceptibility areas over a 20 year period, approximately equivalent to the design life of a low-cost road in Nepal. For the study area this assumption was considered reasonable and, in the absence of information to the contrary, was extended to include probabilities of failure of 0.75, 0.5 and 0.25 in moderate, low and very low susceptibility areas, respectively.

Discussions were held with village and rural development committees and with road authorities to determine the monetary value of the various land uses and engineering structures in the study area. Using these values, a risk map was prepared on the assumption that total value loss would occur if impacted by a future landslide. This is a pessimistic assumption given that most agricultural land can eventually be re-established following inundation by landslide debris. Nevertheless, for the purpose of the exercise, a vulnerability of 1.0 was assumed in all cases (Fig. B2.15).

The illustrations provided in Figures B2.13 and B2.15 are simple manipulations of desk study and field-derived data

Table B2.10. *Typical uses of GIS*

Required data	*Potential data sources*
Planning of new roads	
Land use and infrastructure for route corridor selection	Published small-scale topographical maps if available digitally. If not they may require to be digitized. Large scale remote sensing can be used to update existing maps using GIS. Surveyed cadastral maps can form a separate mapping layer
Small-scale topographical data for route corridor selection	Published small-scale topographical maps if available digitally. If not they may require to be digitized
Large-scale topographical data for alignment design and design of cross-sections	Project-derived photogrammetry or LiDAR and site survey
Geology for route corridor selection	Published small-scale geological maps. Usually not available digitally. Can be digitised into GIS and augmented by remote sensing data
Landslide locations for route corridor selection	Remote sensing can be interpreted on-screen or digitised, scanned and geo-referenced into GIS. Landslide locations from field observations can be geo-referenced into GIS
Detailed information on rock and soil types for alignment design and design of cross-sections	Project-derived field mapping and ground investigation. Not usually digitized, but scanned and geo-referenced into GIS
Detailed information on landslides for alignment design and design of cross-sections	
Drainage details (catchment areas, rivers, streams etc)	Digitized from topographical maps, less commonly from remote sensing
Information on rainfall and seismicity	Can be stored and analysed using GIS
Slope management along existing roads (see Part D)	
Locations of cut and fill slopes, retaining walls and drainage structures	Inventory of earthworks and structures and their condition can be collected and stored and analysed using GIS in both spreadsheet and graphical format
Condition of cut and fill slopes, retaining walls and drainage structures	
Soil and rock types in earthworks	Project-derived field mapping, slope inventory and ground investigation. Not usually digitized, but scanned and geo-referenced into GIS
Landslide locations	Project-derived field mapping, supplemented with remote sensing can be digitized or scanned and geo-referenced into GIS
Landslide and slope stability	Landslide inventories can be stored as spreadsheets, geo-referenced and analysed in GIS
Locations of slope monitoring sites	Monitoring locations can be surveyed and stored in GIS as a separate mapping layer
Results of slope monitoring	Monitoring spreadsheets can be stored, geo-referenced and analysed in GIS
Hazard and risk data for maintenance management	Assessment of hazard and risk can be derived from mapping and inventories, stored and analysed in GIS for prioritizing maintenance

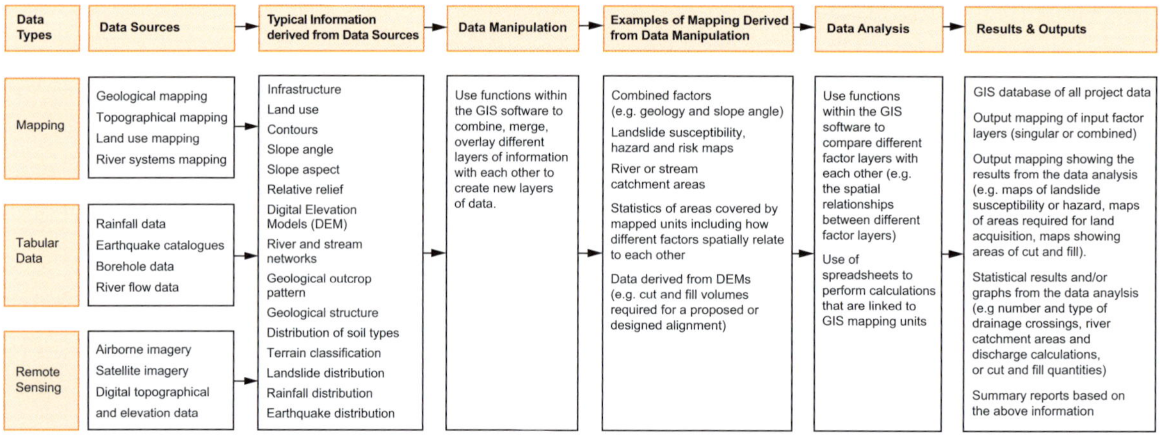

Fig. B2.17. Flowchart of GIS activities for terrain and landslide assessment.

that have been combined with assumptions and generalizations to yield output landslide susceptibility, hazard and risk maps that offer assistance in the selection of route corridors. From the point of view of route corridor comparison, the landslide hazard map illustrated in Figure B2.15 offers the greatest value as not only does it show areas of high susceptibility but it also provides an indication of their potential displacement and runout if and when they do fail. However, several months of work were required to derive these maps and significant levels of uncertainty still remained. Consequently, in the context of the usual short time frame and limited resources available for route corridor comparison for low-cost roads, it will probably prove more cost-effective at present to rely on the conventional methods of remote sensing (Section B2.2) and field reconnaissance surveys (Section B3.1 and Fig. B2.2) together with simple landslide susceptibility mapping where the requisite data are available.

Landslide hazard mapping is discussed further in Section B3.3 in relation to field-based activities.

B2.7 Geographical information systems (GIS) applications

B2.7.1 Introduction

A GIS is a powerful and important computer-based tool for the storage, management and analysis of spatial data (e.g. Chacon *et al.* 2006; Morgenstern & Martin 2008; van Westen 2010). The use of GIS has grown rapidly as the availability of powerful, low-cost computers and the development of user-friendly software systems have also grown. This means that GIS is now a tool that is applicable to a wide range of projects including the route selection, design, construction and maintenance of low-cost roads.

A GIS can be used to digitally represent and analyse the geographical features present on the Earth's surface (e.g. topography, infrastructure, land use, geological units) and attribute data linked to those features such as specific details about the infrastructure, land use, landslides and rock types. The use of a GIS allows the integration of common database operations such as query and statistical analysis with the geographical visualization benefits offered by maps. These abilities distinguish GIS from other databases and make it a valuable tool for collecting, managing, analysing and displaying data in a range of scales and formats.

A GIS can be used to combine data from a variety of sources including:

- published mapping;
- satellite and aerial imagery;
- digital topographical and elevation data; and
- any tabular data that have or can be given a spatial context (e.g. earthquake or rainfall data).

One of the strengths of a GIS is that, once it has been set up, it can be easy to maintain and develop as new data become available.

B2.7.2 Using GIS

Table B2.10 summarizes the issues that need to be addressed when considering using GIS technology for low-cost roads. Figure B2.17 illustrates how various data sources can be managed and analysed in order to yield terrain and landslide assessments using a GIS. These applications are best suited in their entirety to new road construction projects, although parts are also relevant to road improvement schemes and the management of slopes along existing roads (Part D).

References

ACHARYA, G., DE SMEDT, F. & LONG, N. T. 2006. Assessing landslide hazard in GIS: a case study from Rasuwa, Nepal. *Bulletin of Engineering Geology and the Environment*, **65**, 99–107.

ALEOTTI, P. & CHOWDHURY, R. 1999. Landslide hazard assessment: summary review and new perspectives. *Bulletin of Engineering Geology and the Environment*, **58**, 21–44.

ANON 1972. The preparation of maps and plans in terms of engineering geology. Report by the Geological Society Engineering Group Working Party. *Quarterly Journal of Engineering Geology*, **5**, 295–382.

BAYNES, F. 1997. Problems associated with geological characterisation for quantitative landslide risk assessment. *In*: CRUDEN, D. & FELL, R. (eds) *Landslide Risk Assessment*. Balkema, Rotterdam, 153–163.

BAYNES, F. J., FOOKES, P. G. & KENNEDY, J. F. 2005. The total engineering geological approach applied to railways in the Pilbara, Western Australia. *Bulletin of Engineering Geology and the Environment*, **64**, 67–94.

BONACHEA, J., REMONDO, J., GONZALEZ-DIAZ, A., DIAZ DE TARAN, J. R. & CENDRERO, A. 2009. Landslide risk modelling: an experience from northern Spain. *In*: MALET, J. P., REMAITRE, A. & BOGAARD, T. (eds) *Landslide Processes: From Geomorphological Mapping to Dynamic Modelling*. Strasbourg European Centre on Geomorphological Hazards (CERG), Strasbourg, 259–264.

BOOTH, A. M., ROERING, J. J. & PERRON, J. T. 2009. Automated landslide mapping using spectral analysis and high-resolution topographic data: Puget Sound lowlands, Washington, and Portland Hills, Oregon. *Geomorphology*, **109**, 132–147.

BRUNSDEN, D., DOORNKAMP, J. C., FOOKES, P. G., JONES, D. K. C. & KELLY, J. M. H. 1975. Large scale geomorphological mapping and highway engineering design. *Quarterly Journal of Engineering Geology*, **8**, 227–253.

CAMPBELL, J. B. 2007. *Introduction to Remote Sensing*. 4th edn. Taylor and Francis, London.

CHACON, J., IRIGARAY, C., FERNANDEZ, T. & EL HAMDOUNI, R. 2006. Engineering geology maps: landslides and geographical information systems. *Bulletin of Engineering Geology and the Environment*, **65**, 341–411.

CHIGIRA, M., DOSHIDA, S. & NAKAMURA, T. 2010. Morphological features of slope movement identified by the airborne laser scanner. *In*: WILLIAMS, A. L., PINCHES, G. M., CHIN, C. Y., MCMORRAN, T. J. & MASSEY, C. I. (eds) *Geologically Active. Proceedings of the 11th IAEG Congress, Auckland, New Zealand, 5–10 September, 2010*, Taylor and Francis Group, London, 103–110.

CIRIA 1978. Tunnelling – improved contract practices. Construction Industry Research and Information Association, Report 79, London.

COROMINAS, J. 1993. Spatial prediction of landslides. *In*: NEMEC, J., NIGG, J. M. & SICCARDI, F. (eds) *Prediction and Perception of Natural Hazards*. Kluwer Academic Publishers, Dordrecht, Holland, 125–132.

COROMINAS, J. 1996. The angle of reach as a mobility index for small and large landslides. *Canadian Geotechnical Journal*, **33**, 260–271.

COROMINAS, J., PENARANDA, R. & BAEZA, C. 1988. Identificacion de factores que condicionan la formacion de movimientos superficiales en los valles altos del Llobregat y Cardener. *II Symp Taludes y Laderas Inestables*, Andorra La Vella, 195–207.

COTECCHIA, F., LOLLINO, P., SANTALOIA, F., VITONE, C. & MITARITONNA, G. 2009. A research project for deterministic landslide risk assessment in southern Italy: methodological approach and preliminary results. *In*: HONJO, Y., HARA, T., SUZUKI, M. & ZUANG, F. (eds) *Geotechnical Risk and Safety*. Taylor & Francis Group, London, 363–370.

CRUDEN, D. M. & VARNES, D. J. 1996. Landslide types and processes. *In*: TURNER, A. K. & SCHUSTER, R. L. (eds) *Landslide Investigation and Mitigation*. Transportation Research Board, National Research Council, National Academy Press, Washington DC, Special Report, **247**, 36–75.

DAI, F. C., LEE, C. F. & NGAI, Y. Y. 2002. Landslide risk assessment and management: an overview. *Engineering Geology*, **64**, 65–87.

DAS, I., SAHOO, S., VAN WESTEN, C., STEIN, A. & HACK, A. 2010. Landslide susceptibility assessment using logistic regression and its comparison with a rock mass classification system, along a road section in the northern Himalayas (India). *Geomorphology*, **114**, 627–637.

DOWLING, J. W. F. 1968. The classification of terrain for road engineering purposes. *Conference on Civil Engineering Purposes Overseas, Session V, Communications Paper 16*, Institution of Civil Engineers, London.

DUMBLETON, M. J. & WEST, G. 1974. *Guidance on Planning, Directing and Reporting Site Investigations in Britain*. Transport and Road Research Laboratory, **LR625**, Crowthorne, UK.

DUNNING, S. A., MASSEY, C. I. & ROSSER, N. J. 2009. Structural and geomorphological features of landslides in the Bhutan Himalaya derived from Terrestrial Laser Scanning. *Geomorphology*, **103**, 17–29.

ELLIS, L. A., HARRISON, E. & BOWDEN, A. J. 2011. Lanslides on Gault: geomorphological identification and qualitative risk assessment. *Quarterly Journal of Engineering Geology and Hydrogeology*, **44**, 35–48.

EVANS, N. C. & KING, J. P. 1998. The Natural Terrain Landslide Study: Debris Avalanche Susceptibility. Technical Note No TN 1/98. Geotechnical Engineering Office, Hong Kong.

EVANS, S. G. & HUNGR, O. 1993. The assessment of rock fall hazards at the base of talus slopes. *Canadian Geotechnical Journal*, **30**, 620–636.

FANNIN, R. J. & BOWMAN, E. T. 2010. Debris flows – entrainment, deposition and travel distance. *In*: WILLIAMS, A. L., PINCHES, G. M., CHIN, C. Y., MCMORRAN, T. J. & MASSEY, C. I. (eds) *Geologically Active. Proceedings of the 11th IAEG Congress, Auckland, New Zealand, 5–10 September 2010*, Taylor and Francis Group, London, 1111–1118.

FELL, R., GLASTONBURY, J. & HUNTER, G. 2007. Rapid landslides: the importance of understanding mechanisms and rupture surface mechanics. The Eighth Glossop Lecture. *Quarterly Journal of Engineering Geology and Hydrogeology*, **40**, 9–27.

FELL, R., COROMINAS, J., BONNARD, C., CASCINI, L., LEROI, E. & SAVAGE, W. Z. 2008. Guidelines for landslide susceptibility, hazard and risk zoning for land use planning. JTC-1 Joint Technical Committee on Landslides and Engineered Slopes. *Engineering Geology*, **102**, 85–98.

FOOKES, P. G. 1997. Geology for engineers: the geological model, prediction and performance. The First Glossop Lecture. *Quarterly Journal of Engineering Geology*, **30**, 293–424.

FOOKES, P. G. & MARSH, A. H. 1981. Some characteristics of construction materials in the low to moderate metamorphic grade rocks of the Himalayas of east Nepal. 1: Occurrence and geological features. *Proceedings of the Institution of Civil Engineers*, Part 1, **70**, 123–138.

FOOKES, P. G., BAYNES, F. J. & HUTCHINSON, J. N. 2000. Total geological history: a model approach to the anticipation, observation and understanding of site conditions. *Proceedings of the International Conference on Geotechnical & Geological Engineering, GeoEng 2000*, Melbourne, Technomic Publishing, Basel.

FRANKS, C. A. M. 1999. Characteristics of some rainfall-induced landslides on natural slopes, Lantau Island, Hong Kong. *Quarterly Journal of Engineering Geology*, **32**, 247–259.

GRIFFITHS, J. S. & STOKES, M. 2008. Engineering geomorphological input to ground models: an approach based on Earth systems. *Quarterly Journal of Engineering Geology and Hydrogeology*, **41**, 73–91.

GRIFFITHS, J. S., MATHER, A. E., TRUSCOTT, J. B. & HARTLEY, A. 2010. Application of low-cost satellite imagery for reconnaissance mapping and terrain evaluation: a case study from coastal landslides in Chile. *In*: WILLIAMS, A. L., PINCHES, G. M., CHIN, C. Y., MCMORRAN, T. J. & MASSEY, C. I. (eds) *Geologically Active. Proceedings of the 11th IAEG Congress, Auckland, New Zealand, 5–10 September 2010*, Taylor and Francis Group, London, 4145–4152.

GUZZETTI, F., CARRARA, A., CARDINALI, M. & REICHENBACH, P. 1999. Landslide hazard evaluation: a review of current techniques and their application in a multi-scale study, Central Italy. *Geomorphology*, **31**, 181–216.

GUZZETTI, F., REICHENBACH, P., ARDIZZONE, F., CARDINALI, M. & GALLI, M. 2006. Estimating the quality of landslide susceptibility models. *Geomorphology*, **81**, 166–184.

HADLEY, D., HEARN, G. J. & TAYLOR, G. R. 1998. Debris flow assessments for the Foothills Bypass, Hong Kong. *In*: LI, K. S., KAY, J. N. & HO, K. K. S. (eds) *Slope Engineering in Hong Kong. Proceedings of the Annual Seminar on Slope Engineering in Hong Kong*, Hong Hong Institution of Engineers, A. A. Balkema, Rotterdam, 153–162.

HAMMOND, R. & MCCULLAGH, P. S. 1978. *Quantitative Techniques in Geography: an Introduction*. Clarendon Press, UK.

HANSEN, A. 1984. Landslide hazard analysis. *In*: BRUNSDEN, D. & PRIOR, D. B. (eds) *Slope Instability*. J Wiley and Sons, New York, 523–602.

HART, A. B., HEARN, G. J., PETLEY, D. N., TIWARI, S. C. & GIRI, N. K. 2003. Using remote sensing and GIS for rapid landslide hazard assessment: potential public sector uptake in Nepal and Bhutan. *Proceedings of the PIARC (Permanent International Association for Roads Congress) Symposium on Sustainable Slope Management for Roads*, Kathmandu, March 2003.

HART, A. B., GRIFFITHS, J. S. & MATHER, A. E. 2009. Some limitations in the interpretation of vertical stereo photographic images for a landslide investigation. *Quarterly Journal of Engineering Geology and Hydrogeology*, **42**, 21–30.

HEARN, G. J. 1987. *An Evaluation of Geomorphological Contributions to Mountain Highway Design with Particular Reference to the Lower Himalaya*. PhD thesis, University of London.

HEARN, G. J. 1993. Rapid geomorphological reconnaissance survey for road alignment in west Nepal. *Bulletin of the International Association of Engineering Geology*, **48**, 59–68.

HEARN, G. J. 1995. Engineering geomorphological mapping and opencast mining in unstable mountains – a case study. *Transactions of the Institution of Mining & Metallurgy*, **A104**, A1–A18.

HEARN, G. J. 2002*a*. Engineering geomorphology for road design in unstable mountainous areas: lessons learnt after 25 years in Nepal. *Quarterly Journal of Engineering Geology and Hydrogeology*, **35**, 143–154.

HEARN, G. J. 2002*b*. Natural terrain hazard assessment: the art of applied science. *Proceedings of the Conference on Natural Terrain – a Constraint on Development?* Institution of Mining and Metallurgy, Hong Kong Branch, 39–60.

HEARN, G. J. 2004. The role of geology in landslide risk assessment for civil engineering purposes. *In*: JARDINE, R. J., POTTS, D. M. & HIGGINS, K. G. (eds) *Advances in Geotechnical Engineering. The Skempton Conference*. Thomas Telford, London, 1316–1329.

HEARN, G. J. & LAWRANCE, C. J. 2000. *Geomorphological Processes in Eastern Nepal as a Basis for Road Design*. Transport Research Laboratory, Project R6026, PR/INT/186/00, 71pp.

HEARN, G. J. & GRIFFITHS, J. S. 2001. Landslide hazard mapping and risk assessment. *In*: GRIFFITHS, J. S. (ed.) *Land Surface Evaluation in Engineering Practice*. Geological Society, London, Engineering Geology Special Publication, **18**, 43–52.

HSÜ, K. J. 1975. Catastrophic debris streams (sturzstroms) generated by rockfalls. *Geological Society of America, Bulletin*, **86**, 129–140.

HUABIN, W., GANGJUN, L., WEIYA, X. & GONGHUI, W. 2005. GIS-based landslide hazard assessment: an overview. *Progress in Physical Geography*, **29**, 548–567.

HUNGR, O. 1995. A model for the runout analysis of rapid flow slides, debris flows and avalanches. *Canadian Geotechnical Journal*, **32**, 610–623.

HUNGR, O., COROMINAS, J. & EHERHARDT, E. 2005. Estimating landslide motion mechanism, travel distance and velocity. *In*: HUNGR, O., FELL, R., COUTURE, R. & EHERHARDT, E. (eds) *Landslide Risk Management*. Taylor and Francis Group, London, 99–128.

HUNT, T. 1979. Geotechnical aspects of road design in Libya. *Ground Engineering*, October, **12**, 7, 15–19.

HÜRLIMANN, M., RICKENMANN, D., MEDINA, V. & BATEMAN, A. 2008. Evaluation of approaches to calculate debris-flow parameters for hazard assessment. *Engineering Geology*, **102**, 153–163.

JAISWAL, P., VAN WESTEN, C. J. & JETTEN, V. 2010. Quantitative assessment of landslide risk along transportation lines in southern India. *In*: WILLIAMS, A. L., PINCHES, G. M., CHIN, C. Y., MCMORRAN, T. J. & MASSEY, C. I. (eds) *Geologically Active. Proceedings of the 11th IAEG Congress, Auckland, New Zealand, 5–10 September, 2010*, Taylor and Francis Group, London, 1185–1193.

KASPERSKI, J. & VARREL, E. 2010. Characterisation of the active zone of the Sechilienne landslide using terrestrial laser scanning. *In*: WILLIAMS, A. L., PINCHES, G. M., CHIN, C. Y., MCMORRAN, T. J. & MASSEY, C. I. (eds) *Geologically Active. Proceedings of the 11th IAEG Congress, Auckland, New Zealand, 5–10 September, 2010*, Taylor and Francis Group, London, 1981–1988.

KO KO, C., FLENTJE, P. & CHOWDHURY, R. 2003. Quantitative landslide hazard and risk assessment: a case study. *Quarterly Journal of Engineering Geology and Hydrogeology*, **36**, 261–272.

LAWRANCE, C. J., BYARD, R. J. & BEAVAN, P. J. 1993. *Terrain Evaluation Manual*. State of the Art Review 7. Transport Research Laboratory, HMSO, London.

LEE, E. M. 2009. Landslide risk assessment: the challenge of estimating the probability of landsliding. *Quarterly Journal of Engineering Geology and Hydrogeology*, **42**, 445–458.

LILLESAND, T. M., KIEFER, R. W. & CHAPMAN, J. W. 2008. *Remote Sensing and Image Interpretation*. 6th edn. John Wiley and Sons, Chichester.

MEXE. 1969. Field Trials of a Terrain Classification System. Military Engineering Experimental Establishment, Report 995, Christchurch, Hants.

MINER, A., VAMPLEW, P., WINDLE, D. J., FLENTJE, P. & WARNER, P. 2010. A comparative study of various data mining techniques as applied to the modelling of landslide susceptibility on the Bellarine Peninsula, Victoria, Australia. *In*: WILLIAMS, A. L., PINCHES, G. M., CHIN, C. Y., MCMORRAN, T. J. & MASSEY, C. I. (eds) *Geologically Active. Proceedings of the 11th IAEG Congress, Auckland, New Zealand, 5–10 September, 2010*, Taylor and Francis Group, London, 1327–1336.

MITCHELL, C. W. 1973. *Terrain Evaluation*. Longmans, London.

MORGENSTERN, N. R. & MARTIN, C. D. 2008. Landslides: seeing the ground. *In*: CHEN, Z., ZHANG, J.-M., HO, K., WU, F.-Q. & LI, Z.-K. (eds) *Landslides and Engineered Slopes. From the Past to the Future. Proceedings of the 10th International Symposium on Landslides and Engineered Slopes, 30 June–4 July 2008, Xi'an, China*, Taylor & Francis Group, London, 3–23.

NAGARAJAN, R., ROY, A., VINOD KUMAR, R., MUKHERJEE, A. & KHIRE, M. V. 2000. Landslide hazard susceptibility mapping based on terrain and climatic factors for tropical monsoon regions. *Bulletin of Engineering Geology and the Environment*, **58**, 275–287.

PETLEY, D. N. 2010. Landslide hazards. *In*: ALCANTARA-AYALA, I. & GOUDIE, A. (eds) *Geomorphological Hazards and Disaster Prevention*. Cambridge University Press, Cambridge, 63–73.

PHIPPS, P. J. 2001. Terrain systems mapping. *In*: GRIFFITHS, J. S. (ed.) *Land Surface Evaluation for Engineering Practice*. Geological Society, London, Special Publication, **18**, 59–61.

RODRIGUEZ ORTIZ, J. M. & PRIETO, C. 1979. A proposal for quantitative terrain evaluation and highway construction. *Quarterly Journal of Engineering Geology*, **12**, 139–146.

ROSSER, N. J., PETLEY, D. N., DUNNING, S. A. & ALLISON, R. J. 2005. Terrestrial laser scanning for monitoring the process of hard rock coastal cliff erosion. *Quarterly Journal of Engineering Geology & Hydrogeology*; **38**, 363–375.

SCHEIDEGGER, A. F. 1973. On the prediction of the reach and velocity of catastrophic landslides. *Rock Mechanics*, **5**, 231–236.

SELBY, M. J. 1993. *Hillslope Materials and Processes*. 2nd edn. Oxford University Press, Oxford.

SOETERS, R. & VAN WESTEN, C. J. 1996. Slope instability, recognition, analysis and zonation. *In*: TURNER, K. & SCHUSTER, R. L. (eds) *Landslides, Investigation and Mitigation*. Special Report 247, Transportation Research Board, National Academy Press, Washington, DC, 129–177.

SORALUMP, S. 2009. Landslide mitigation in Thailand – from research to practice. *International Symposium on Geotechnical Engineering, Ground Improvement and Geosynthetics for Sustainable Mitigation and Adaption to Climate Change, Including Global Warming*, 3–4 December 2009, 399–409.

SOUSA, J. & VOIGHT, B. 1991. Continuum simulation of flow failures. *Géotechnique*, **41**, 515–535.

TRL. 1997. *Principles of Low Cost Road Engineering in Mountainous Regions*. Overseas Road Note 16, Transport Research Laboratory, Crowthorne.

VAN WESTEN, C. J. 2010. GIS for the assessment of risk from geomorphological hazards. *In*: ALCANTARA-AYALA, I. & GOUDIE, A. (eds) *Geomorphological Hazards and Disaster Prevention*. Cambridge University Press, Cambridge, 205–219.

VAN WESTEN, C. J., VAN ASCH, T. W. J. & SOETERS, R. 2005. Landslide hazard and risk zonation – why is it still so difficult? *Bulletin of Engineering Geology and the Environment*, **65**, 167–184.

VARNES, D. J. 1984. Landslide hazard zonation. A review of principles and practice. International Association of Engineering Geology, Commission on Landslides and Other Mass Movements on Slopes, UNESCO, Paris.

VERSTAPPEN, H. TH. 1983. *Applied Geomorphology: Geomorphological Surveys for Environmental Development*. Elsevier, Amsterdam.

WALLER, A. W. & PHIPPS, P. 1996. Terrain systems mapping and geomorphological studies for the Channel Tunnel rail link. *In*: CRAIG, C. (ed.) *Advances in Site Investigation Practice*. Thomas Telford, London, 25–38.

WASOWSKI, J., LAMANNA, C. & CASARANO, D. 2010. Mapping mid-slope road landslides and instability factors with high resolution satellite imagery. *In*: WILLIAMS, A. L., PINCHES, G. M., CHIN, C. Y., MCMORRAN, T. J. & MASSEY, C. I. (eds) *Geologically Active. Proceedings of the 11th IAEG Congress, Auckland, New Zealand, 5–10 September 2010*, Taylor and Francis Group, London, 2563–2570.

WHITWORTH, M. C. Z., GILES, D. P. & MURPHY, W. 2005. Airbourne remote sensing for landslide hazard assessment: a case study on the Jurassic escarpment slopes of Worcestershire, UK. *Quarterly Journal of Engineering Geology and Hydrogeology*, **38**, 285–300.

B3 Field mapping

G. J. Hearn

URS Scott Wilson Ltd, Scott House, Alençon Link, Basingstoke, Hampshire RG21 7PP, UK
(e-mail: gareth.hearn@scottwilson.com; garethhearn@talktalk.net)

B3.1 Reconnaissance surveys

Reconnaissance surveys are usually carried out to establish the main topographical, geological and engineering criteria that will influence the selection and design of a new alignment (e.g. Brunsden *et al.* 1975) or the key issues that need to be addressed in the case of a road improvement project (environmental and social impact reviews also form critical elements of these surveys). Reconnaissance surveys also allow validation of the desk study interpretations, and provide field information that can then be used to calibrate the desk study outputs.

B3.2 Reference condition mapping

The use of reference conditions to assist in classifying ground conditions originated as an aid to the management

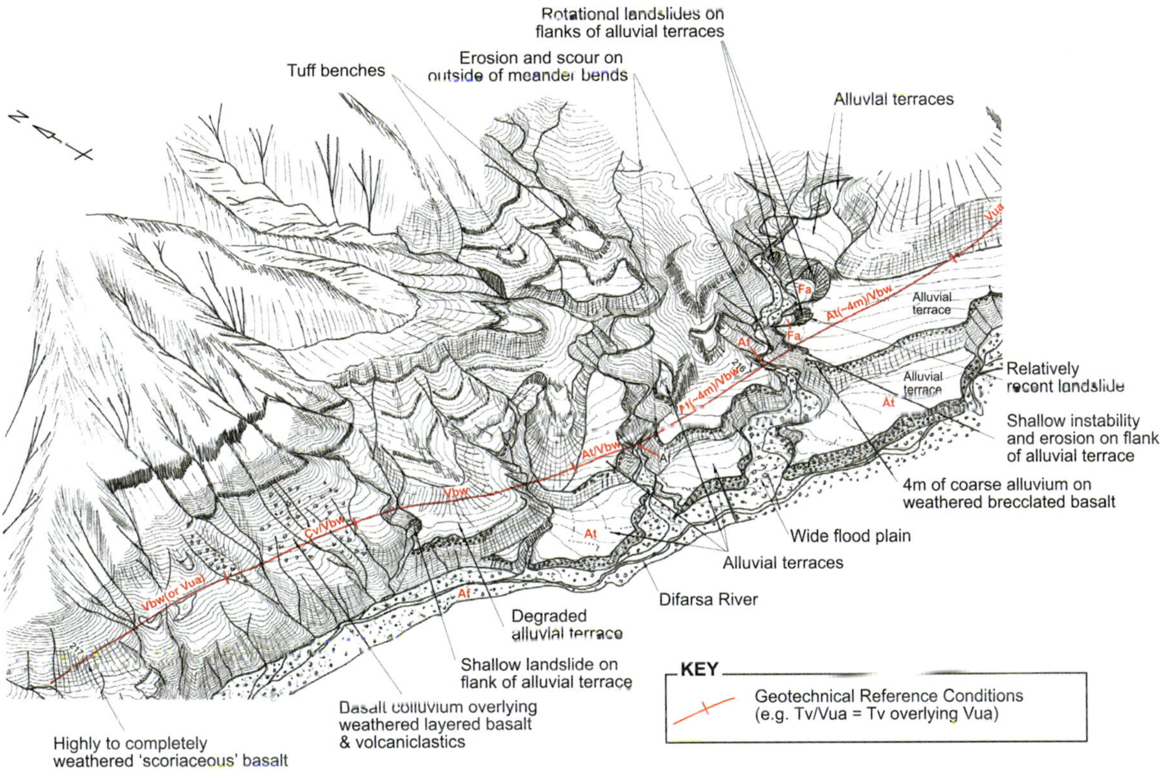

Fig. B3.1. Perspective map showing reference condition distribution along the Blue Nile gorge road alignment, Ethiopia (mapped by D. Wise).

From: Hearn, G. J. (ed.) *Slope Engineering for Mountain Roads*. Geological Society, London, Engineering Geology Special Publications, **24**, 103–116.
DOI: 10.1144/EGSP24.7 0267-9914/11/$15.00

Table B3.1. *Reference condition summary for* in situ *rock and transported soil units in the Blue Nile gorge, Ethiopia*

Reference condition	*Lithology*	*Landforms*	*Materials*	*Associated engineering geological considerations*	*Preferred cross-section*	*Embankment foundation suitability*	*Cut slope stability*	*Anticipated maximum cut slope angles V : H*	*Expected Excavation Techniques*	*Specific retaining, slope protection or drainage measures anticipated*
IN SITU										
Vbw	Highly to completely weathered basalt	Basalt plateau, moderate slopes	Spheroidal weathering of basalt creates gravels, cobbles and small boulders with a clayey and sandy matrix.	Both rock and soil type behaviour. Differential weathering, ravelling. Low material strength promoting circular failures through rock mass and relict discontinuities.	Full cut	Variable and unreliable, dependent on weathering profile and depth	May be controlled either by rock structure or soil type behaviour.	1:1 (<5m) 1:1.75 (5-10m) These weathering grades not expected beyond 10m depth.	Conventional excavation.	Removal of loose boulders. Rock catch ditches below slope.
Vua	Undiff-erentiated, interbedded basalt lava and tuff ('Ashangi Basalt' formation)	Moderate to steep slopes	Subhorizontal layers of jointed and fractured basalt, fine to coarse tuff and volcaniclastic sandstone/siltstone and volcaniclastic breccia.	Variable strengths, differential weathering and erosion, ravelling of undermined layers, structurally controlled failures in stronger basalt and tuff layers. Possible thin clay layers, including paleosols.	Cut or retained fill	Very good.	Generally good, subject to local structural control and rock types. Some ravelling in breccia and agglomerates expected. Buried paleosoils and clayey horizons may encourage shear failures.	Nominal angle of 1:1 suitable. Refer to individual reference conditions on exposure during construction. Multi-sloped design may be required where considerable variations in materials are encountered with depth.	Variable, generally conventional excavation and ripping, with some blasting of locally stronger materials.	Rock catch ditches below slopes. Consider anchored wire netting on slopes with particularly variable and ravelling slopes. Consider slope drainage
TRANSPORTED										
At	Alluvium	Terraces and alluvial fans, flat to very gently sloping, elevated surfaces	Uncemented sands, gravels, cobbles and boulders of variable lithologies with a minor silt/clay content.	Fines content highly prone to erosion. Terrace and fan flanks subject to river scour, oversteepening and landsliding.	Fill	Generally acceptable for low embankments, though some variable ground conditions.	Moderate. Controlled by groundwater, fines content and plasticity. Vertical variations in particle size and density may warrant multi-sloped design.	1:1.5 (<5m) 1:2 (5-20m) 1:2.5 (20-30m)	Conventional excavation.	Controlled slope drainage. For cut slopes in areas of higher groundwater, toe walls with additional drainage measures may be required.
Af	As above	Flood plains, flat infill on valley floors		Flood plain, mobilization of fine to coarse debris.	Fill	Potentially troublesome as some materials compressible and highly prone to erosion and scour.	Avoid due to groundwater/water bearing fines content and plastic fines.		Conventional excavation. Dewatering or stabilisation may be required during excavation.	Scour protection from existing watercourses and floods.
Cv	Colluvium derived from volcanic rocks (basalt, tuff and breccia)	Gentle to moderately steep, concave slopes, often steeply incised	Uncemented well graded clay, silt, sand, gravels, cobbles and occasional boulders of volcanic rock origin.	Soil type failure. Broad stratigraphic layering occasionally parallel to ground surface promotes translational landslides. Possibly subject to creep. Circular failures through fine-grained materials. Highly erodible and transportable on steep slopes (debris flows) Otherwise stable.	Balanced cut and fill in low angle slopes.	Potentially troublesome as fine-grained matrix may be plastic and prone to shrink/swell and creep. Residual (low strength) shear surfaces may be present and may be remobilised upon loading.	Controlled by soil behaviour though adversely oriented bedding fabric may initiate planar failures in cut slopes in side long ground.	1:1.5 (<5-15m) 1:1.75 (15-20m) 1:2 (20-30m)	Conventional excavation. Some larger boulders will be encountered.	Masonry or gabion retaining walls likely to be required in cuts >10m to ensure cut slope stability. Controlled slope drainage.
Tv	Talus derived from volcanic debris (from basalt, tuff and breccia rock sources)	Planar talus slopes close to or at limiting slope angle, developed beneath ravelling exposures or from rockfall failures in cliffs formed in volcanic rocks.	Uncemented basalt cobbles and boulders with a low clay/silt content	Subject to creep and shallow circular or planar failures through fine-grained components. Potentially subject to further rockfall and rock avalanche.	Full cut in steep slopes. Cut/fill in shallower natural slope angles.	Poor and unreliable due to variable materials on steep slopes close to their limiting angle of stability, potentially subject to creep.	Controlled by soil behaviour. Troublesome due to loose, unconsolidated deposits. Excavation may initiate new shallow plane failures which regress upslope.	1:1.5 (<5m) 1:1.75 (5-10m) Reference condition not expected beyond 5m depth. Retaining structures required if cut slopes need to be steepened to daylight.	Generally conventional excavation.	Masonry road retaining wall and cut slope retaining walls founded into underlying bedrock. Ditches and gabion barriers on upslope side of road to contain rockfalls.
Fa	Alluvium	Translational/ rotational landslides on flanks of alluvial terraces	Uncemented sands, gravels, cobbles and boulders of variable lithologies with a minor silt/clay content.	Base of slope may be subject to river scour, potentially destabilizing slopes and enlarging landslides.	Avoid or cross in full cut or balanced cut and fill.	Not recommended for embankment as additional loading may reactivate movements along residual shear surfaces leading to gradual creep or sudden failure.	Variable ground conditions possible. High groundwater table may be encountered. Stability controlled by residual shear strengths, groundwater, fines content and plasticity.	Site specific assessment and design required for excavations into this Reference Condition.	Conventional excavation.	Gabion retaining walls to support failing slopes. Scour protection. Controlled slope drainage.

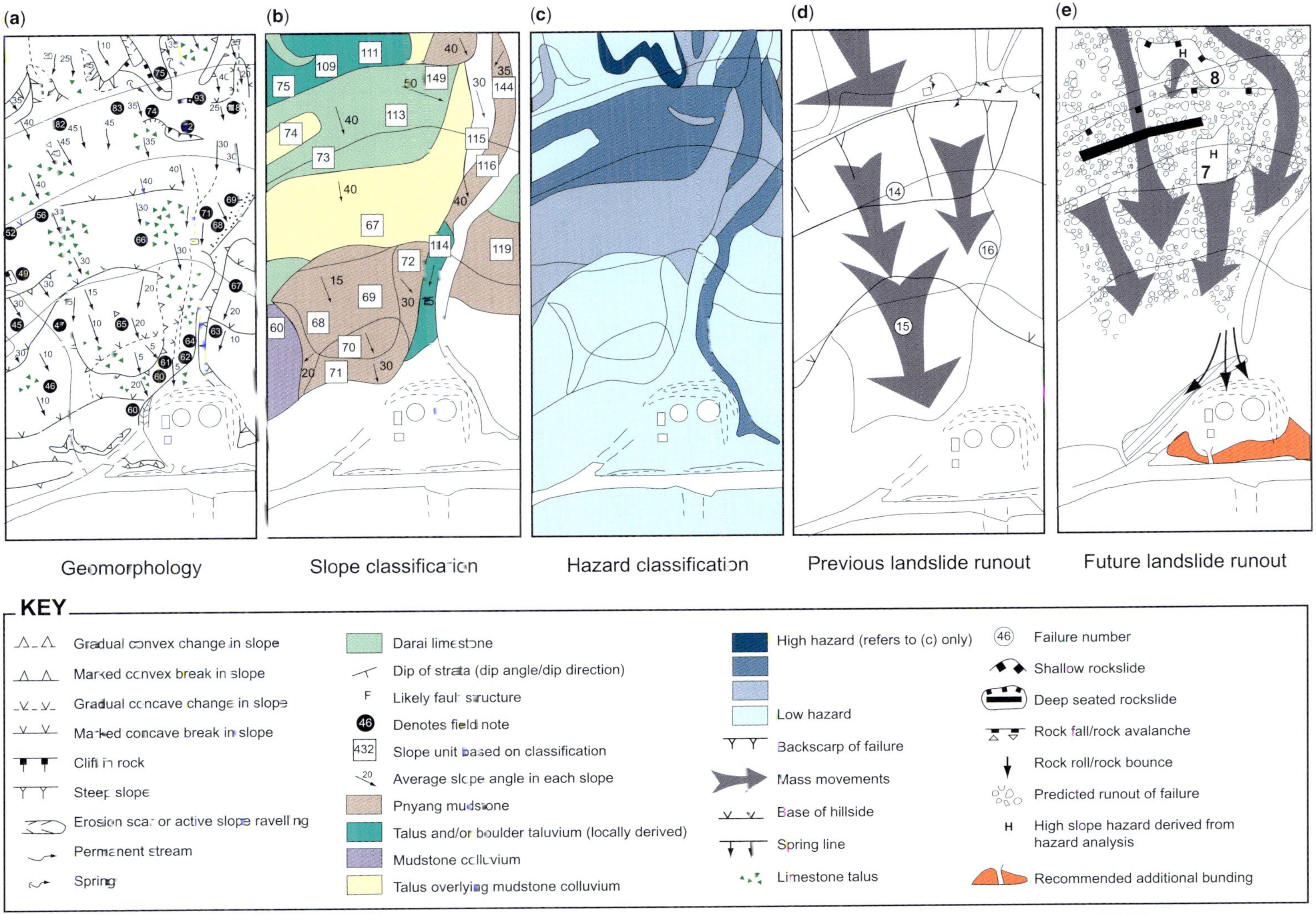

Fig. B3.2. Landslide hazard and runout mapping, Papua New Guinea (modified from Hearn 1995 with permission of Maney Publishing).

Engineering geological proforma rock slopes

1: LOCATION

Failing slopes above road, south east of bridge	Weather	Dry, sunny	Date of report	31/3
	Operator	Gareth Hearn	Reference No	R187

2: GEOMORPHOLOGY

Eastern valley side slope of river composed of upper cliffs in limestone with scree and mudflow on slopes below, Part of a much larger older failure embayment caused by failure of rock avalanche terrace overlying mudstone

3: LANDUSE

Haul road, water supply and slurry pipe on outside edge

4: DRAINAGE/GROUNDWATER ETC: at time of survey

Limestone cliffs dry. Probable seepage at limestone/mudstone contact though very little at time of survey

5: SLOPE DESCRIPTION (see map Plan A)

	Slope length (m)	Ave. (°) angle	Max angle (°) & Height (m)	Aspect (°)	Slope height (m)	%Veg.	%Rock outcrop
Cut slope	None						
Natural Slope	450 - 600	35-40	-	230	260	10	20

6: ROCK TYPE

Moderately fractured limestone, open jointed and locally disturbed due to undercutting and loss of support in mudslide heads below. Pnyang mudstone - highly fractured weak - very weak, locally softened by seepage

7: SOIL TYPE DEPTH

Local limestone colluvium/taluvium (<2m) remaining on 'unfailed' slopes below limestone cliffs

8: DISCONTINUITIES

Discontinuities	1	2	3	4	5
Dip angle (°)	60	46	50		
Dip direction (°)	200	200	360		
Persistence (m)	1	2	bedding		
Width of opening (m)	closed	closed	closed		
Spacing (m)	?	0.2	0.2		
Roughness	3	3	?		

9: ROCK ASSOCIATIONS

Weathering grade	IB-III
Rock strength	3 (lmst) - 5 (mst)
Structures/bedrock associations	

Darai limestone overlying Pnyang mudstone

10: CONDITIONS DOWNSLOPE

Moderately fractured limestone, open jointed and locally disturbed due to undercutting and loss of support in mudslide heads below. Pnyang mudstone - highly fractured weak - very weak, locally softened by seepage.

11: GEOMORPHOLOGICAL FEATURES

Mudslides/mudflows on slopes above road with local undercutting and rockfall from limestone cliffs above. Old quasi stable mudslide below road with reactivated failure in toe and in head immediately below road.

12: STABILITY (see Hazard Matrix)

Small slides and topple failures from 5-10m high disturbed limestone cliffs above eroding mudstone slopes. Active rock falls onto mudflow below with breakup of previously failed rock from cliff face. River scour.

Hazard Matrix

	HAZARD TYPE	SIZE (mxm)	DEPTH	STABILITY	CAUSE	CURRENT IMPACT	POSSIBLE IMPACT	INFERRED RATE
a)	Shallow (<2m) channel erosion							
b)	Deep (>2m) channel erosion							
c)	Surface rill erosion							
d)	Undercutting/lateral scour	-	-	Active	Scour by river	None	Could reactivate mudslide (k1)	2-5m/yr
e)	Flooding							
f)	Sedimentation							
g)	Debris fall							
h)	Rock fall	10x10	3?	Active	Seepage, loss of support	None	Could eventually regress back and cause cliff above to fail (R190)	<0.5m/yr
i)	Debris slide							
j)	Debris flow							
k1)	Mudslide/flow (Below road)	80x300	4?	Quasi stable	Seepage, toe erosion	None	Could regress back to road	mm/ month
k2)	Mudslide/flow (Above road)	40x100	3?	Active	Seepage, undrained loading, steep slope, softened/ weathered mudstone	Periodic flows onto road	continued failure causing h)	cm/day
m)	Block slide							
n)	Rotational slide (slump)							
o)	Talus							
p)	Settlement							
q)	Solution (Karst)							

Performance of existing stabilisation/protection measures	Comments/recommendations	Operator site hazard and risk classification					
d) None h) None k1) None k2) Berm alongside road ok	Slope above road may evolve to quasi stable angle as in H186 & H188. Review stability of cliff face above road & mudslide erosion scar below road every 3 months. Carry out engineering geological mapping within 5 years.	d) h) k1) k2)	H = 2 H = 1 H = 3 H = 2	P = 2 P = 2 P = 2 P = 3	Vu= 1 Vu= 1 Vu= 2 Vu= 1	Va= 3 (haul road) Va= 3 NB refer to 190 Va= 3 (haul road) Va= 3 (haul road)	R = 12 R = 6 R = 36 R = 18

Plan A

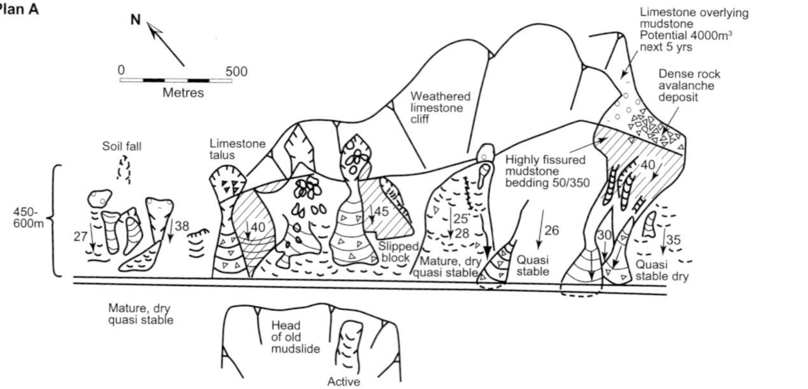

Fig. B3.3. Hazard matrix and risk classification applied to mine access road field investigations, Papua New Guinea.

Fig. B3.4. Failure surface daylights in cut slope.

of construction contracts in complex/unknown ground conditions (CIRIA 1978). The technique can be used to define the expected engineering parameters and extent of each difficult ground condition, which can then be built into the contract documents. If the ground conditions encountered during construction are different at any particular location to those anticipated, a price adjustment can be made based on the rates provided in the contract for the reference condition actually found.

More recently, the technique has been used to assist in the classification of anticipated ground conditions for project feasibility study and design purposes (Baynes *et al.* 2005; Fookes & Baynes 2008), and is a logical follow-on from terrain modelling and classification (Sections B2.4 & B2.5). By grouping together geological or combined geological-terrain units with approximately similar ground conditions, the technique can be used in the prediction of soil and rock profiles for earthworks schedules and preliminary estimates of quantities. The ground conditions exposed in excavations during construction can be highly variable and sometimes confusing in mountain terrain, and reference condition mapping developed from terrain modelling and classification can help to explain and clarify these variations. The technique can therefore also provide a rationale for the observation and interpretation of ground conditions by site supervision staff during construction.

The reference condition approach is illustrated using the example of the Blue Nile gorge road in Ethiopia. This mapping was undertaken in conjunction with terrain modelling (Figs B2.7 & B2.8) during the initial design review phase of the construction period because existing geological and geotechnical data were largely absent. The reference conditions were defined on a section-by-section basis for the entire alignment in the gorge, and these were shown on perspective geomorphology plans (illustrated in Fig B3.1). Engineering geological and geotechnical descriptions and design parameters were assigned and tabulated for each reference condition (those relevant to Fig B3.1 are shown in Table B3.1). The cutting angles were derived from stability charts and soil and rock descriptions.

As with terrain modelling, specialist skills are required to derive these outputs, but the time required to do so is relatively short and inexpensive in comparison to normal design and construction periods and budgets. Nevertheless, predictions of subsurface conditions from the ground surface should be validated using exposures and ground investigation data (Section B4); relationships between morphology and weathering profile, for example, are seldom clear-cut in the humid tropics and subtropics (e.g. Hencher & Lee 2010 and Fig. A3.3).

B3.3 Landslide hazard mapping

Desk study-based landslide susceptibility and hazard mapping for route corridor selection is discussed in Section

Fig. B3.5. Identifying principal landslide morphology for mapping purposes (for discussion see text).

B2.6. However, this mapping is rarely based on desk study alone and varying degrees of field mapping and investigations are required in its compilation and verification. In the case of existing roads, field-based landslide hazard mapping can be undertaken to help draw attention to sections of alignments and engineering structures at risk. This can be used to design movement monitoring schemes and implement risk management during maintenance (Part D).

Figure B3.2 is an extract from a landslide hazard and runout prediction map developed for a mining township in Papua New Guinea (Hearn 1995). The mapping was based entirely on field observations and on the use of geomorphological mapping (Section B3.4), in particular, to derive the required data. A slope materials classification was developed and a hazard rating was assigned in terms of the potential for each of the units to fail, based on the evidence of previous slope failures. The slope geometry, volumes and runout distances of previous failures were approximated from field observations, and an empirical relationship between these parameters was determined. From the geomorphology, the volume of anticipated future failures in each hazard zone was estimated and the expected runout of each was computed using the same empirical relationships. Mitigation measures were then proposed accordingly.

Methods of rockfall and landslide runout modelling are also referred to in Section B2.6 and, for example, in publications by Scheidegger (1973), Hsü (1975), Corominas *et al.* (1988), Corominas (1993, 1996), Evans & Hungr (1993), Hearn (1995), Hungr (1995), Evans & King (1998), Hadley *et al.* (1998), Franks (1999), Fell *et al.* (2007), Hürlimann *et al.* (2008) and Fannin & Bowman (2010).

The Papua New Guinea study extended over 3 km^2 and took 10 man-weeks to complete in very difficult, densely forested terrain. This intensity of mapping was warranted because of the proximity of the occupied township at potential risk and is probably inappropriate for the majority of low-cost mountain roads, except where high-value assets are at risk or

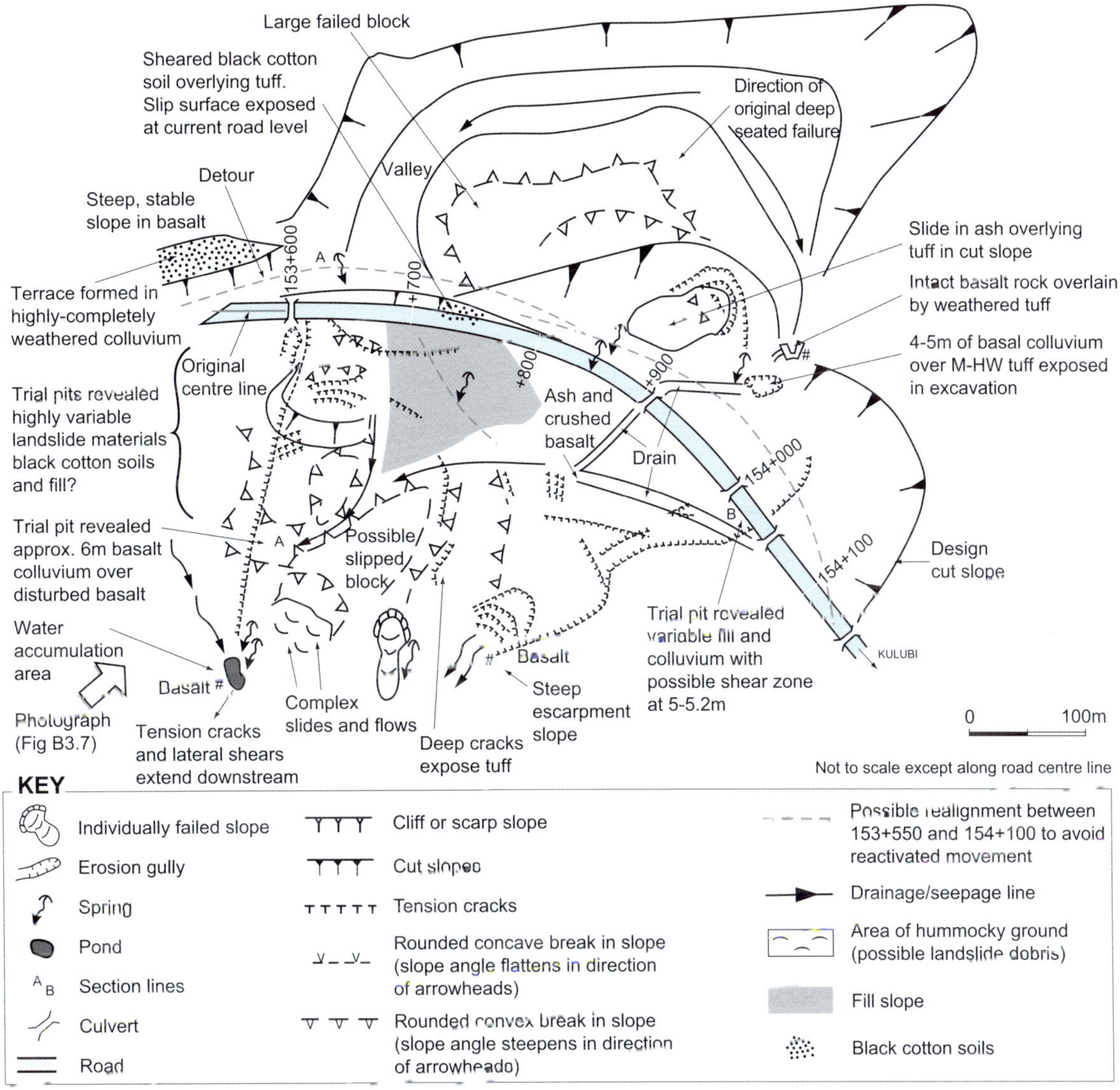

Fig. B3.6. Geomorphological map of deep-seated landslide.

Fig. B3.7. Mapped slope showing subtle landslide features (dotted lines are back scarps, dashed lines are landslide deposits (also in orange), solid line is road; photograph taken in direction of arrow on Fig. B3.6).

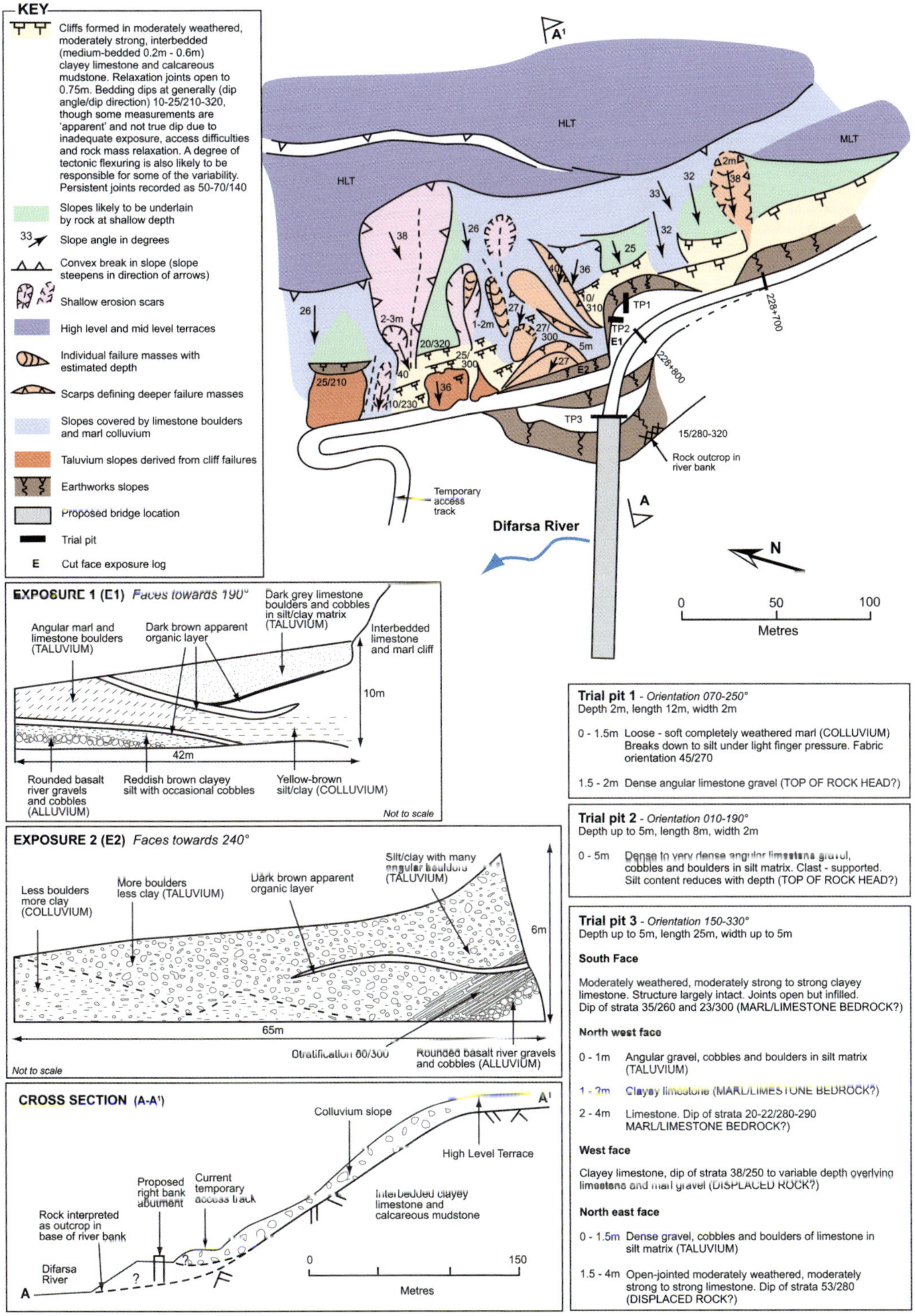

Fig. B3.8. Engineering geological investigations for bridge abutment stability assessment.

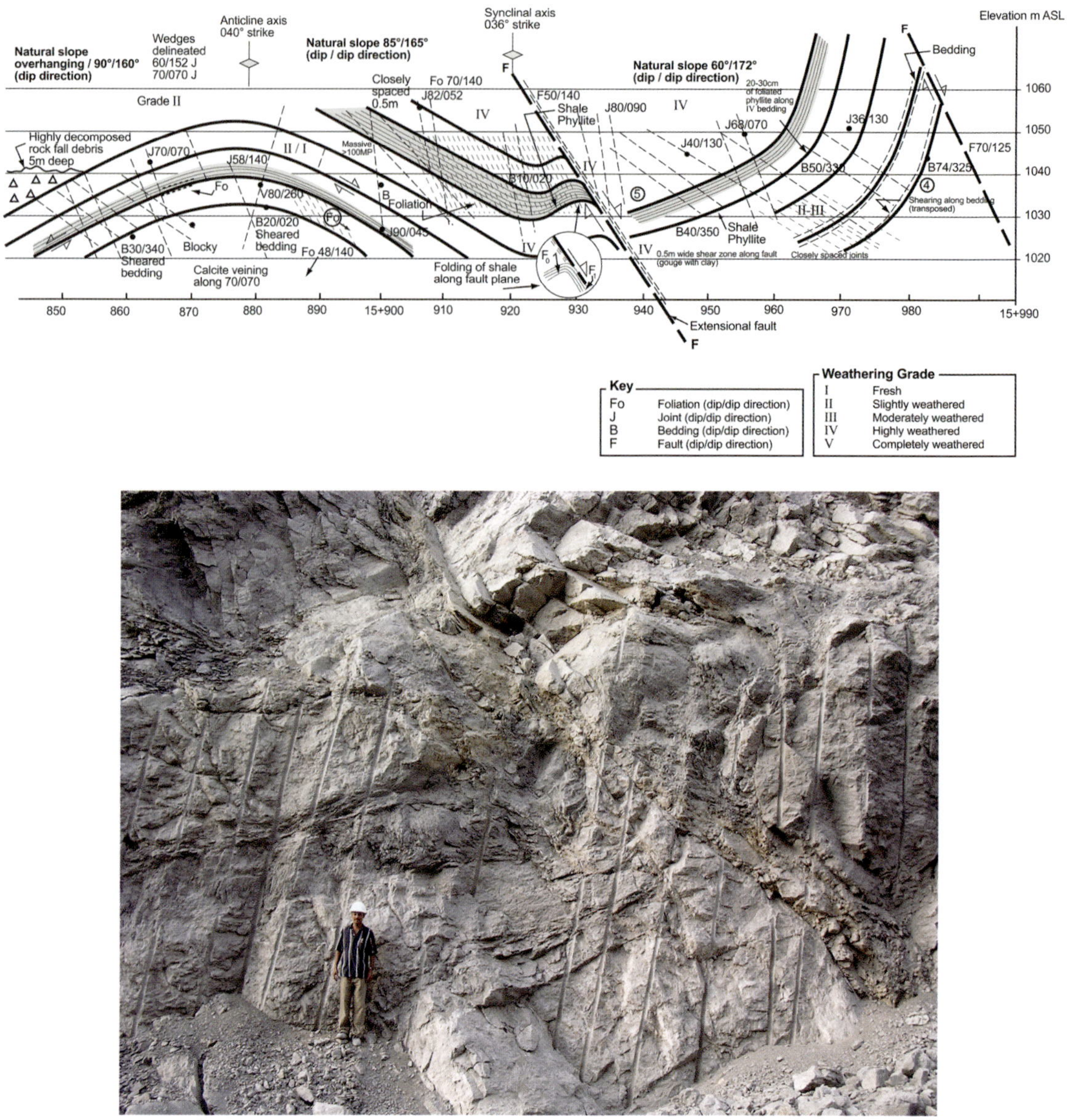

Fig. B3.9. Structural geological rock slope stability mapping (mapped by C. Massey).

where there are recurrent landslide problems that result in regular road blockage. Even then, the uncertainty over runout predictions compounds the difficulties in determining where landslides are most likely to occur in the first place (Section B2.6). Simpler approaches to landslide susceptibility and hazard mapping are more likely to be appropriate for the average conditions found on most mountain roads. These might be based on slope inventories and hazard ranking from a simple matrix of observations, as illustrated in Figure B3.3 from the same project in Papua New Guinea.

B3.4 Engineering geological mapping

B3.4.1 Recognition of landslide features

'The geomorphology of any site is precious and fragile, the end result of an interplay of thousands to millions of years between solid and Quaternary geology, hydrogeology, climate, process and the nature of the ground as controlled by its physical properties. All too often, however,

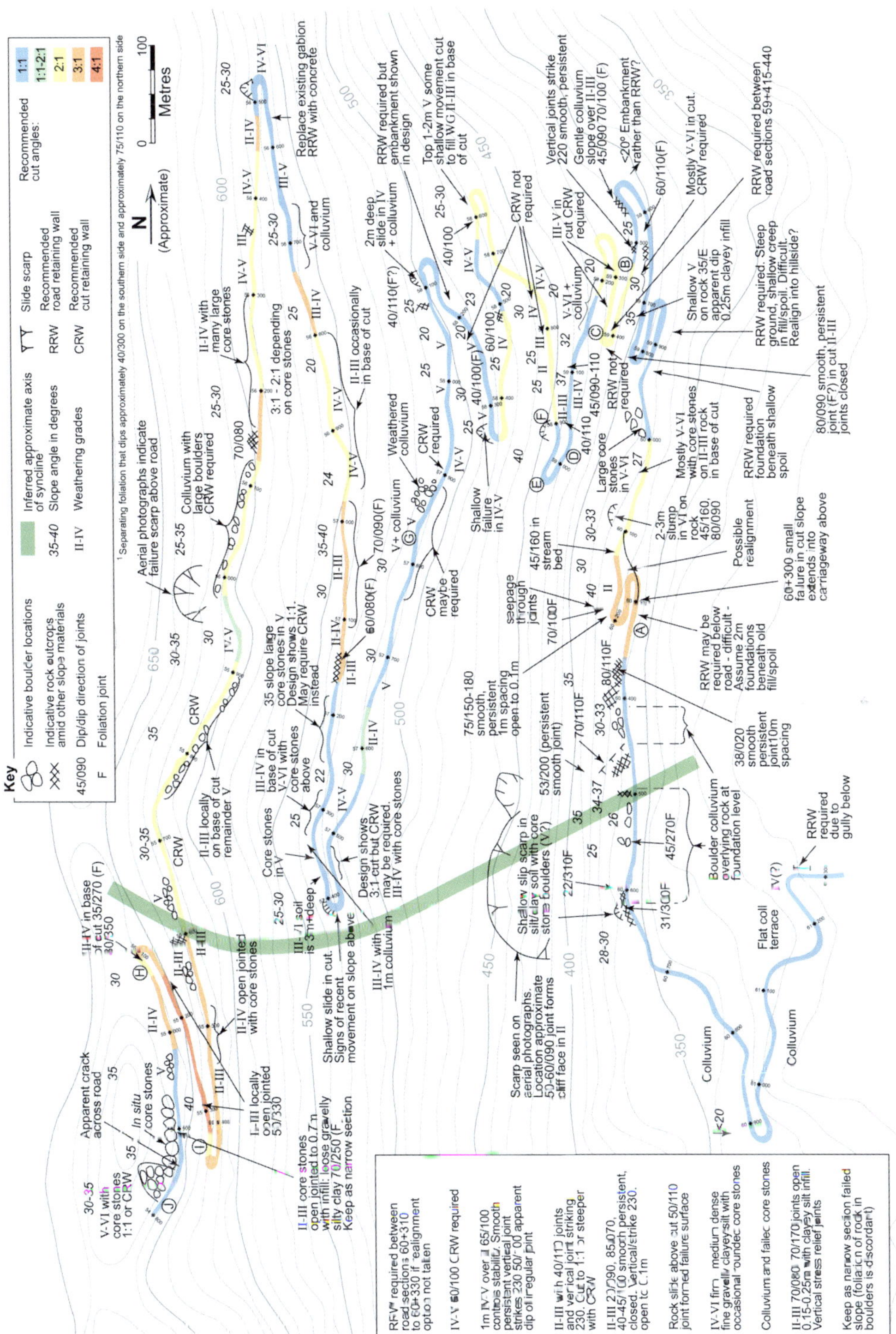

Fig. B3.10. Engineering geological and geomorphological map for road widening and slope protection on a hairpin stack in Sri Lanka.

Table B3.2. *Common features indicative of landslides and potential landslide locations*

Indicators of landslides and potentially unstable slopes	*Description and comments*	*Method*				
		*Optical satellite Imagery**		*Airborne imagery (air photos or LiDAR†)*		*Field*
		Resolution		*Scale*		
		Low	High	1:40k	1:20k	
Active Landslides						
Tension cracks	Often orientated in an arc (segment of a circle) with vertical displacement on the downslope side		✓		✓	✓
Slip scarps	Steps across terraces and other slopes		✓	✓	✓	✓
Disturbed/displaced terracing	Lines of vertically/laterally displaced terracing often mark the margins of ground movement		✓	✓	✓	✓
Hummocky ground	Slope surface is irregular and often formed by a series of low amplitude hummocks		✓		✓	✓
Cracking to structures and paved surfaces	Can be due to settlement of fill and foundations, supporting evidence is required, unless extensive				✓	✓
Dislocation of drainage structures	Either directly observed or seen as seepages		✓		✓	✓
Springs and seepages	Creating marshy ground		✓	✓	✓	✓
Trees leaning or with curved trunks	Wind, steep slopes and slope movement can cause leaning tree trunks, careful interpretation required					✓
Relict Landslides						
Spoon-shaped landforms	Steep upper scarp often semi-circular in section, lower-angled, possibly tongue-shaped deposit		✓	✓	✓	✓
Chaotic debris forming landslide deposits	Boulders often protrude above the surface		✓	✓	✓	✓
Hummocky ground	Slope surface is irregular and often formed by a series of low amplitude hummocks		✓		✓	✓
Immature soil profile, indicates disturbed ground	Normal weathering profile is replaced by a structureless, and usually loose taluvium soil					✓
Disturbed or uncharacteristic vegetation pattern	Could be related to land use, so needs to be interpreted with care		✓		✓	✓
Colluvium/taluvium vulnerable to future movement						
Chaotic debris forming landslide deposits	Boulders often protrude above the surface		✓	✓	✓	✓
Hummocky ground	Slope surface is irregular and often formed by a series of low amplitude hummocks		✓	✓	✓	✓
Immature soil profile, indicates disturbed ground	Normal weathering profile is replaced by a structureless, and usually loose taluvium soil					✓
Waterlogged ground and marshy areas	Water is seen to collect, either from surface water or groundwater seepage		✓		✓	✓
Future first time failures						
Slopes underlain by adverse‡ geological structures and rock types prone to failure	Smooth and persistent discontinuities often form slopes and may form potential failure surfaces if exposed by excavation or by river downcutting		✓	✓	✓	✓
Slopes with high groundwater tables or ground saturation in deep low density soils	Slopes where water is seen to collect, either from surface water or groundwater seepage				✓	✓
Slopes prone to river or stream scour at their base	Should be directly observable		✓	✓	✓	✓
Debris flows from upstream						
Flow deposits border main channel indicating future events are also possible	These deposits lack stratification, are predominantly boulders and often have low amplitude levees parallel to the direction of flow		✓	✓	✓	✓

*See Table B2.3 for explanation.

†High density LiDAR data can be displayed on screen at a range of resolutions, according to area of interest.

‡*Adverse* in this context means where discontinuities within the rock mass dip out of the slope (Fig A3.18), thus forming planes of weakness, either singularly or in combination, along which slope failure can take place. These discontinuities usually comprise bedding in sedimentary rocks and foliation in metamorphic rocks, as well as tectonically-derived joints, faults and thrusts in all rock types.

the site morphology is ignored, sometimes even destroyed in a day by the bulldozing of access tracks . . . . and its Quaternary geology neglected. These factors are believed to be responsible for the tendency of initial site appraisal to lag behind the impressive developments in other areas of geotechnics and to be the source of many of our worst mistakes' (Hutchinson 2001, pp. 7–8).

This quotation emphasizes the importance in recording and understanding the geomorphology of a site or area for engineering purposes. In particular, during all project phases it is important to be able to recognize features indicative of landslides and slope instability. This is critical to engineering geological mapping, but also as a means of monitoring the development of slope instability during operation and maintenance by trained inspectors (Part D). If these features are identified early enough then timely intervention measures can be implemented. Table B3.2 lists the common features of landslide recognition, and differentiates between those that are usually identifiable from optical satellite imagery and aerial photography and those that can only be effectively identified or assessed in the field.

B3.4.2 Landslide mapping

Engineering geological mapping enables geological and geomorphological information to be recorded for engineering purposes including rock and soil exposures, drainage conditions, landslide back scarps, toe bulges, tension cracks and failure surfaces exposed in excavations. In some cases, the evidence for slope instability can be quite distinct while in others it can be very subtle, and in some cases non-existent. For example, at the location where the photograph in Figure B3.4 was taken, the only evidence for 'incipient' ground movement was the shear surface exposed in the cut slope excavated during road widening.

Procedures for engineering geological mapping are given in Fookes (1969), Anon (1972), Dearman & Fookes (1974), IAEG (1979) and GEO (2004). Case histories in the use of engineering geological mapping are presented in Griffiths (2002) while GEO (2007) describes the use of engineering geological mapping for landslide and slope stability assessment in Hong Kong. The description and classification of soils and rocks for engineering purposes are given in Anon (1972, 1977, 1981, 1995), Martin & Hencher (1996), GCO (1988), Matheson (1989), Bell (1995), Dearman (1995), BSI (1999) and Norbury (2010). GEO (2007) describes the engineering classification of rocks, transported soils and *in situ* weathered soils developed in the tropical climate of Hong Kong. However, for slope stability assessment especially, classification is no substitute for description, particularly where geological structures and fracture spacing exert major influences on mass strength (Hencher & Lee 2010).

In areas of subtle topography dominated by soil slopes, mapping will focus on the careful identification and recording of slope geomorphology. Figure B3.5 illustrates how the morphology of a failed slope can be defined in terms of the main back scarp, areas of previous or intermittent ground movement and areas of active movement. The delineation of the back scarps (shown in black) helps to define the full extent of failed ground (shown in orange). Areas of active movement (in red) are determined from the evidence of bare ground in back scarps and visible evidence of ongoing ground movement, such as tension cracks. Although the production of maps from these observations is based on the recording of factual information, it can require a significant degree of interpretation by the mapper; the recognition of even subtle features can significantly influence the final interpretation of ground conditions.

Figure B3.6 shows the principal geomorphological features identified when mapping a large landslide area in Ethiopia. In this example, there were very few exposures of rock on the slope and so the mapping focused on the morphology of the landslide (some of which was extremely subtle). The mapping helped to identify the extent of the failure and allowed the interpretation of its mechanism and likely depth. The failed hillside and the road across it are shown in Figure B3.7.

Figure B3.8 illustrates the use of engineering geological mapping and investigation to record geological and geomorphological data for purposes of slope stability assessment at the location of a bridge abutment in Ethiopia. Maximum use is made of landform interpretation, geological outcrop and structural observations, soil exposure mapping in temporary cut slopes and trial pit excavations (Section B4) in order to interpret the ground conditions to the fullest extent possible.

In steeper terrain, where bedrock is close to or at the slope surface, mapping will focus on structural geology, outcrop pattern and rock weathering grade, as it is these factors that will have the greatest influence on slope stability. Figure B3.9 shows a detailed structural geological strip map of a rock excavation on a mountain road in Tajikistan. The map is used to identify the main discontinuities (joints and faults) that control the stability of the slope. Measurements taken of the orientation of these discontinuities were then used for rock slope stability analysis and slope design (Section C4).

Figure B3.10 shows how engineering geological and geomorphological mapping are used to assist in the design of road improvement and widening on a hairpin stack in Sri Lanka. The key issues in deciding upon cut slope design and slope protection are the weathering grade of the underlying slope materials and the orientation of foliation and other discontinuities in the gneiss bedrock, where exposed.

References

ANON 1972. The preparation of maps and plans in terms of engineering geology. Report by the Geological Society Engineering Group Working Party. *Quarterly Journal of Engineering Geology*, **5**, 293–382.

ANON 1977. The logging of rock cores for engineering purposes. *Quarterly Journal of Engineering Geology*, **10**, 45–52.

ANON 1981. Rock and soil description and classification for engineering geological mapping. Report of the International

Association of Engineering Geology, Commission on Engineering Geological Mapping. *Bulletin of the International Association of Engineering Geology*, **24**, 235–274.

ANON 1995. The description and classification of rocks for engineering purposes. Geological Society Engineering Group Working Party Report. *Quarterly Journal of Engineering Geology and Hydrogeology*, **28**, 207–242.

BAYNES, F. J., FOOKES, P. G. & KENNEDY, J. F. 2005. The total engineering geological approach applied to railways in the Pilbara, Western Australia. *Bulletin of Engineering Geology and the Environment*, **64**, 67–94.

BELL, F. G. 1995. *Engineering Geology*. Blackwell Science, Cambridge.

BRUNSDEN, D., DOORNKAMP, J. C., FOOKES, P. G. F., JONES, D. K. C. & KELLY, J. M. H. 1975. Large scale geomorphological mapping and highway engineering design. *Quarterly Journal of Engineering Geology*, **8**, 227–253.

BSI 1999. *Code of practice for site investigations*. **BSI 5930**, British Standards Institution, HMSO, London. First published in 1981 (Amendment 2, 2010).

CIRIA 1978. Tunnelling – improved contract practices. Construction Industry Research and Information Association, Report 79, London.

COROMINAS, J. 1993. Spatial prediction of landslides. *In*: NEMEC, J., NIGG, J. M. & SICCARDI, F. (eds) *Prediction and Perception of Natural Hazards*. Kluwer Academic Publishers, Amsterdam, 125–132.

COROMINAS, J. 1996. The angle of reach as a mobility index for small and large landslides. *Canadian Geotechnical Journal*, **33**, 260–271.

COROMINAS, J., PENARANDA, R. & BAEZA, C. 1988. Identificacion de factores que condicionan la formacion de movimientos superficiales en los valles altos del Llobregat y Cardener. *In*: *II Symposium Taludes y Laderas Inestables*, Andorra La Vella, 195–207.

DEARMAN, W. R. 1995. Description and classification of weathered rocks for engineering purposes: the background to the BSI5930:1981 proposals. *Quarterly Journal of Engineering Geology*, **28**, 267–276.

DEARMAN, W. R. & FOOKES, P. G. 1974. Engineering geological mapping for civil engineering practice in the United Kingdom. *Quarterly Journal of Engineering Geology*, **7**, 223–256.

EVANS, N. C. & KING, J. P. 1998. *The Natural Terrain Landslide Study: Debris Avalanche Susceptibility*. Technical Note No TN 1/98. Geotechnical Engineering Office, Hong Kong.

EVANS, S. G. & HUNGR, O. 1993. The assessment of rock fall hazards at the base of talus slopes. *Canadian Geotechnical Journal*, **30**, 620–636.

FANNIN, R. J. & BOWMAN, E. T. 2010. Debris flows – entrainment, deposition and travel distance. *In*: WILLIAMS, A. L., PINCHES, G. M., CHIN, C. Y., MCMORRAN, T. J. & MASSEY, C. I. (eds) *Geologically Active. Proceedings of the 11th IAEG Congress, Auckland, New Zealand, 5–10 September, 2010*, Taylor and Francis Group, London, 1111–1118.

FELL, R., GLASTONBURY, J. & HUNTER, G. 2007. Rapid landslides: the importance of understanding mechanisms and rupture surface mechanics. The Eighth Glossop Lecture. *Quarterly Journal of Engineering Geology and Hydrogeology*, **40**, 9–27.

FOOKES, P. G. 1969. Geotechnical mapping of soils and sedimentary rock for engineering purposes with examples of practice from the Mangla Dam Project. *Géotechnique*, **19**, 52–74.

FOOKES, P. G. & BAYNES, F. J. 2008. Successful use of the Total Geology Approach, using Reference Conditions for the design and construction of heavy duty railways in the Pilbara, Western Australia. *Ground Engineering*, March 2008, 39–45.

FRANKS, C. A. M. 1999. Characteristics of some rainfall-induced landslides on natural slopes, Lantau Island, Hong Kong. *Quarterly Journal of Engineering Geology*, **32**, 247–259.

GCO 1988. *Guide to Rock and Soil Descriptions*. Geoguide 3. Geotechnical Control Office, Hong Kong.

GEO 1997. *Model Specification for Prestressed Ground Anchors*. 2nd edn. Geospec 1. Geotechnical Engineering Office, Hong Kong (1989), reprinted 1997.

GEO 2004. Guidelines on geomorphological mapping for natural terrain hazard studies. *Technical Guidance Note* **22 (TGN22)**, Geotechnical Engineering Office, Civil Engineering Development Division, Hong Kong.

GEO. 2007. *Engineering Geological Practice in Hong Kong*. Publication No. 1/2007. Geotechnical Engineering Office, Civil Engineering and Development Department, The Government of Hong Kong Special Administrative Region.

GRIFFITHS, J. S. (compiler). 2002. *Mapping in Engineering Geology. Key Issues in Earth Sciences*, Vol. **1**, Geological Society, London.

HADLEY, D., HEARN, G. J. & TAYLOR, G. R. 1998. Debris flow assessments for the Foothills Bypass, Hong Kong. *In*: LI, K. S., KAY, J. N. & HO, K. K. S. (eds) *Slope Engineering in Hong Kong. Proceedings of the Annual Seminar on Slope Engineering in Hong Kong*, Hong Hong Institution of Engineers, A. A. Balkema, Rotterdam, 153–162.

HEARN, G. J. 1995. Engineering geomorphological mapping and opencast mining in unstable mountains – a case study. *Transactions of the Institution of Mining & Metallurgy*, **A104**, A1–A18.

HENCHER, S. & LEE, G. 2010. Landslide mechanisms in Hong Kong. *In*: CALCATERRA, D. & PARISE, M. (eds) *Weathering as a Predisposing Factor to Slope Movements*. Geological Society, London, Engineering Geological Special Publication, **23**, 77–103.

HSÜ, K. J. 1975. Catastrophic debris streams (sturzstroms) generated by rockfalls. *Geological Society of America, Bulletin*, **86**, 129–140.

HUNGR, O. 1995. A model for the runout analysis of rapid flow slides, debris flows and avalanches. *Canadian Geotechnical Journal*, **32**, 610–623.

HÜRLIMANN, M., RICKENMANN, D., MEDINA, V. & BATEMAN, A. 2008. Evaluation of approaches to calculate debris-flow parameters for hazard assessment. *Engineering Geology*, **102**, 153–163.

HUTCHINSON, J. N. 2001. Reading the ground: morphology and geology in site appraisal. The Fourth Glossop Lecture. *Quarterly Journal of Engineering Geology and Hydrogeology*, **34**, 7–50.

IAEG 1979. Report on the commission on engineering geological mapping. *Bulletin of the International Association of Engineering Geology*, **19**, 364–371.

MARTIN, R. P. & HENCHER, S. R. 1986. Principles for description and classification of weathered rock for engineering purposes. *In*: HAWKINS, A. B. (ed.) *Site Investigation Practice: Assessing BSI 5930*. Geological Society, London, Engineering Geology Special Publication, **2**, 299–308.

MATHESON, G. D. 1989. The collection and use of field discontinuity data in rock slope design. *Quarterly Journal of Engineering Geology*, **22**, 19–30.

NORBURY, D. 2010. *Soil and Rock Description in Engineering Practice*. Whittles Publishing, CRC Press, Taylor & Francis Group, Caithness, Scotland.

SCHEIDEGGER, A. E. 1973. On the prediction of the reach and velocity of catastrophic landslides. *Rock Mechanics*, **5**, 231–236.

B4 Ground investigation

T. Hunt & G. J. Hearn*

URS Scott Wilson Ltd, Scott House, Alençon Link, Basingstoke, Hampshire RG21 7PP, UK
**Corresponding author (e-mail: gareth.hearn@scottwilson.com; garethhearn@talktalk.net)*

B4.1 Purpose of a ground investigation

The term *ground investigation* refers specifically to the investigation of subsurface soil, rock and groundwater conditions, either through intrusive methods (principally boreholes and trial pits and the associated soil/rock sampling and testing) or surface and borehole geophysics. Ground investigation methods are described in numerous textbooks including more recently Clayton *et al.* (1995), Simons *et al.* (2002), Cornforth (2005) and Bond & Harris (2008), for example.

Ground investigations for low-cost roads can often be low on the priority list. They are sometimes seen to be costly and time consuming, providing little information of any value. However, if planned and implemented correctly, they can yield valuable information when compared to the cost of design.

Ground investigations are undertaken for new roads to determine:

- typical soil and rock profiles to calibrate and augment the terrain models, terrain classifications and field mapping referred to earlier (Sections B2 & B3);
- specific soil and rock profiles in the case of deep cuts for site-specific design;
- foundations for structures, such as bridge piers and abutments and large retaining walls;
- depth and geotechnical composition of existing landslides and the design of remedial works; and
- material type and depth for borrow areas.

Ground investigations are undertaken for existing roads when examining:

- landslides that were not stabilized during construction; and
- new landslides and road failures that have occurred since construction.

In all cases, desk studies and field mapping (Sections B2 & B3) are essential prerequisites to ensure that the ground investigation is planned effectively and that it is able to yield the maximum information.

As with field mapping techniques, the description of soils and rocks for engineering purposes forms a key element of ground investigation and should follow recognized engineering standards, such as ISRM (1981), GCO (1988), BSI (1999) and BSI (EN) (2003). Norbury (2010) provides a useful guide to engineering geological logging and classification methods. Fookes *et al.* (2005) reproduce a number of useful standard descriptions and classifications.

B4.2 Scope of investigation

The required scope and sophistication of a ground investigation will be dictated by the complexity of the ground conditions and the level of geotechnical risk posed to the performance of the investment. In the case of low-cost roads, it is usual to employ the simpler and more routine methods to derive the required data.

Usually a ground investigation comprises trial pits and boreholes (or drillholes in the case of hard rock investigation), together with sampling and *in situ* and laboratory testing. In the case of drilling investigations, the standard of practice in many countries can be quite poor, and the results obtained should be viewed with considerable caution. The equipment used may be inappropriate or poorly maintained, and the operators may lack the necessary skills and experience to carry out drilling and testing to the appropriate specified standard. Usually in landslide investigations it is desirable to identify the shear surface on which movement has occurred, and this can be very difficult even when using sophisticated drilling techniques. The heterogeneity of materials found on many mountain slopes may seriously limit the applicability of the test results. As a general rule, therefore, it will be better value for money to carry out a larger number of simple tests than a smaller number of complex tests.

Knowledge of the depth to the actual or potential failure surface is necessary both for the purpose of stability analysis (Section C3) and the design of remedial works. In the latter case, retaining walls (Section C5) often form important components of slope stabilization or the reinstatement of sections of failed road and the depth to stable ground beneath the failure surface needs to be determined before the wall can be designed. However, if the remedial works simply

From: Hearn, G. J. (ed.) *Slope Engineering for Mountain Roads*. Geological Society, London, Engineering Geology Special Publications, **24**, 117–126.
DOI: 10.1144/EGSP24.8 0267-9914/11/$15.00

involve reconstructing a hillside fill slope or cutting back a cut slope to a shallower slope angle, then a ground investigation may not be necessary if the types of material are already known (e.g. engineered fill) or can be seen in excavations. Nevertheless, in the case of engineered fill, a ground investigation may be necessary to determine fill thickness and to check material type.

Other considerations relevant to carrying out ground investigations along mountain roads are summarized below and discussed in Section B4.3.

- If machine access is required, then a track may need to be constructed. This can be difficult in mountainous terrain, and may cause instability and erosion. Either skid-mounted, track-mounted or man-portable equipment will provide a satisfactory solution for most situations. Helicopter rig transfers may also need to be considered.
- Some basic considerations are necessary for the depth of investigation. To what depth is information required at a particular location? Would a machine-dug or hand-dug trial pit reach the required depth? Would pits be stable to that depth? Is it necessary to schedule a borehole/drillhole? Would a probe from existing ground level provide the relevant information?
- The required sampling and testing will also dictate the method of investigation used. A decision will need to be made as to what the test results will be used for and whether disturbed or undisturbed samples will be required.

B4.3 Investigation methods

Ground investigation techniques commonly comprise:

- trial pits and trenches;
- augered holes;
- boreholes/drillholes;
- probing;
- sampling (disturbed and undisturbed);
- *in situ* testing in soil and rock;
- geophysical investigations; and
- the use of boreholes to facilitate groundwater monitoring (Section B4.3.7) and ground movement monitoring (Section B5).

Table B4.1 gives some suggestions for the techniques required in a typical ground investigation. The design of deep excavations and the stabilization of large and deep-seated slope failures may require more sophisticated investigation.

B4.3.1 Trial pits and trenches

Trial pits can be hand dug or machine dug. If there is good access, machine-dug trial pits are far more preferable for the following reasons:

- speed of excavation;
- depth of excavation;
- safety (as personnel are not required to enter the trial pit);
- ability to break through obstructions such as boulders and dense gravels; and
- ability to excavate weathered rock, thus proving outcrop far more effectively.

Other safety precautions must be followed regardless of the pitting method. If trial pits are located where they present a danger to traffic, traffic warning signs and barriers need to be erected. In any event, all trial pits should be properly backfilled and compacted as soon as they have been logged and should not be left open and unprotected overnight. Great care must be taken when pitting in taluvium/colluvium or other failed or otherwise unstable ground, particularly during the wet season, or if groundwater is encountered. Access of personnel into unshored trial pits should not be permitted and is counter to Health and Safety regulations in many countries.

Machine-dug trial pitting also has some disadvantages:

- the speed of excavation may sometimes not be compatible with the need to log the trial pit faces carefully as excavation proceeds;
- the ripping action of the bucket teeth can cause significant disturbance to trial pit faces, frequently resulting in undercutting and collapse in loose taluvial soils; and
- in soils containing clay the side of the bucket can smear exposures on trial pit faces, possibly obscuring important detail.

The main advantage of trial pitting over borehole and drillhole investigations is that the structure and stratification of the underlying ground can be seen and logged in three dimensions. Disturbed samples can be taken during excavation, but if undisturbed sampling is required this can only be done in cohesive soils and then with some difficulty. Trial pits are very useful in determining the nature and composition of the near-surface soils, shallow foundation conditions for existing or new retaining walls, the detection of shallow slip surfaces and seepages. Text box B4.1 illustrates this from a road improvement project in Ethiopia.

The main disadvantage of trial pitting is the restriction on depth. It may be possible to excavate machine-dug trial pits to 4–6 m but advancing through wet soils, bedrock or large boulders can be extremely difficult.

B4.3.2 Augering

Augering can be carried out by hand or machine. Hand augering is suitable in self-supporting (firm to stiff) fine-grained soils down to a depth of a few metres, but is unable to penetrate hard obstructions such as cobbles, boulders or rock. Machine augering, particularly using hollow-stem augers, can be used to reach greater depths and to extract samples through the stem, but will also encounter similar difficulties in heterogeneous and large-sized material.

B4.3.3 Boreholes and drillholes

Borehole and drillhole slope investigations along mountain roads are uncommon, especially in the case of low-volume

Table B4.1. *Typical range of subsurface techniques for the investigation of slopes along low-cost roads*

Type of project	*Project element*	*Investigation technique*	*Purpose*
Design of new road or road realignment/ widening where no apparent instability is occurring/has occurred.	New cut slope	Borehole and sampling	To determine subsoil profile and soil/rock strengths from laboratory tests on samples obtained in order to design cut slope angle(s) and to estimate soil/rock excavation quantities
		Piezometer installation	To determine depth and variation in groundwater level over time.
	New fill slope	Trial pit and sampling	To assess strata and geotechnical parameters for stability against sliding and bearing failure
	New retaining wall	Borehole/trial pit and sampling	To determine a suitable founding level from field descriptions and soil strengths derived from representative samples obtained. To determine groundwater level (if appropriate)
		Probe	To determine a suitable founding level from inferred soil strength
Design of new road or road realignment/ widening where instability is apparent or suspected, or to rectify/prevent slope instability along an existing road.	Unstable or failed cut or fill slope	Borehole/trial pit and sampling	To determine depth to slip surface, groundwater level and subsoil profile and soil strengths from samples obtained
		Piezometer installation	To determine depth and variation in groundwater level over time
	Failure/ distress through/ beneath road	Borehole/trial pit	To determine depth to slip surface
		Piezometer installation	To determine depth and variation of groundwater level over time
		Inclinometer	To determine rate and depth of movement(s)
		Slip indicator	To determine depth to slip surface
		Surface movement monitoring	To determine rate of surface movements
	Unstable or failed retaining wall	Borehole/trial pits	To determine existing/replacement founding conditions and groundwater level
		Probe	To determine a suitable founding level from soil strengths
		Monitoring points	To monitor any existing/replaced wall movements
	New retaining wall	Borehole/trial pits	To determine existing/replacement founding conditions and groundwater level
		Probe	To determine a suitable founding level from soil strengths

roads, and are often limited to investigating the depth to rock head beneath landslide or colluvial cover in order to assess the feasibility of founding retaining walls. Even where the depth to rock head proves to be beyond that of a practicable gravity wall foundation (usually a maximum of 10 m), knowledge of the actual depth to *in situ* material beneath this level will assist in decision making over the need for other options, such as realignments into the hillside or earthworks solutions (Section C3.3).

Boreholes may be advanced with or without the use of a rotational drill bit, depending on the competence of the underlying strata. Borehole and drillhole rigs may be truck, trailer or skid-mounted, depending on the availability and practicalities of access.

B4.3.3.1 Cable percussion boring (boreholes)

Referred to as 'shell and auger boring' in some countries, this method advances the borehole by percussive effort using a clay cutter for cohesive soils, a shell (or bailer) for granular soils and a chisel for cobbles and boulders or *in situ* rock. Shell and auger methods are uncommonly used outside the UK and their future use in UK is likely to

Text box B4.1. Trial pitting investigation of a landslide in Ethiopia

The use of trial pitting in landslide investigations can be illustrated from the Hirna to Kulubi road in Ethiopia (Hearn & Massey 2009). For each of the 26 slope failures that occurred in 2003 an engineering geological or geomorphological map was produced, allowing a programme of trial pitting investigations to be defined. Borehole or drillhole investigation equipment was not available to the project, and therefore all investigation had to be carried out from surface observations and trial pitting. There were no restrictions on access to land for pitting, and there was generally good trafficability between sites. The majority of the soils investigated comprised black cotton soil (vertisol residual clay; Table A3.1) and clayey colluvium, overlying closely jointed basalt which, in turn, usually overlaid completely weathered tuff. Most of the identified failure surfaces passed through the weathered tuff. Although the majority of trial pits reached 4–5 m in depth, the deepest was 9 m and the greatest depth of excavation for ground investigation purposes was of the order of 12 m. This was achieved by progressively lowering the ground surface by dozer over a wide area, logging exposures with increasing depth and completing the exercise with a trial pit in the base of the excavation. The excavation was carried out during the dry season and, for safety reasons, the slopes were constantly monitored for any signs of movement. In the more critical areas, all soil and rock horizon boundaries were surveyed and referenced to a benchmark outside the landslide area. Striated shear surfaces were evident in most of the trial pits and the dip and orientation of the striations were used to help establish the configuration of the failure surface in the ground model (Section C3.2.1). An undisturbed block sample containing part of the shear surface was removed from one of the trial pits and transported to a laboratory for strength testing. The photograph illustrates the failure surfaces typically exposed during the trial pitting investigation. The failure surface is in clay between overlying (failed) basalt and underlying (*in situ*) completely weathered tuff.

It is common on many mountain slopes to encounter landslide materials that are much coarser than those illustrated from Ethiopia, and failure surfaces can be less well-defined. Failure often occurs either within poorly defined zones of heterogeneous debris or along the rockhead surface beneath residual soil or taluvium/colluvium. In many instances, this may be deeper than trial pitting investigations are able to reach.

diminish as they do not conform with Eurocode 7 requirements for undisturbed sampling (BSI EN 2007).

Cable percussion boring can be effective in soils containing few boulders and in highly to completely weathered rock, but is unlikely to be successful in soils where boulders make up more than 30% of the mass. Cable percussion boring will be difficult in rock that is moderately weak (UCS 5–12.5 MPa) or stronger. The action of cable percussion boring will disturb the ground immediately below the base of the hole, and the quality of 'undisturbed' 100 mm diameter (U100) samples taken at that level for strength testing may be compromised.

B4.3.3.2 Rotary drilling (drillholes)

This method advances the hole by rotating a drill bit. The drilling fluid normally used is water which is pumped down through a pipe string, around the drill bit or casing shoe and then allowed to rise back up between the drill string and the casing. Drilling muds, such as bentonite and synthetic polymers, often provide the best recovery in weak rocks but can be expensive, depending upon availability.

Rotary drilling, although the most expensive of ground investigation techniques, is generally preferred for investigating stiff clays and unweathered intact rock, particularly if Standard Penetration Tests (SPTs) are to be performed or cores taken since drilling tends to cause least disturbance to the intact material at the base of the hole. Undisturbed core is recovered from the core barrel for inspection and logging at the ground surface. Wire-line drilling allows drilling core to be recovered quickly from the hole for logging and strength testing purposes, thus reducing the opportunity for disturbance to the core. Anon (1970), Bieniawski (1989), BSI (1999), Blackbourn (2009) and Norbury (2010), for example, provide a description of industry-standard core logging procedures.

It is extremely difficult to recover granular soils, soft clays and some types of highly weathered rock by rotary drilling, and this is usually not attempted (the drill operator will advance through these materials by *open-hole* drilling with no core recovery). Recovery of good-quality undisturbed samples of cohesive soils for strength testing is usually carried out by thin wall or piston sampling (hydraulically advanced) at the base of the hole.

B4.3.3.3 Wash boring

This method utilizes a chisel to break up the ground at the bottom of the hole, with water pumped down the pipe string to bring up the broken material to the surface. As with cable percussion and rotary drilling, casing is sometimes used to support the sides of the hole to prevent collapse.

In general, wash boring is not recommended even if the purpose of the borehole is only to give an indication of the strata penetrated; the material brought up to the surface may not be representative of the material at the base of the hole. In addition, the action of the water will disturb the material and this may render sampling and *in situ* testing ineffective.

B4.3.4 Probing

Probing can sometimes be used to estimate the strength of near-surface soils. It is carried out from an existing ground surface or an excavated foundation level, and is useful for indicating foundation bearing capacity for proposed retaining walls (Section C5). Probing includes hand-operated penetration equipment such as the dynamic cone penetrometer (e.g. Jones 2004) and the static cone penetrometer (e.g. Brouwer 2008), which are usually truck-mounted. However, the latter can only be used in relatively uniform soft or loose soils and is probably inappropriate for most landslide situations on mountain slopes. In any event, probing in gravelly or bouldery soils is unlikely to produce meaningful data. A range of cone types and techniques are available and some of these are described, for example, in Mayne (2007). The Mackintosh probe (e.g. Fakher *et al.* 2006) is used principally for determining the depth of soft soils and can provide a means of assessing approximate undrained shear strength (c_u). As with other probing methods, its application is limited on mountain slopes due to the coarse-grained and heterogeneous nature of the soils commonly encountered.

In the case of dynamic probes, the results are very difficult to interpret if the basic soil composition (gravel, sand, silt, clay) is unknown. Probing should therefore always be carried out in conjunction with trial pitting or boreholes.

B4.3.5 Sampling and testing of soils and rocks

Sampling can be categorized as disturbed or undisturbed.

B4.3.5.1 Disturbed sampling

This includes SPT split-spoon samples, auger and borehole arisings and most trial pit sampling. Disturbed sampling is useful in enabling a good visual description of the subsoil material and for laboratory classification testing (see Table B4.2 and, for example, BSI 1990; Simons *et al.* 2002; Craig 2004; ASTM 2005*a*; Head 2008). However, the intact structure of the material is lost and any descriptions and testing can only apply to the disturbed material. As a minimum, disturbed samples are usually taken at each change in soil composition or at 1 m depth intervals.

B4.3.5.2 Undisturbed sampling

This is, by its very nature, considerably more difficult and expensive than disturbed sampling. A decision must be made as to whether such testing is likely to provide useful and useable information, since a significant number of tests may be necessary to fully represent the characteristics of the material under investigation. For instance, Hencher & Lee (2010, p. 84) note that 'Properties of weathered rocks can be difficult to ascertain because sampling and testing of very weak and sensitive weathered materials without disturbance can be impossible ... The difficulty of conducting high quality laboratory tests that provide realistic properties is one of the main reasons for employing a material weathering classification linked to simple index tests.' In the humid tropics and subtropics, most natural hillside slopes will have a graded weathering profile with depth and each weathering grade will have its own shear strength characteristics that can be approximated from index tests.

Landslides will usually comprise heterogeneous and predominantly coarse-grained materials such as taluvial soils and landslide debris. Therefore, unless the boundaries of these materials are well understood, the determination of the material characteristics (e.g. shear strength parameters) of a very small sample, in comparison with the volume and complexity of the remainder of the landslide mass, may only be of very limited value. Nevertheless, there may be circumstances where undisturbed sampling is desirable or necessary, for example in the investigation of slope failures where the material through which failure has occurred is known (Section C3) and can be sampled.

Undisturbed sampling includes block sampling from trial pits, thin and thick wall sampling within boreholes and cores from core drilling (e.g. Clayton *et al.* 1995). Thick-wall sampling (for instance the U100 sampling often used in the UK) is usually hammer driven, and is now regarded as not

Table B4.2. *Potential range of laboratory tests on soils for slopes affecting mountain roads*

Category	*Test*	*Remarks*
Classification tests (disturbed or undisturbed samples)	Moisture content	Together with plasticity can be helpful in estimating the undrained strength of cohesive materials.
	Liquid and plastic limits (Atterberg tests)	For classifying the fine-grained fraction of soils, can be helpful for estimating both drained and undrained strength of cohesive materials.
	Bulk density	For undisturbed samples only, to use in the calculation of forces exerted by the soil.
	Particle size distribution	
	a) Sieving	a) to determine the grading of a soil coarser than silt size; with description of particle angularity can be used to estimate friction angle of granular soils.
	b) Sedimentation	b) to determine the relative proportions of silt and clay.
Compaction-related tests (disturbed samples)	Dry density	To determine the mass of solids per unit weight of soil.
	Standard compaction tests	To determine Maximum Dry Density and Optimum Moisture Content at which fill materials should be placed. Results used in the control of earthworks, e.g. embankments.
Soil strength tests (undisturbed samples)	Triaxial compression	To determine the strength of cohesive soils.
	a) Unconsolidated undrained	a) undrained tests to assess undrained shear strength (c_u).
	b) Undrained with measurement of pore pressure c) Drained	b) and c) undrained or drained tests with the measurement of pore pressure to assess shear strength parameters in terms of effective stress (c' and φ').
	d) Multi-stage	d) useful if lack of samples, but single stage tests usually more reliable.
	Unconfined compression	Only suitable for saturated, uniform fine-grained soils.
	Laboratory vane shear	Only for soft and firm clays. May not give representative results for remoulded samples; usually better to take *in situ* vane shear measurements if possible.
	Direct shear box	Cheaper alternative to undrained triaxial test. Drainage conditions cannot be controlled during testing. Samples can be orientated, residual strength can be determined.
	a) Multiple reverse shear box b) Ring shear	To determine the residual shear strengths of cohesive soils. a) and b) use undisturbed or remoulded samples for use in the absence of a pre-formed shear surface.
	c) Triaxial test with pre-formed shear surface d) Shear box test with pre-formed shear surface	Preparation and alignment of undisturbed samples for c) and d) test specimens can be very difficult.
Soil deformation tests (undisturbed samples)	Consolidation test (oedometer)	To determine the magnitude and rate of settlement of soft soil under loading.

producing good enough quality samples for strength testing. Thin-wall sampling (ASTM 2008), preferably hydraulically driven, is much preferred although is only suitable for clay/silt/sand soils. Coring can only really be described as undisturbed if a triple tube core barrel is used and the drilling operation is of a very high quality. In the vast majority of cases the drill operator will use a double tube core barrel and simple drilling techniques, with the result that any seams of weaker material will be washed out and the percentage core recovery (the length of core recovered as a percentage of the total distance advanced down the hole) will be significantly less than 100%. The weaker material might well be the key factor in the assessment of slope stability. Sampling of the shear surface can be difficult to achieve, except perhaps in trial pits. Even then, it may be more practical to obtain undisturbed samples close to the suspected location of the shear surface and to perform residual shear tests on these in the laboratory (see Table B4.2).

B4.3.5.3 In situ *testing*

Standard Penetration Testing (SPT) records the number of blows of a drop hammer for a cone or split spoon to penetrate 300 mm into the base of a borehole (e.g. Clayton 1990; ASTM 1999, 2005*b*). The SPT is a simple test to carry out and there are many published correlations between the SPT blow count and other material properties. Of all the tests that can be undertaken in a landslide investigation, it still remains one of the easiest to do and most useful in indicating the relative strength of the underlying soils. As with all simple tests, the results have to be used with extreme caution especially in soils containing cobbles and boulders, such as taluvium or landslide debris, or *in situ* weathered soils containing corestones. High water tables can also cause spurious results.

The *in situ* shear strength of clay soils can be measured directly using a shear vane in the sides or base of trial pits or boreholes or by hand-held penetrometer in trial pits or on recovered undisturbed samples.

B4.3.5.4 Laboratory testing

Soils. Table B4.2 gives some details of the range of laboratory testing on soils that may be appropriate for slopes affecting mountain roads. Further information on these and other tests can be found in a number of references and codes of practice (e.g. BSI 1990; Simons *et al.* 2002; Craig 2004; ASTM 2005*a*, 2007; Head 2008).

Soil tests commonly undertaken on disturbed samples include natural moisture content determination, Atterberg limits, particle size distributions and compaction testing of materials to be used as fill. If a detailed investigation is deemed necessary, more specialized testing such as triaxial or shear box testing on undisturbed samples may be appropriate. While an assessment of the strength of all strata will be required, in the case of an unfailed slope this will focus on identifying the weakest strata through which failure might take place (often not recovered in boreholes); in a failed slope this focus will be the failure surface itself.

Where embankments are to be constructed on compressible clay soils, located on valley floors for example, then oedometer tests to determine consolidation parameters may also be required to enable the magnitude and rate of settlement to be estimated.

An important consideration in the humid tropics and subtropics is the fact that aggregation of clay particles can occur in many residual soils. During sieve analyses these particles are recorded as silt-sized, but they can break down under loading or dynamic testing (such as Atterberg limit testing) into their constituent clay particles and behave quite differently (Fookes 1997).

Rocks. In the case of rock slopes and landslides that have occurred as a result of failure through rock strata, sampling of the rock core can be undertaken for laboratory determination of compressive strength and other parameters. These tests are predominantly outside the scope and requirements of investigations undertaken for landslides affecting low-cost roads, and are therefore not discussed here. Further information can be obtained from standard publications such as those of Hoek & Bray (1981), Bieniawski (1989), Wyllie & Mah (2004) and Jaeger *et al.* (2007). However, rock slope behaviour is seldom controlled by material strength (Section C4) so correlation from inspection or simple tests is usually sufficient. It may be necessary to carry out strength tests on any joint infill if this is considered to be critical to stability (Section A3.2).

B4.3.6 Geophysical investigations

Surface geophysics can be used to detect variations in seismic velocity and other physical properties of soil and rock due to groundwater fluctuations or changes in strata and slope materials (e.g. Donnelly *et al.* 2005; Reynolds 2011). The strength of the technique is its ability to detect anomalies in an otherwise constant profile as illustrated, for example, in the detection of corestones using seismic resistivity in the tropical soils of Brazil by Taioli *et al.* (2010). The technique becomes less clear-cut when there are multiple sources of anomaly as would be the case, for example, where a soil profile contains a groundwater table, corestones and a landslide failure surface.

The main advantage of these techniques is their ability to provide continuous profiling of subsurface conditions using portable equipment and testing that can be carried out from the surface at relatively low-cost. However, conventional borehole or drillhole information will be required to confirm and calibrate the readings, and the techniques can be unreliable where ground conditions are complex. One of the more common applications of the technique is the detection of distinct boundaries, such as the depth and configuration of the rockhead surface beneath landslide material and colluvium or taluvium deposits (e.g. Glade *et al.* 2005).

Seismic refraction is generally the most frequently used geophysics technique in engineering applications but will be reliable only if:

- seismic velocity is constant, or increases with depth (this is not always the case in landslide materials);
- there is a distinct boundary and a marked difference in velocity between the layers being investigated;
- boulders are absent from the profile (the presence of large boulders complicates the interpretation of results); and
- trained field operators and specialist interpretation are utilized.

Nevertheless, compressional wave velocity associated with seismic refraction is reported to have been used, for example, by Latham *et al.* (2010) to identify the depth of shear in a rockslide in N Carolina, USA and by Larson (1995) to define shear surfaces in colluvium and saprolite in Puerto Rico.

Although geophysical surveys are infrequently used on low-cost road projects, they offer some potential in the following applications:

- the detection of the boundary between transported soil or landslide debris and an underlying unweathered rockhead surface;
- the definition of distinct layers in slope deposits; and
- the detection of underground cavities and other karst features developed in calcareous rocks and deposits.

Further details can be found in Anon (1988), McDowell *et al.* (2002), Waltham (2009) and Reynolds (2011).

B4.3.7 Groundwater observations

The presence of groundwater has a profound effect on slope stability. Groundwater levels and pore pressures will respond to rainfall and groundwater flow patterns, and therefore the presence of groundwater in a borehole or trial pit is only indicative of the water condition in a slope at that point in time. However, observations of depths to these water levels during the course of the ground investigation will at least provide some basic indication of groundwater conditions. A particular point to note when excavating trial pits or performing borehole investigations in clay soils, is that the absence of any observed water does not necessarily mean that the trial pit or borehole is above the water table. Due to the very low permeability of clay soils, the trial pit or borehole will have to be left open for some considerable time before the true water table can be ascertained. The ingress of rainfall or surface water runoff during this period may prevent meaningful observations from being made. A more reliable method of recording groundwater levels is the use of piezometers.

Long-term monitoring of groundwater can be undertaken using a standpipe piezometer installed in a borehole prior to backfilling. The main drawback with a simple open standpipe is that the response time (i.e. the time delay for the water in the pipe to respond to the water pressure in the ground) is slow in fine-grained soils. In addition, if zones of different permeabilities have been encountered, the readings may not be representative of actual groundwater levels. The main advantages of a standpipe are its simplicity and reliability. Water levels may be measured manually using dipmeter probes from the ground surface or by installing submersible water level logging probes. The logging probes use measurements of temperature and pressure to calculate water levels. The logged data may be transmitted automatically using a telemetry-based system, or may be regularly downloaded from data logger units by connection to a computer. These logging systems are usually utilized in relation to high-risk landslide sites and are seldom used on low-cost roads.

Vibrating wire electrical and pneumatic (pressure sensor) piezometers are also used in landslide investigations to provide real-time measurement of pore pressures (e.g. Reid *et al.* 2008). The former measures the deflection of a sensitive diaphragm by the surrounding pore-water pressure while the latter operates via a sealed tip with a pressure-sensitive valve connected to tubes leading back to the ground surface. Issues such as response time, reliability, ruggedness and cost will need to be considered when deciding upon the best approach to be adopted. For most low-cost road investigations, an open standpipe installed in a borehole or drillhole is usually the most appropriate but peak water levels may be missed if observations are not made frequently enough. Alternatively, peak water levels can be detected by lowering a series of small buckets into the standpipe. When withdrawn after a few days or months, the highest filled bucket will indicate the peak level reached since the last reading. This system would be useful in remote locations that can only be accessed infrequently.

B4.4 Supervising a ground investigation

The investigation should be carried out by an approved and reliable contractor, with day-to-day supervision by an experienced and qualified engineering geologist. The key issues to consider are:

- good core recovery if drilling through rock;
- close attention paid by the driller to depth of progress, water strikes and material changes and removal of all disturbed materials in the base of the hole prior to sampling/SPT testing;
- an appropriate sampling/testing programme for the materials encountered and an approved laboratory in which to carry out the tests;
- engineering geological description and logging to an approved standard (e.g. USBR 1998, BSI 1999) of all materials and samples obtained from the trial pitting, boring/drilling and sampling;
- laboratory test scheduling in accordance with international procedures; and

- routine inspection of laboratory testing procedures to ensure tests are carried out according to specification.

The results of the field and laboratory investigations and testing should be examined in relation to the engineering geological interpretation of the site from surface observations. This should include, for example, a comparison of the SPT values and borehole log descriptions and, for landslide investigations, a review of the observed or inferred depth to the failure surface compared to that indicated by the geological and geomorphological evidence on the slope itself. If there are any uncertainties or inconsistencies, then a supplementary ground investigation may be necessary.

References

ASTM 1999. *Standard Test Method for Penetration Test and Split-barrel Sampling in Soils*. American Society for Testing and Materials, ASTM **D1586-99**.

ASTM 2005*a*. *Standard Test Methods for Liquid Limit, Plastic Limit, and Plasticity Index of Soils*. American Society for Testing and Materials, Annual Book of ASTM Standards, ASTM Manual, ASTM **D4318-05**.

ASTM 2005*b*. *Standard Test Method for Energy Measurement for Dynamic Penetrometers*. American Society for Testing and Materials, Annual Book of ASTM Standards, ASTM Manual, ASTM **D4633-05**.

ASTM 2007. *Standard Test Method for Unconsolidated-Undrained Triaxial Compression Test on Cohesive Soils*. American Society for Testing and Materials, Annual Book of ASTM Standards, ASTM Manual, ASTM **D2850-03a**.

ASTM. 2008. *Standard Practice for Thin-Walled Tube Sampling of Soils for Geotechnical Purposes*. American Society for Testing and Materials, Annual of ASTM Standards, ASTM, **D1587-08**.

ANON 1970. The logging of core for engineering purposes. Report of the Geological Society Engineering Group Working Party. *Quarterly Journal of Engineering Geology*, **3**, 1–24.

ANON 1988. Engineering geophysics. *Quarterly Journal of Engineering Geology and Hydrogeology*, **21**, 207–271.

Bieniawski, Z. T. 1989. *Engineering Rock Mass Classifications. Technology and Engineering*. John Wiley Ltd, New York.

Blackbourn, G. A. 2009. *Cores and Core Logging for Geoscientists*. 2nd edn. Whittles Publishing, Caithness, Scotland.

Bond, A. & Harris, A. 2008. *Decoding Eurocode 7*. Taylor & Francis, London.

Brouwer, J. J. M. 2008. *In-situ Soil Testing*. Lankelma, East Sussex, UK.

BSI 1990. Methods of test for soils for civil engineering purposes. **BSI 1377-1**, British Standards Institution.

BSI 1999. Code of practice for site investigations. **BSI 5930**, British Standards Institution, HMSO, London. First published in 1981. (Amendment 2, 2010).

BSI (EN) 2003. Geotechnical investigation and testing. Identification and classification of rock. Part 1: Identification and description. British Standards Institution, EN ISO **BSI 14689-1**: 2003.

BSI (EN) 2007. Eurocode 7: Geotechnical Design (Part 2 Ground Investigation and Testing). British Standards Institution EN ISO **BSI 1997-2**: 2007.

Clayton, C. R. I. 1990. SPT energy transmission: theory, measurement and significance. *Ground Engineering*, **23**, 10, 35–43.

Clayton, C. R. I., Matthews, M. C. & Simmons, N. E. 1995. *Site Investigation*. Blackwell Science, Oxford.

Cornforth, D. H. 2005. *Landslides in Practice. Investigation, Analysis and Remedial/Preventative Options in Soils*. John Wiley and Sons Ltd, Chichester.

Craig, R. F. 2004. *Soil Mechanics*, 7th edn. Taylor and Francis, London.

Donnelly, L. J., Culshaw, M. G., Hobbs, P. R. N., Flint, R. C. & Jackson, P. D. 2005. Engineering geological and geophysical investigations of a slope failure at Edinburgh Castle, Scotland. *Bulletin of Engineering Geology and the Environment*, **64**, 119–137.

Fakher, A., Khodaparast, M. & Jones, C. J. F. P. 2006. The use of the Mackintosh Probe for site investigation in soft soils. *Quarterly Journal of Engineering Geology & Hydrogeology*, **39**, 189–196.

Fookes, P. G. (ed.) 1997. *Tropical Residual Soils*. Geological Society, London, Engineering Group Working Party Report.

Fookes, P. G., Lee, E. M. & Milligan, G. 2005. *Geomorphology for Engineers*. Whittles Publishing, Caithness, UK.

GCO 1988. *Guide to Rock and Soil Descriptions*. Geoguide 3. Geotechnical Control Office, Hong Kong.

Glade, T., Stark, P. & Dikau, R. 2005. Determination of potential landslide shear plane using seismic refraction – a case study in Rheinhessen, Germany. *Bulletin of Engineering Geology and the Environment*, **64**, 151–158.

Head, K. H. 2008. *Manual of Soils Laboratory Testing*. 3rd edn. Whittles Publishing, Caithness, UK.

Hearn, G. J. & Massey, C. I. 2009. Engineering geology in the management of roadside slope failures: contributions to best practice from Bhutan and Ethiopia. *Quarterly Journal of Engineering Geology and Hydrogeology*, **42**, 511–528.

Hencher, S. & Lee, G. 2010. Landslide mechanisms in Hong Kong. *In*: Calcaterra, D. & Parise, M. (eds) *Weathering as a Predisposing Factor to Slope Movements*. Geological Society, London, Engineering Geology Special Publication, **23**, 77–103.

Hoek, E. & Bray, J. 1981. *Rock Slope Engineering*. Revised 3rd edn. Institution of Mining and Metallurgy, London.

ISRM 1981. Brown, E. T. (ed.) *Suggested Methods of Rock Characterisation, Testing and Monitoring*. Pergamon Press, Oxford.

Jaeger, J. C., Cook, N. G. W. & Zimmerman, R. W. 2007. *Fundamentals of Rock Mechanics*. 4th edn. Blackwell Publishing, Oxford.

Jones, C. 2004. Improved Measurement of Pavement Strength by Dynamic Cone Penetrometer. R8157, Transport Research Laboratory, UK.

Larson, M. C. 1995. Use of seismic refraction techniques for investigating recent landslides in a tropical rainforest in Puerto Rico. *In*: Miller, R. L., Escalante, J. A., Reinemund, J. A. & Bergin, M. J. (eds) *Energy and Mineral Potential of the Central American – Caribbean Regions*. Springer, Berlin, 411–414.

Latham, R. S., Wooten, R. M. *et al*. 2010. The Hunters Crossing weathered rockslide, Haywood County, North Carolina, USA. *In*: Calcaterra, D. & Parise, M. (eds) *Weathering as a Predisposing Factor to Slope Movements*. Geological Society, London, Engineering Geology Special Publication, **23**, 149–165.

Mayne, P. W. 2007. *Cone Penetration Testing: A Synthesis for Highway Practice*. National Co-operative Highway Research Programme. Transportation Research Board, Washington, DC.

McDowell, P. W., Barker, R. D. et al. 2002. *Geophysics in Engineering Investigations*. Geological Society of London and CIRIA.

Norbury, D. 2010. *Soil and Rock Description in Engineering Practice*. Whittles Publishing, CRC Press, Taylor & Francis Group, Caithness, Scotland.

Reid, M. E., Baum, R. L., LaHusen, R. G. & Ellis, W. L. 2008. Capturing landslide dynamics and hydrological triggers using near real-time monitoring. *In*: Chen, Z., Zhang, J.-M., Ho, K., Wu, F.-Q. & Li, Z.-K. (eds) *Landslides and Engineered Slopes. From the Past to the Future. Proceedings of the 10th International Symposium on Landslides and Engineered Slopes*, 30 June–4 July 2008, Xi'an, China, Taylor & Francis Group, London, 179–191.

Reynolds, J. M. 2011. *An Introduction to Applied and Environmental Geophysics*. 2nd edn, John Wiley and Sons Ltd, Chichester.

Simons, N., Menzies, B. & Matthews, M. 2002. *A Short Course in Geotechnical Site Investigation*. Thomas Telford, London.

Taioli, F., Marchioreto, A., Machado, R. & Gallas, J. D. F. 2010. Indirect detection of boulders in tropical regions. *In*: Williams, A. L., Pinches, G. M., Chin, C. Y., McMorran, T. J. & Massey, C. I. (eds) *Geologically Active. Proceedings of the 11th IAEG Congress, Auckland, New Zealand, 5–10 September, 2010*, Taylor and Francis Group, London, 2511–2516.

USBR. 1998. *Engineering Geology Field Manual*. 2nd edn. United States Bureau of Reclamation, Department of the Interior, Washington DC.

Waltham, T. 2009. *Foundations of Engineering Geology*. 3rd edn. Spon Press, London.

Wyllie, D. C. & Mah, C. W. 2004. *Rock Slope Engineering*. 4th edn, Taylor and Francis Group, Suffolk.

B5 Slope movement monitoring

G. J. Hearn* & T. Hunt

URS Scott Wilson Ltd, Scott House, Alençon Link, Basingstoke, Hampshire RG21 7PP, UK
**Corresponding author (e-mail: gareth.hearn@scottwilson.com; garethhearn@talktalk.net)*

B5.1 Purpose

Slope movement monitoring enables rates of ground movement to be measured and assessed, allowing:

- confirmation as to which slopes are undergoing movement during the early stages of a road construction project;
- confirmation of the areal extent and depth of movement (if inclinometers and slip indicators are used; see Section B5.4);
- a decision to be made as to when investment in stabilization measures is required;
- the performance of implemented stabilization measures to be assessed and whether any further works are required; and
- the development of a monitoring database that relates recorded ground movement to rainfall patterns in order to be able to assess future risks.

Desk studies and field observations (Sections B2 & B3) will help define the more obvious landslide areas, but it is possible that other failed slopes may not be so readily identifiable and may be subject to slow or periodic ground movements. Slope monitoring can be used to help confirm whether movement is occurring in these areas. The sooner slope monitoring schemes are put in place, the greater the length of time available to collect a reliable dataset (e.g. discussion in Baynes *et al.* 2010).

B5.2 Monitoring methods

Approaches to slope monitoring can be subdivided into methods based on observation (qualitative) and measurement (quantitative). The more common of these are outlined below.

B5.2.1 Observational methods

B5.2.1.1 Remote sensing

Historically, the use of remote sensing as a possible means of landslide monitoring was limited to the interpretation of slope changes between successive flyovers of aerial photography. Text box B5.1 illustrates the use of successive aerial photography for projects in Nepal and Hong Kong. In the latter case the aerial photograph record was comprehensive and was able to provide the required data, while in the former case the exercise proved unsuccessful for the following reasons:

- limited sets of successive photographs and occasional incomplete cover;
- shade, relief distortion and cloud effect (common in mountain terrain); and
- small scale and limited resolution in some of the photography.

Liu & Wang (2008) review some of the methods of landslide monitoring from remote sensing. Any satellite or airborne imagery capable of providing a digital elevation model (DEM) (Section B2.3) can be used for monitoring purposes if the changes in elevation and slope morphology over a given monitoring period are significant. Singhroy (2005) and Henderson *et al.* (2011), for example, describe the use of InSAR (Interferometric Synthetic Aperture Radar) to provide high-accuracy slope displacement information. However, as with any remote sensing, it should be noted that InSAR data provide information on changes in elevation only, that is, not absolute ground movement. There can also be problems with the use of the technique in areas of dense vegetation cover. Cascini *et al.* (2009) and Plank *et al.* (2010) describe the use of differential radar interferometry (DInSAR) – both satellite and airborne – to detect landslide movements. Plank *et al.* (2010) describe how various imaging issues and complications can be rectified by using GIS correction factors for land cover and shadow effects in the radar image. InSAR techniques can be expensive and complex to use, and probably have limited potential application to low-cost roads at present. Repeat airborne imagery, such as LiDAR, can allow changes in slope morphology to be detected to an accuracy of *c.* 0.5 m horizontally and 0.1–0.25 m vertically. Chigira *et al.* (2010) describe how LiDAR was used to map landslides and monitor ground movements taking place beneath a dense canopy in Japan while Yin *et al.* (2010) describe its use in combination with surface movement monitoring in China. Dense vegetation cover, low cloud and atmospheric haze will however complicate the

From: Hearn, G. J. (ed.) *Slope Engineering for Mountain Roads*. Geological Society, London, Engineering Geology Special Publications, **24**, 127–133.
DOI: 10.1144/EGSP24.9 0267-9914/11/$15.00

Text box B5.1. Mapping landslide movements from aerial photographs

Nepal

A study was undertaken in east Nepal (Hearn 2003) whereby all available aerial photography was used to assess the development of landsliding over time. The photography ranged in age from 1954 to 1990 but coverage and scale were the limiting factors in both the period and degree of detail that could be determined for monitoring purposes. Although some useful results were obtained, they were largely inconclusive.

Hong Kong

Stereo aerial photograph coverage in Hong Kong dates back to 1924 and has comprised regular flyovers since then, including annual updates in recent years (e.g. Parry & Ng 2010). The map below shows the distribution and record date (when they first became apparent in the aerial photography) of landslides mapped from this photographic record as part of a landslide hazard study undertaken for a proposed residential development in the New Territories. The mapping allowed an assessment to be made of landslide frequency, thus contributing to the assessment of hazard. Hong Kong is one of a very small number of areas in the world where such a long and continuous photographic record exists. Nevertheless, the example illustrates the value in using historical data in this way to assess rates of change in the landscape for planning and engineering purposes.

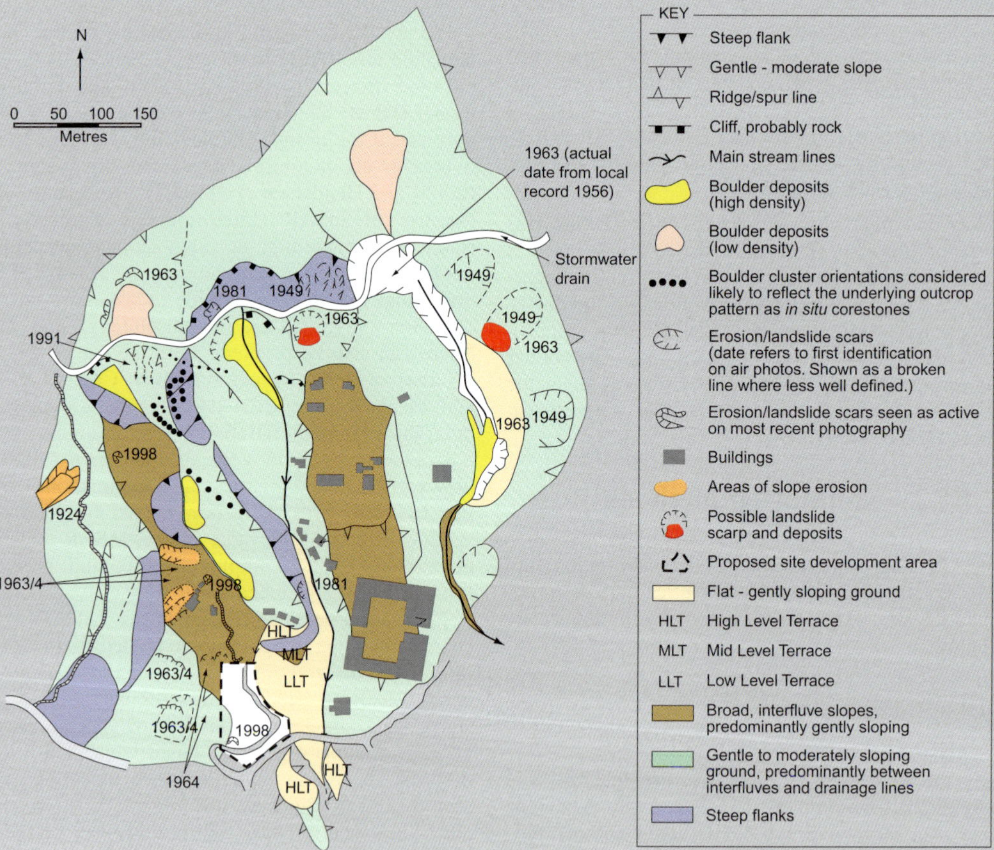

interpretation, and these measurement accuracies may not always be achievable.

Terrestrial laser scanning can also provide the means of monitoring changes caused by ground movement. The technique has been used to map and monitor movements in cliff faces and in landslide areas (e.g. Rosser *et al.* 2005; Kasperski & Varrel 2010), and offers potential application to the structural geological and geomorphological assessment of landslides along mountain roads (Dunning *et al.* 2009).

B5.2.1.2 Field inspections

Basic visual observation (usually on foot or by vehicle or helicopter) is the cheapest and, in many ways, the most effective approach available to identify ground movements and related effects such as damage and deflection to retaining walls, cracking to road surfaces and erosion beneath culverts. However, there is clearly a minimum extent of movement that is detectable from visual observation alone, and recourse may have to be made to conventional ground survey methods. Even small ground movements can sometimes have important implications for the stability of a road and its structures.

B5.2.1.3 Line-of-sight monuments

Usually wooden pegs are driven into the ground surface in a straight line perpendicular to the steepest slope or the direction of observed movement. Those stakes located within failing ground become displaced and the extent of movement can be judged from the line-of-sight (Fig. B5.1). The technique does not allow accurate measurements to be made, but serves as a rapid and inexpensive means of monitoring less critical areas.

B5.2.2 Measurement methods

The following methods are regularly used in landslide monitoring schemes, and have potentially useful application to low-cost roads:

- repeat survey of surface monuments using total station methods (Fig. B5.1);
- tension crack displacement measurements (Fig. B5.1);
- inclinometers (Section B5.4); and
- slip indicators (Section B5.4).

B5.2.2.1 Repeat survey of surface monuments

Monuments should be constructed in concrete and be of sufficient size that they are easily visible and will not become subsequently disturbed by site operations. If concrete is impracticable, then painted steel bars driven deeply (preferably >1 m) into the ground may suffice. A minimum of three control points are required in stable ground for accurate triangulation outside the existing and expected future limits of landslide movement. It is always preferable to fix permanent control points on stable rock outcrops where these are present.

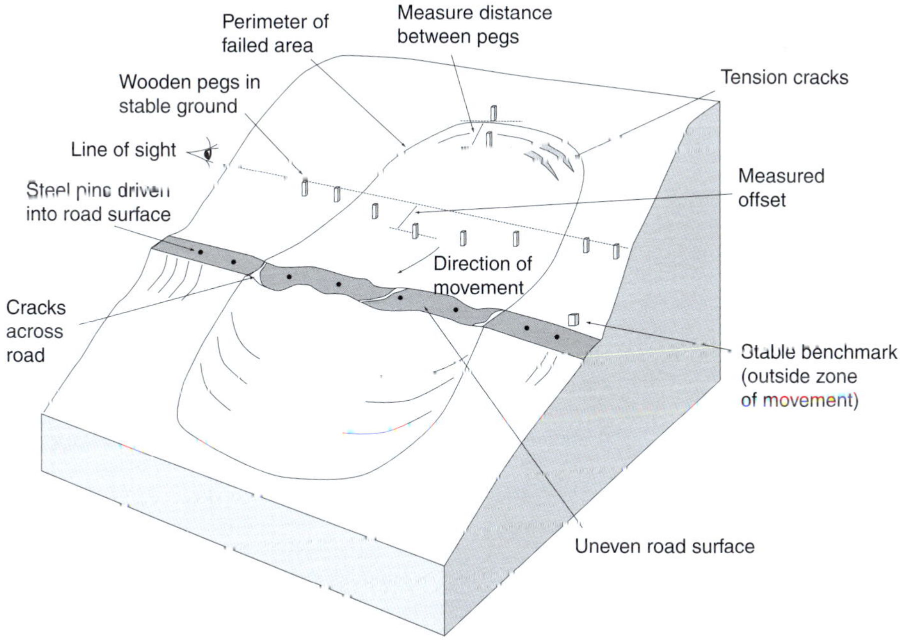

Fig. B5.1. Common simple methods of surface movement monitoring.

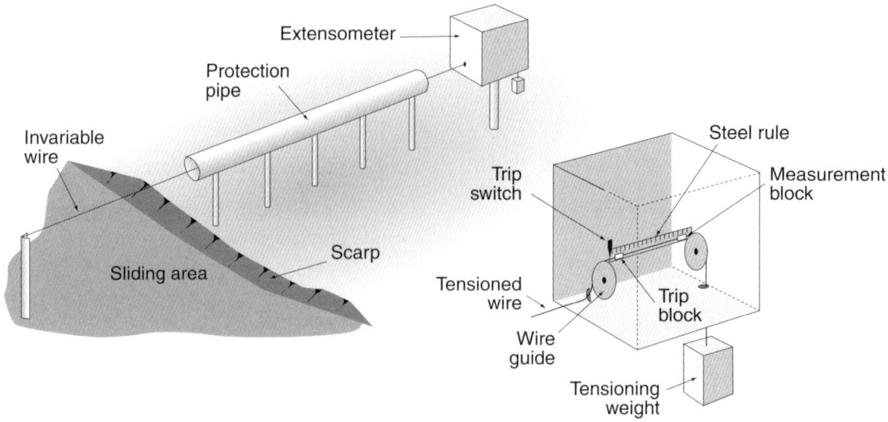

Fig. B5.2. Typical layout of crack extensometer.

B5.2.2.2 Tension crack measurements

The location and dimensions of tension cracks should be documented during site inspections, particularly those cracks identified within or directly adjacent to the road and its earthworks slopes. The simplest form of crack monitoring involves the installation of pairs of markers on opposite sides of a crack with regular measurement of their horizontal and vertical offsets. It is important that monitoring personnel are trained to differentiate between tension cracks and ground desiccation cracks which may occur in clay or mud deposits. In high-risk locations, extensometers can be used to allow continual monitoring of tension cracks (Fig. B5.2). The monitoring data can be relayed remotely via a transmitter to a receiving centre.

Text box B5.2. Inclinometers and slip indicators

Inclinometers are placed in vertical boreholes through a landslide mass and suspected slip plane into the underlying undisturbed ground to monitor deformation and tilt of the borehole casing. Successive measurements down the length of the tubing enable subsurface motion to be determined as a function of depth and as a function of time. The top of the casing should be accurately surveyed relative to stable survey monuments in case of significant lateral displacement.

There are essentially two types of inclinometer.

- Portable probe inclinometer: a portable probe instrument is the standard device for surveying the inclinometer casing. It is manually inserted into longitudinal grooves in the casing and lowered and raised, and measurements of inclination are made by the sensor at fixed increments. Two axes of grooves in the casing allow measurements in two perpendicular planes.
- In-place inclinometer strings: sensors are permanently installed at intervals along the borehole axis and inclination readings are collected continuously or at fixed time intervals by a remote data logger. Readings can be regularly downloaded in the field for subsequent processing or can be automatically transmitted for near-real-time remote monitoring. The combination of automatic data acquisition with telemetry or data transmission systems (e.g. by a fibre-optic network) provides the significant advantages of near-real-time processing; alarm systems can be put in place for monitoring ground deformation in critical locations. These systems are more relevant to high-risk landslide sites than to general movement monitoring on low-cost roads.

Although not inclinometers as such, slip indicators can provide a simple method for locating the depth to failure surfaces. Metal rods (generally an upper and a lower rod) are used to regularly probe a tube inserted in a borehole. The probe may be stopped by deformation in the tubing which potentially indicates the depth of the shear surface but does not provide information on rate or direction of movement. A light lowered down the tube can also indicate the depth of movement when the depth to which it can no longer be seen is recorded. This method is likely to be less accurate and other effects may lead to misinterpretation of the results.

Text box B5.3. The use of inclinometer readings to assess movement depth in conditions of complex failure

The use of inclinometers as a means of helping to determine failure surface depths over short monitoring periods is illustrated by the Halsema Highway rehabilitation project in the Philippines (Text box A3.5). Significant ground movement took place in five main areas following typhoons that occurred between 2002 and 2004. While most of the reinstatement design was undertaken using engineering geological mapping and trial pitting, in these five areas it was decided that a design could not be finalized without confirmation of where slip planes were located at depth. A drillhole investigation proved inconclusive and so inclinometers were installed to ascertain movement depths. One of these critical areas is described below.

Typhoon Igme occurred during the summer of 2004 and caused the entire failure of the road formation over a length of 50 m. This failure was the result of progressive landslide movements that had been taking place over several years at this location. The area is shown below in the geomorphological sketch map prepared prior to the ground investigation. The mapping led to the conclusion that the failure affecting the road might daylight on the slope up to 40 m vertically below road level, but extend only as far as the cut slope on the inside edge of the road. Two drillholes were put down and one inclinometer was installed along the outside edge of the road. The drill core revealed alternating sequences of strong and weak andesite, varying from Weathering Grade II to Weathering Grade IV. Very weak and disturbed material in the core that might represent failure surfaces or shear zones was recorded at 7.3–7.5, 13.3–13.6, 34 and 38 m below the level of the outside edge of the road. However, the core was so variable and its quality so poor that a single shear surface could not be positively identified. The inclinometer record, spanning a period of only two wet season months, was used to assist in slip zone identification. A marked deflection was recorded at 3.5–4.5 m and a 'kink' at 13–15 m. On a cross-section, a straight line drawn between the interpreted toe of the ground movement on the mapping and the inside edge of the road intersected the drillholes on the outside edge of the road at *c.* 10 m below ground level. It was concluded that the bulk of the movement was probably taking place at 13–15 m below the level of the outside edge of the road, and the design was finalized accordingly. Although a longer inclinometer record would have provided more conclusive evidence of failure depth, this short record proved effective given the extent of movement that occurred during the recording period.

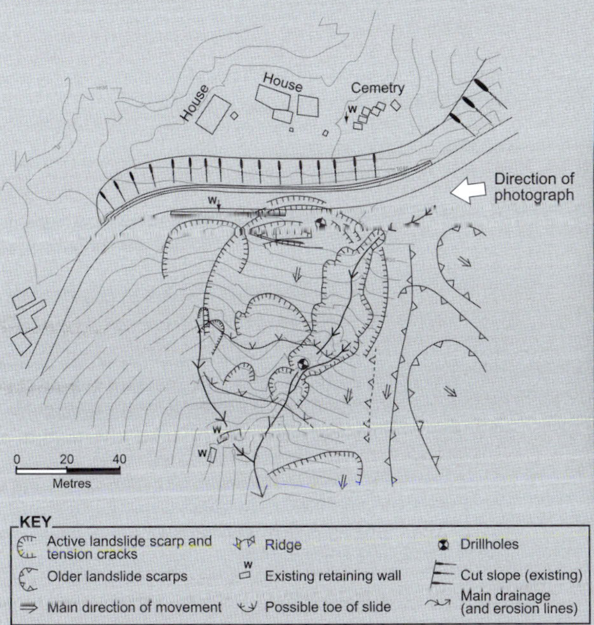

B5.3 Interpretation of surface monitoring data

It is usual to interpret slope movement monitoring data in relation to an independent variable, for example the pattern of ground movement plotted against daily or monthly rainfall or the progress of a deep excavation during construction. The plot of cumulative rainfall against cumulative movement can be particularly helpful in the examination of ground movements. Chang *et al.* (2005) for example describe correlations between surface and subsurface landslide monitoring and rainfall on a mountain road in Taiwan.

When comparing ground movements against rainfall, it is preferable to use raingauges that have been established over several decades in order to be able to carry out meaningful magnitude against frequency analysis (TRL 1997). However, since the nearest raingauge may be some distance from the slope concerned, there may be no choice but to establish a raingauge on site in conjunction with slope monitoring. While this will provide useful comparative information from one day, month or year to the next it may not allow statistical analysis of slope movement v. recorded rainfall to be carried out until at least a 10-year record is established. Recorded slope movement data can also be compared to groundwater levels, determined from standpipes or piezometers (Section B4.3.7).

Interpretation of monitoring data will indicate whether movement:

- is accelerating and, if so, whether it would be prudent to increase the prioritization for remedial works in order to avert total slope failure; or
- is of such a small magnitude, or is slowing or stopping, that remedial works can be delayed or postponed indefinitely.

Slope movement monitoring can also enable the performance of stabilization works to be assessed. Usually this is done through visual monitoring, although the survey of surface monuments and the use of inclinometers (Section B5.4) are often used at high-risk locations. One of the main considerations, and usually the ultimate limitation, is whether or not the monitoring period includes rainfall and groundwater levels that will provide the design level test of the works.

B5.4 Assessing depth of slope movement from monitoring data

Movement depth may be determined using either or both of the following methods:

- inclinometers (Text boxes B5.2 & B5.3) to detect and measure displacement rates with depth below ground level; and
- probing open standpipes or using slip indicators (Text Box B5.2) to determine depth of internal deformation beyond which probing is prevented.

Further information on field instrumentation with respect to landslide monitoring can be found in Mikkelsen (1996) and Hunt (2007). Singer *et al.* (2010) and Lin & Chung (2010) describe the use of Time Domain Reflectometry (TDR) to monitor movement on discrete (well-defined in this context) landslide slip surfaces. The technique uses electrical pulses which are sent through a coaxial cable to record the depth of rupture and can be as little as 20% of the cost of conventional inclinometer monitoring. Dixon *et al.* (2010) describe the use of acoustic emission monitoring to detect displacements on shear surfaces in boreholes.

References

Baynes, F. J., Moon, A. T. & Gough, J. A. 2010. Investigating the Kapit Landslide. *In*: Williams, A. L., Pinches, G. M., Chin, C. Y., McMorran, T. J. & Massey, C. I. (eds) *Geologically Active. Proceedings of the 11th IAEG Congress, Auckland, New Zealand, 5–10 September, 2010*, Taylor and Francis Group, London, 35–43.

Cascini, L., Fornaro, G., Peduto, D., Ferlisi, S. & Di Nocera, S. 2009. A new approach to the use of DInSAR data to study slow-moving landslides over areas. *Proceedings of the Fringe Workshop.* Frascati, Italy. *Advances in the Science and Applications of SAR Interferometry, ESA ESRIN.* November/December 2009.

Chang, M., Chiu, Y., Lin, S. & Ke, T.-C. 2005. Preliminary study on the 2003 slope failure in Woo-wan-chai area. Mt Ali Road, Taiwan. *Engineering Geology*, **80**, 93–114.

Chigira, M., Doshida, S. & Nakamura, T. 2010. Morphological features of slope movement identified by the airborne laser scanner. *In*: Williams, A. L., Pinches, G. M., Chin, C. Y., McMorran, T. J. & Massey, C. I. (eds) *Geologically Active. Proceedings of the 11th IAEG Congress, Auckland, New Zealand, 5–10 September, 2010*, Taylor and Francis Group, London, 103–110.

Dixon, N., Spriggs, M. P., Meldrum, P., Ogilvy, R., Haslam, E. & Chambers, J. 2010. Development of a low cost acoustic emission early warning system for slope instability. *In*: Williams, A. L., Pinches, G. M., Chin, C. Y., McMorran, T. J. & Massey, C. I. (eds) *Geologically Active. Proceedings of the 11th IAEG Congress, Auckland, New Zealand, 5–10 September, 2010*, Taylor and Francis Group, London, 1803–1810.

Dunning, S. A., Massey, C. I. & Rosser, N. J. 2009. Structural and geomorphological features of landslides in the Bhutan Himalaya derived from terrestrial laser scanning. *Geomorphology*, **103**, 17–29.

Hearn, G. J. 2003. Landslide risk warning for mountain communities and rural infrastructure planning. *ISDR Highlights*, **11**, 2–3.

Henderson, I. H. C., Lauknes, T. R., Osmundsen, P. T., Dehls, J., Larsen, Y. & Redfield, T. F. 2011. A structural, geomorphological and InSAR study of an active rock slope failure development. *In*: Jaboyedoff, M. (ed.) *Slope Tectonics.* Geological Society, London, Special Publications, **351**, 185–199.

HUNT, R. E. 2007. *Geotechnical Investigation Methods. A Field Guide for Geotechnical Engineers*. CRC Press, Taylor and Francis Group, Florida.

KASPERSKI, J. & VARREL, E. 2010. Characterization of the active zone of the Séchilienne landslide using terrestrial laser scanning. *In*: WILLAMS, A. L., PINCHES, G. M., CHIN, C. Y., MCMORRAN, T. J. & MASSEY, C. I. (eds) *Geologically Active*: *Proceedings of the 11th IAEG Congress, Auckland, New Zealand, 5–10th September, 2010*, Taylor and Francis Group, London, 1981–1988.

LIN, C. P. & CHUNG, C. C. 2010. *In*: WILLIAMS, A. L., PINCHES, G. M., CHIN, C. Y., MCMORRAN, T. J. & MASSEY, C. I. (eds) *Geologically Active. Proceedings of the 11th IAEG Congress, Auckland, New Zealand, 5–10 September, 2010*, Taylor and Francis Group, London, 2087–2094.

LIU, S. T. & WANG, Z. W. 2008. Choice of survey methods for landslide monitoring. *In*: CHEN, Z., ZHANG, J.-M., HO, K., WU, F.-Q. & LI, Z.-K. (eds) *Landslides and Engineered Slopes. From the Past to the Future. Proceedings of the 10th International Symposium on Landslides and Engineered Slopes*, 30 June–4 July 2008, Xi'an, China, Taylor & Francis Group, London, 1211–1216.

MIKKELSEN, P. E. 1996. Field instrumentation. *In*: TURNER, A. K. & SCHUSTER, R. L. (eds) *Landslides – Investigation and Mitigation*. Transportation Research Board Special Report **247**, 278–316.

PARRY, S. & NG, K. C. 2010. The assessment of landslide risk from natural slopes in Hong Kong: an engineering geological perspective. *Quarterly Journal of Engineering Geology & Hydrogeology*, **43**, 307–320.

PLANK, S., SINGER, J., THURO, K. & MINET, CH. 2010. The suitability of the differential radar interferometry method for deformation monitoring of landslides – a new GIS based evaluation tool. *In*: WILLIAMS, A. L., PINCHES, G. M., CHIN, C. Y., MCMORRAN, T. J. & MASSEY, C. I. (eds) *Geologically Active. Proceedings of the 11th IAEG Congress, Auckland, New Zealand, 5–10 September, 2010*, Taylor and Francis Group, London, 2345–2352.

ROSSER, N. J., PETLEY, D. N., DUNNING, S. A & ALLISON, R. J. 2005. Terrestrial laser scanning for monitoring the process of hard rock coastal cliff erosion. *Quarterly Journal of Engineering Geology and Hydrogeology*, **38**, 363–375.

SINGER, J., FESTL, J. & THURO, K. 2010. Application of Time Domain Reflectometry (TDR) as a monitoring system for subsurface deformations. *In*: WILLIAMS, A. L., PINCHES, G. M., CHIN, C. Y., MCMORRAN, T. J. & MASSEY, C. I. (eds) *Geologically Active. Proceedings of the 11th IAEG Congress, Auckland, New Zealand, 5–10 September, 2010*, Taylor and Francis Group, London, 2459–2465.

SINGHROY, V. 2005. Remote sensing of landslides. *In*: GLADE, T., ANDERSON, M. & CROZIER, M. (eds) *Landslide Hazard and Risk Journal*. J. Wiley & Sons Ltd, Chichester, 469–492.

TRL 1997. *Principles of Low Cost Road Engineering in Mountainous Regions*. Overseas Road Note 16, Transport Research Laboratory, Crowthorne.

YIN, Y., ZHENG, W., LIU, Y. & ZHANG, J. 2010. Integration of GPS and InSAR techniques for the monitoring of a landslide in Danba, Sichuan, China. *In*: WILLIAMS, A. L., PINCHES, G. M., CHIN, C. Y., MCMORRAN, T. J. & MASSEY, C. I. (eds) *Geologically Active. Proceedings of the 11th IAEG Congress, Auckland, New Zealand, 5–10 September, 2010*, Taylor and Francis Group, London, 1567–1576.

Part C: Design and Construction

C1 Route corridor and alignment selection

G. J. Hearn* & T. Hunt

URS Scott Wilson Ltd, Scott House, Alençon Link, Basingstoke, Hampshire RG21 7PP, UK

*Corresponding author (e-mail: gareth.hearn@scottwilson.com; garethhearn@talktalk.net)

C1.1 Controlling factors

As outlined in Section A2, there are several factors that determine the selection of route corridors and the design of the alignment within them. These include road network planning and traffic forecasts, construction and maintenance costs and engineering, socio-economic, environmental and political considerations. The following discussion focuses on route corridor and alignment design in relation to topography and geometric standards, and describes how these considerations interface with those of slope stability and ground conditions.

C1.2 Route corridor identification and selection

The selection of the route corridor is the most critical element in any road construction project in hilly or mountainous terrain. It affects route length, construction quantities, cost and alignment stability, as well as environmental and social impact. In many circumstances the selection of the route corridor may be straightforward, for example for roads linking villages and towns on ridge tops or valley floors. In other cases, however, several alternatives will present themselves. These will be determined, *inter alia*, by topography, hydrology and river crossings, land use and the location of landslides and other difficult ground conditions. The availability of suitable naturally occurring construction materials may also be a consideration in route corridor selection in some circumstances.

C1.2.1 Topography

Usually the objective will be to avoid the steepest areas and to develop an alignment that follows the lineaments or *grain* of the topography, as controlled by the underlying geology. In fold mountain belts, for example, there is usually a distinct topographic grain; in the Himalayas it is essentially east–west and in the Andes it is north–south. Where there is a choice, alignments that follow this grain will normally require less rise and fall. Earthworks quantities and slope stability hazards will also tend to be less if, for example, a ridge or valley floor alignment can be followed. Where roads are required to cross the grain of the topography complex alignments will usually result, comprising some or all of the following:

- valley floor (floodplain and river terraces);
- valley side traverses on sloping ground (side-long ground);
- high level terraces and midslope benches;
- climbing sections (hairpin stacks) to connect sections of alignment at different elevations; and
- ridge lines.

While there will be good engineering and economic reasons to follow the more gently sloping terrain in any landscape, this type of terrain could reflect an underlying weak geology that is prone to slope instability. It also may reflect areas of failed ground (Fig. C1.1 and Section C1.2.3) and, while construction costs might be lower, the cost of maintaining a road in this terrain in the longer term could be higher. Conversely, alignments that follow the more prominent features of the landscape, such as steep-sided ridges and steep slopes, are likely to involve greater earthworks costs but may well encounter stronger and more stable materials that pose fewer problems for road maintenance. The desk study and field mapping techniques described in Sections B2 and B3 are critically important in identifying these areas and in contributing to these decisions.

Geometric standards and, in particular, ruling gradients in relation to topography are also of critical importance in the identification of route corridor options. However, if the corridor is wide enough and the topography is not extreme, it will usually be possible to design an alignment conforming to the required standards (Section C1.3).

C1.2.2 River crossings

The location and number of large river crossings will also influence the selection of the route corridor and the detailed location of the alignment within it. Hill and ridge routes will have substantially fewer river crossings than lower valley side and valley floor routes. Flooding, scour and sediment

From: Hearn, G. J. (ed.) *Slope Engineering for Mountain Roads*. Geological Society, London, Engineering Geology Special Publications, **24**, 135–144.
DOI: 10.1144/EGSP24.10 0267-9914/11/$15.00

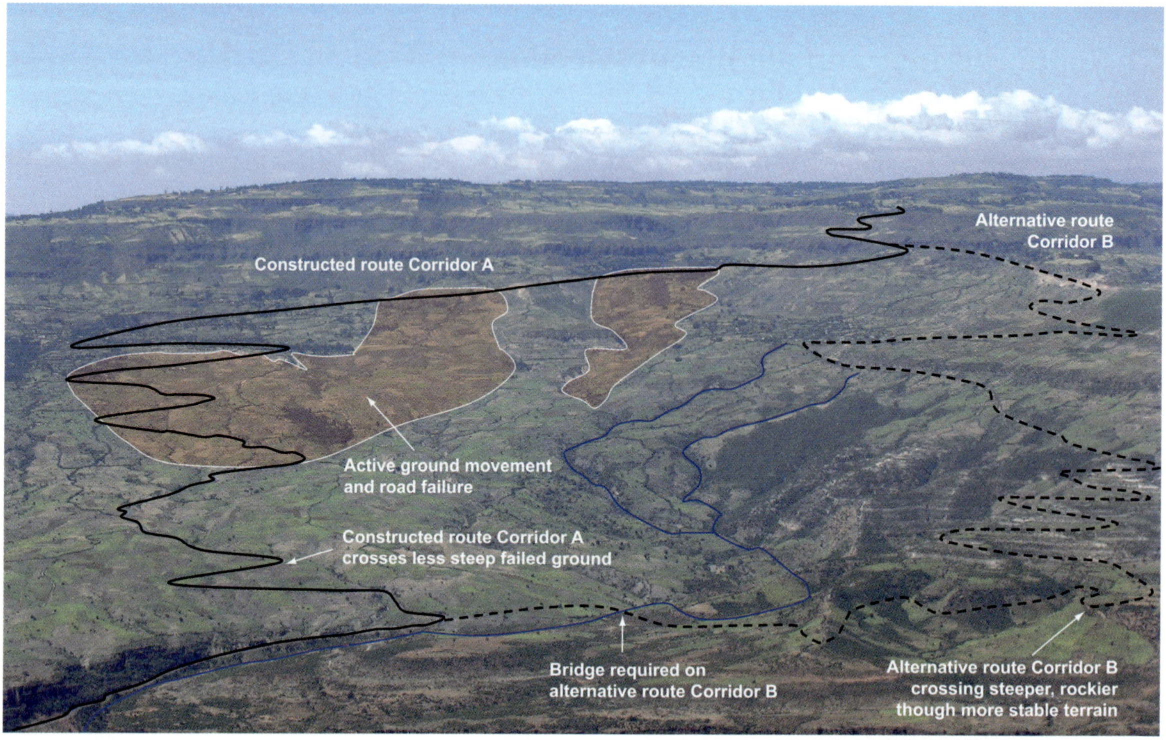

Fig. C1.1. Choice between gentle, unstable ground (Corridor A) and steep and rocky (although largely stable) ground (Corridor B).

transport and deposition are common hazards in most mountain areas, but their size and potential consequences are usually most severe along alignments located in the lower portions of the landscape. River meanders also pose a significant hazard. Meanders migrate downstream, sometimes within engineering timescales, and this can cause scour of road structures and instability on adjacent valley sides. Factors that affect the rate of meander migration include the magnitude of flood flows in any given period, the susceptibility or resistance to erosion of the bed and bank material and the presence of hard points (either geological or man-made) that exert some control on downstream migration. While rates are variable, meander progression of the order of 1–2 m/year can be considered quite normal in some environments; much higher rates have been observed around the world. River confluences can also be problematic locations for alignments. Alluvial fans from tributaries can build up by several metres during single floods (Fig. C1.2) and can force the main channel against the opposite river bank (Fig. A3.19), encouraging scour and failure of the adjacent valley side. In Figure C1.2 the constructed alignment across a tributary fan is shown in solid white. A safer alignment might have been as shown by one of the two broken white line options (see Section C7.6 for further consideration of fan crossings).

C1.2.3 Slope stability

The desk study and mapping techniques described in Sections B2 and B3 should allow landslide zones and route corridors to be defined that avoid hazard areas to the greatest extent possible. Sources of rockfall and areas of taluvium and colluvium that might become unstable naturally or due to road construction effects can also be identified by these techniques and avoided wherever possible. In addition, some landforms and land uses pose potential implications for slope stability and construction difficulty, and these should be considered in the identification and comparison of route corridors. Table C1.1 lists some of these features and their potential implications.

A road will experience long-term problems if constructed over the following unstable ground conditions:

- landslides, or areas of taluvium and colluvium, that remain active or become reactivated due to:
  - high water tables during the wet season;
  - stream or river erosion; or
  - construction effects; and
- slopes comprising talus, taluvium and landslide debris at or near their limiting angles for stability (usually 35–38°, but much less for fine-grained materials including colluvium).

Fig. C1.2. Bridge and bridge approaches buried by tributary fan deposition.

Where alignments cross landslides and unstable ground, periodic or catastrophic ground movements can cause road blockages, progressive road subsidence or sudden breaches in road access (Section A3.6). The cost of keeping a road open in these circumstances may become excessive and may not even be practicable in the long term. Figure C1.3, for example, shows the initial earthworks for a hairpin stack that was eventually abandoned due to slope failures taking place in the underlying highly disturbed quartzite rock and taluvium. Excavated material was dumped onto sections of road below creating slope instability, uncontrolled runoff and erosion, resulting in the worst conceivable combination of ground conditions and engineering practice. Figure C1.4, by contrast, shows a hairpin stack that has been designed and constructed to fit the topography as much as possible, thus reducing earthworks and minimizing slope disturbance. A balance of cut and fill (Section C2.5) has been achieved through very careful alignment design and the use of retaining walls to support road fill and minimize excavations. Drainage has been controlled using lined channels and cascades (note the cascade that discharges side drain water over the retaining wall down to river level at the base of the stack). Vegetation has also been preserved, to an extent, although a significant degree of regrowth has taken place due to the underlying stability of the roadside slopes.

C1.2.4 Land use

Land use is also an important factor in the selection of an alignment (Table C1.1). In mountainous areas, roads that cross irrigated farmland may suffer drainage and instability problems caused by high water tables while at the same time consuming valuable farmland. In densely populated regions where land is at a premium, areas that remain covered by forest or jungle are often either too steep or too unstable to cultivate. Road construction in these areas may also prove very difficult. Forests are also protected in many areas, and road construction may in any case be prohibited.

C1.2.5 Comparison of alternatives

Although it can rarely be proven, the cost of the most stable corridor for a low-volume road is likely to be the lowest in the long term. Where this might not apply is where the most stable alignment involves lengthy departures from the shortest distance route or where it requires costly additional bridges, major rock excavation or a large number of retaining structures. Figure C1.1 illustrates this comparison. The constructed Corridor A has required reconstruction over several sections due to recurring slope failure. This is likely to have proven more costly in the longer term than had Corridor B been constructed, even though Corridor B would have been more expensive to construct, with its additional bridge and rock excavation.

The higher the AADT of a road, the more important alignment length is likely to become. On motorways and expressways it is usual to find geotechnical problems 'engineered out' rather than have the length of alignment increased to avoid them.

Table C1.1. *Landform and land use features, slope stability and construction difficulty*

Alignment type	*Landform and land use features*	*Typical problems encountered*	*Likelihood of existing landslide*	*Potential for road construction to cause landslide*
Ridge top	Rounded relief	Deeply weathered soils likely; some erosion potential	Unlikely	Possibly
	Sharp relief	Rock at surface; costly and difficult rock excavation possible	Unlikely	Unlikely
	Irregular relief	Difficult alignment along ridge top between high points and low points	Possibly	Possibly
	Asymmetric relief	Joint-controlled slopes will influence stability of alignments and cut slopes	Possibly – check for failed debris downslope	Possibly on dip slope
	Ridge lines generally Ridge lines generally	May be subject to greater rainfall than valley sides May be more affected by seismicity (topographic amplification)	Possibly Possibly	Possibly Possibly
Valley side	Slopes are steeper than 40°	Probably underlain by rock, therefore alignment likely to be more costly to construct but less costly to maintain	Unlikely	Possibly, depends on jointing
	Slopes are 35–40°	Potential to be shallow taluvium on rock	Possibly	Possibly
	Slopes are 20–35°	Potential to be deep taluvium, colluvium or failed slope	Possibly	Possibly
	Continuous rock slopes with persistent jointing approximately parallel to slope	Depending on strength of rock mass this joint set could be problematic in excavations and foundations	Possibly – check for failed debris downslope	Likely
	Embayments	Either erosional in origin or formed by landslide(s)	Probably	Possibly, reactivated movements in landslide debris
	Areas of irrigated paddy field	Drainage problems likely; soils possibly taluvial/colluvial in origin and potentially unstable locally	Possibly, but mass as a whole may be stable	Possibly
	Forest/jungle areas on otherwise cultivated hillside	Possibly areas of wet ground, steep slopes, instability that cannot be cultivated	Possibly	Possibly
	Rounded spurs	Probably formed in residual soils and stable	Unlikely	Unlikely
	Elongated mid-slope benches	Either ancient river terraces or rock benches; both stable and 'easy' for road construction	Unlikely	Unlikely
	Local mid-slope benches	Could be as above, or part of deep-seated landslide	Possibly	Unlikely
Valley floor	Steep slopes forming margins of river channel (i.e. no river terrace)	Possibly unstable; difficult for road alignments, especially on meander bends; possible flood risk and high water table	Likely	Possibly
	Steep slopes forming valley side margins to river terrace	Possible ancient landslides and high water table	Possible	Possibly
	River terrace	Possible flood risk, soft soils and terrace edge scour; high water table	Unlikely, except at terrace edges	Unlikely
	Tributary streams	Possibly active debris flows and debris fan deposition causing scour and blockage/damage to road structures; possible flood risk and high water table	Debris flows only	Debris flows only

Fig. C1.3. Hairpin stack construction abandoned due to slope instability.

Fig. C1.4. Hairpin stack constructed on stable slope with minimal earthworks disturbance.

When more than one corridor option is identified, it may be quite clear as to which is preferred in terms of topography, engineering, stability, environmental considerations and cost. However, each option may encounter areas of difficult topography and existing or potential slope instability, and it will be necessary to ascertain the level of hazard posed by each. As part of the cost estimation exercise (which at the feasibility stage is probably only possible to an accuracy of about $\pm 20\%$ at most) the lengths of the alignment crossing landslides or potentially unstable ground can be measured (Section B2.5) and appropriate factors applied to the assumed per metre road construction costs. In a recent example from Ethiopia, the per metre construction cost at landslide sites was approximately five times that across the remaining stable sections of alignment due to the need for increased earthworks, deeply-founded retaining structures and slope drainage.

C1.3 Alignment and carriageway design

Once a route corridor is selected it will be necessary to identify a preferred alignment within it based on alignment length, topography and geometric standards, construction quantities and cost, preferred locations for river crossings, slope stability (and other difficult ground conditions), socio-economic, environmental impact and land use considerations. In some terrain it may be a fairly easy task to identify the optimum alignment; more than one option may need to be considered in other cases. A case study presented in Section C1.4 illustrates this.

C1.3.1 Alignment

The opportunity to avoid difficult topography and unstable slopes will often depend on the geometric standards that need to be followed. This relates principally to allowable road gradients, but the minimum horizontal curve radius and road width (Section C2.1) are also essential factors to take into consideration during detailed alignment design. Text box C1.1 provides some background information and discussion on geometric standards for low volume roads.

The length of road at maximum gradient is another important consideration. Most national standards will specify a maximum length of road at steepest gradient, to be followed by a minimum recovery length at a shallower gradient. In Nepal for instance, the maximum gradient of 12% for low-volume roads is only permitted for a length of 300 m which is then followed by a maximum gradient of 4% for at least 150 m.

With regard to minimum horizontal curve radius, one consideration will be the need to maintain sufficient space for the swept paths of larger vehicles and to allow drivers to manoeuvre when approaching other vehicles. Special requirements for curve widening may have to be adopted depending on the type of vehicle for which access is required, as would be the case for a road providing access to a hydropower plant or a mine, for example. Strategic roads, such as those constructed for reasons of national security, are often designed to a higher standard than that required for traffic volumes alone.

A relaxation of alignment standards may be necessary in order to avoid major earthworks and thereby achieve an

Text box C1.1. Recommended geometric standards for low-volume roads

The following has been adapted from Overseas Road Note 6 (TRRL 1988) but with some changes to the standards. Geometric standards should take into account:

- the function of the road;
- the volume of traffic and traffic safety requirements; and
- the terrain through which the road is passing.

The function of a road can be considered to fall into one of three categories:

- arterial, where the road connects national centres from which traffic is mainly derived;
- collector, where the road connects rural areas to the urban centres or the arterial network; and
- access, where the road provides access to small settlements.

In rural areas the AADT (average annual daily traffic) is often less than 100. However, road traffic can grow at significantly high rates in response to economic development and increased mobility, and an average annual growth rate of 10% or more might not be unrealistic. Due to the difficulties in improving the geometric standards of existing roads in mountainous areas to take into account increased traffic flows, some authorities design for predicted traffic flows 10 years after the road is expected to be completed.

Terrain is usually taken into consideration through side slope categorization, that is, level: defined as 0–5° side slope; rolling: 5–15° side slope; hilly: 15–30° side slope; mountainous: greater than 30° side slope. The table below shows the recommended carriageway and shoulder widths, maximum gradient and design speeds for the various types of road and terrain.

Recommended geometric standards

Road function	*AADT*	*Minimum width (m)*		*Max gradient (%)*	*Maximum design speed (km/h)*			
		Carriageway	*Shoulder*		*Level*	*Rolling*	*Hilly*	*Mountainous*
Arterial	<1000	5.5	1.0	10	85	70	60	60
Collector or access	100–400	5.0	1.0	10	70	60	50	40
Access	20–100	3.0	1.5	15	60	50	40	20
Access	<20	2.5–3.0	Passing places	15/20	n/a	n/a	n/a	n/a

Once the design speed has been determined, the minimum horizontal curve radii given in the table below are recommended.

Minimum horizontal curvature

Design speed (km/h)	*Minimum horizontal curve radius (m)*
85	210
70	130
60	85
50	60
40	30
20	15

Table C1.2. *Typical national standards for low-volume roads in mountainous terrain*

Attribute	*Nepal*	*Vietnam*	*Laos*	*Ethiopia*
AADT	100–400	50–300	100–300	100–200
Carriageway width (m)	3.5	3.5	5.5	7.0
Shoulder width (m)	0.5	1.5	0.5	0.0
Max grade (%)	12	12	9	9
Design speed (km/h)	n/a	25	20	40
Minimum horizontal curve radius (m)	12.5	15	20	50

acceptable level of return on the investment. Significant savings may arise from the inclusion of a short section of lower standard road where steep side-long ground or slope instability is encountered. Marwa & Kimaro (2005), for example, describe the use of single lane access with passing places in Tanzania to avoid excessive cut and fill in areas of steep and unstable side slopes. Road safety cannot be compromised and *whole life costing* considerations of fuel consumption, vehicle operating costs and journey time will need to be taken into account. However, a reduction in design class by one level will often have little effect on vehicle operating costs and road traffic safety, and even two levels may be acceptable if accompanied by the appropriate traffic warning signs. This reduction in design class can make a significant cost saving and can result in less slope disturbance if, for instance, it enables a large retaining wall to be omitted or a major rock cut to be avoided. Relaxation of alignment standards could also include the omission of a shoulder for short distances.

Design classes are usually set by each national road authority according to traffic volume (AADT). Table C1.2 illustrates how national standards can vary significantly. In Nepal, where most of the road network is located in hilly and mountainous terrain, the design standard for low-volume roads is, by necessity, considerably lower than in countries such as Ethiopia for example, where the majority of the road network is located on flat or gently rolling terrain. Further discussion of geometric design standards for low-volume roads is given, for example, in Giummarra (2003).

Fig. C1.5. Blue Nile and Difarsa River confluence showing Option 1.

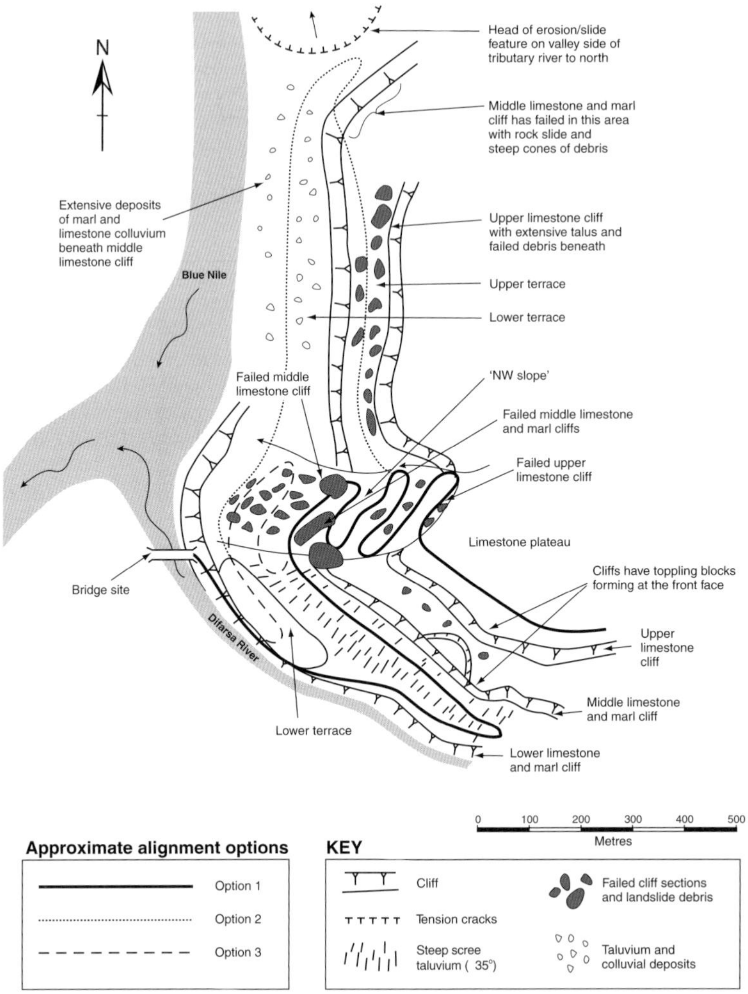

Fig. C1.6. Summary geomorphology of alignment options.

C1.3.2 Camber and crossfall

A camber of 3% is normally recommended for a sealed road (up to 5 or 6% for gravel roads). In mountainous terrain this camber is usually replaced by a crossfall towards the hillside of 3%. Some commentators (e.g. Kojan 1978; Keller & Sherar 2003) recommend a crossfall away from the hillside in order to avoid the need for roadside drains and to preserve rainfall-runoff patterns that existed before road construction. However, this is not recommended for roads in areas where rainfall intensities are high and where there is an opportunity for uncontrolled runoff to become channelized onto the slopes below rather than into natural gullies. It is virtually impossible, even when a road is sealed, to create an even and continuous discharge of rainwater over a road edge. Instead, runoff will become concentrated, causing erosion and gullying that may eventually lead to slope failure. It is recommended that, for these reasons, the crossfall should normally be directed towards the hillside and runoff collected in a roadside drain for eventual discharge into a cross culvert (Section C6).

C1.3.3 Superelevation

For traffic safety reasons, superelevation with a resultant crossfall away from the hillside will usually be required on re-entrant curves. In these cases, provision will need to be made to prevent uncontrolled road runoff from causing erosion below the road. This is done by constructing a

Fig. C1.7. Option 2 (bend out of sight).

kerb or a drain along its outer edge, leading the runoff to a suitable discharge point.

C1.4 Case study

The Gundewein to Mekane Selam Road in Ethiopia crosses the gorge of the Blue Nile River and is required to descend and ascend through almost 1000 m in the process. The terrain is composed of an approximately horizontally bedded sequence of basalt and pyroclastic rock overlying sandstone, limestone and marl. This sequence has been eroded as the Blue Nile has cut down to form a series of benches. Large volumes of failed material mantle each bench and tension cracks extending 10–20 m behind cliff faces are indicative of ongoing instability. The marl crops out on the slopes adjacent to the Blue Nile and its tributaries, and landslides and ground movements are occurring in this material and the overlying limestone. On the eastern side of the gorge the steepness of the terrain and the presence of unstable cliffs, taluvium deposits and landslides posed significant problems for alignment selection and design. The situation was further complicated by the need to find a suitable crossing of the Difarsa River (Fig. C1.5) in order to gain access to the proposed site of the Blue Nile bridge.

Three options were considered for the final alignment descent to the Difarsa bridge site from the eastern side of the gorge (Figs B2.8 & C1.6). Option 1 involves a hairpin stack that descends from an upper limestone plateau through an area of failed cliffs at the top of the *NW slope* (Fig. C1.6). This is followed by a hairpin loop (Fig. C1.5) across steep taluvial slopes and a final traverse on slopes above the Difarsa River to the proposed right bank abutment location of the Difarsa bridge. Option 2 (Figs B2.8, C1.6 & C1.7) avoids most of the upper stack across the failed cliffs of the NW slope by adopting an extended hairpin loop to the north that requires a deep box cut through 20 m high limestone and marl cliffs before returning to the Difarsa right bank abutment across relatively gentle taluvial and colluvial slopes. Option 3 (Figs B2.8 & C1.6) adopts the hairpin stack alignment of Option 1 but continues to descend through the entire failed NW slope before traversing steep side-long ground to the proposed bridge site.

Option 3 was quickly discounted on grounds of slope instability and uncertainty over founding conditions for retaining walls. This uncertainty would ordinarily have been reduced by the use of ground investigation, but poor access and the unavailability of suitable drilling equipment precluded this. Option 3 would have involved an extended hairpin section of alignment on a failed slope and would require long lengths of retaining wall founded within suspected landslide debris. Consequently, only Options 1 and 2 were taken forward for detailed comparison.

Table C1.3. *Comparison of alignment alternatives for a section of new road, Ethiopia*

Comparison parameter	*Alignment option*	
	Option 1	*Option 2*
Cost (US$ million)	3.0	9.1
Length (km)	5.4	5.9
Length ≥9% gradient, but <12% (km)	1.27	0.77
Length at 12% (km)	0.41	1.15
'Substandard' horizontal curve radii	20 m @ 227 + 378 20 m @ 227 + 408 26 m @ 225 + 750 26 m @ 225 + 791 Others are ≥29 m	25 m @ 226 + 261 28 m @ 225 + 775 25 m @ 225 + 824
Length of potential slope instability above alignment – cut slope hazard (km)	2.35	1.35 (assumes 50% of box cut alignment across upper terrace encounters slipped/failed taluvium, and that the outline design for box cut side slopes are stable)
Length of potential slope instability below/ through alignment – formation hazard (km)	0.25	0.15 (assumes that the only section susceptible to failed/failing ground below the road is in the vicinity of the bend beneath the box cut exit)
Soil cut to fill ($\times 10^3$ m^3)	57.2	199
Rock cut to spoil ($\times 10^3$ m^3)	108	545
Boulders excavation class A to spoil ($\times 10^3$ m^3)	Assumed small	414
Boulders excavation class C to spoil ($\times 10^3$ m^3)	66	0
Soft cut to spoil ($\times 10^3$ m^3)	474	90.5
RC retaining wall concrete ($\times 10^3$ m^3)	0.68	0
RC retaining wall rebar (tonnes)	53.46	0
Masonry retaining wall ($\times 10^3$ m^3)	0.42	0
Foundations for retaining walls	Low–mod bearing capacities	According to the outline design, no walls required
Drainage considerations	Normal	Normal plus drainage of 1.4 km of box cut and erosion control structures over fill slope at box cut exit on cliff
Uncertainty of ground conditions	Low–mod	Mod–high

Table C1.3 summarizes the various factors and parameters that were considered when comparing the two options. Although Option 1 involved the traverse of a failed slope and steep taluvium, it was selected in preference to Option 2 on the grounds of less geotechnical uncertainty and lower cost. While this selection may appear fairly clear-cut from Table C1.3, it is emphasized that a formal comparison of options in this way is the only means by which an objective assessment can be made.

References

GIUMMARRA, G. J. 2003. Establishment of a road classification system and geometric design and maintenance standards for low-volume roads. *In: Transportation Research Record*, 1819, Paper no. LVR8-1090, Transportation Research Board, National Research Council, Washington DC, 132–140.

KELLER, G. & SHERAR, J. 2003. *Low Volume Roads Engineering*. United States Department of Agriculture, Forest Service.

KOJAN, E. 1978. *Engineering Geological Evaluation of Landslide Problems, Western Hills Road Project: Godavari to Dadeldhura, Nepal*. Department of State Agency for International Development, Washington DC.

MARWA, E. M. M. & KIMARO, D. N. 2005. Geomorphological mapping for route selection and road construction at lower Kihansi in Tanzania. *Bulletin of Engineering Geology and the Environment*, **64**, 333–339.

TRRL 1988. *A Guide to Geometric Design*. Overseas Road Note 6. Transport and Road Research Laboratory, Crowthorne, Berkshire.

C2 Earthworks

G. J. Hearn* & T. Hunt

URS Scott Wilson Ltd, Scott House, Alençon Link, Basingstoke, Hampshire RG21 7PP, UK

*Corresponding author (e-mail: gareth.hearn@scottwilson.com; garethhearn@talktalk.net)

C2.1 Choice of cross-section

The choice of cross-section on mountain roads is of critical importance. Even small increases in road width on steeply sloping ground can have a major impact on earthworks volumes and the need for retaining structures. An increase of road pavement width from 5 to 6 m, for example, can increase construction costs by as much as 50%; formation widths commonly adopted for roads in lowland terrain may be difficult to justify economically in hilly and mountainous areas. Furthermore, the greater the road width the greater the disturbance to the natural hillside and usually the larger the volumes of spoil required to be disposed of. These outcomes may exacerbate slope stability problems and necessitate a higher maintenance commitment in later years.

Many road projects involve the upgrading of existing roads through a combination of widening and improvements to the horizontal and vertical geometry. Widening into the cut slope will require slopes to be completely reformed and will invariably reactivate or trigger slope instability and erosion. Widening into the hillside can also be quite disruptive to traffic if the road is to be kept open at the same time. By contrast, widening on the outside edge of the road may encounter difficulties with the stability of fill slopes and the suitability of foundations for retaining structures (see discussion in Section A4.3).

For a new road in hilly or mountainous terrain there are essentially three main choices of cross-section: full-cut, part-cut/part-fill and full-fill, as shown in Text box C2.1. The costs shown undoubtedly represent a simplification, and assume that a stable cut slope can be formed without the need for slope retaining structures and stabilization measures. Nevertheless, the figures clearly indicate that a full cut section in stable terrain is significantly less costly to construct than a full-fill section and still appreciably cheaper than a half-cut, half-fill section, even when taking the cost of spoil disposal into account. The comparisons are not particularly surprising given the advantages cited in Table C2.1, where the full-cut section minimizes the need for compaction of the road formation and removes the requirement for fill retaining walls.

Even in moderately-sloping terrain (>25°), fill slopes will almost always need to be supported by a retaining wall unless rock fill or reinforced fill is used. In steep terrain full-cut usually predominates with the use of retaining walls across embayments and other topographic low points (Fig. C2.1). In extreme cases, where neither cut nor retained fill are practicable, the use of bridging structures, such as that illustrated in Figure C2.2, might be the only solution. In this example, the columns are embedded into rock to a depth of between 3 and 7 m due to the steeply sloping ground.

In riverside locations, steep and potentially unstable slopes above a proposed alignment might lead to a preference towards a full-fill or retained-fill solution. However, where river scour poses a major hazard (Section C7.5), a part-cut/part-retained fill cross-section might be the most effective compromise. Even if river flooding were to remove all protection works, retaining walls and fill, that part of the cross-section constructed in rock cut should remain (thus allowing the road to remain open prior to and during reinstatement).

In practice, the vertical and horizontal alignment constraints will impose a significant control on the choice of cross-section at any one location. To prevent an alignment from becoming unduly sinuous and to avoid large volumes of cut and fill, all three cross-sections in Table C2.1 are likely to be adopted at different locations along an alignment during detailed design. A solution in which the excavation generated from cuts can be incorporated into properly constructed fills within a distance of one or two kilometres (i.e. a balanced cut and fill) is much preferred in order to minimize haul distances and spoil disposal requirements.

There are other important factors to consider in the choice of cross-section that relate to the stability of hillsides and the location of alignments on them. Some of these are summarized in Table C2.2 and illustrated in Figure C2.3.

C2.2 Design of cut slope angle

Cut slope instability is most commonly the result of:

- cut slope angles not taking into account the underlying geology (this can include adverse bedding/foliation or

From: Hearn, G. J. (ed.) *Slope Engineering for Mountain Roads*. Geological Society, London, Engineering Geology Special Publications, **24**, 145–163.
DOI: 10.1144/EGSP24.11 0267-9914/11/$15.00

Text box C2.1. Choice of cross-section

The table gives a cost comparison of the three cross-sections shown for a 5 m formation width and for varying natural slope angles. For a 10° natural slope angle, a profile of 2.0 m of soil overlying 1.5 m of rippable/excavatable rock on top of rock requiring blasting is assumed. On slopes in excess of 60° it is assumed that rock requiring blasting will be encountered at the surface. Cut slopes are expected to vary from 45° to 85° with increase in natural slope angle. Below-road retaining walls are assumed to be necessary for half-cut/half-fill and full-fill sections when the hillside slope exceeds 25° (see Table C2.1). Typical costs per m^3 were averaged from projects in Laos, Nepal and Ethiopia; the work items comprising bulk excavation in soil, excavation in rippable/excavatable rock, blasting in rock, fill and compaction in embankments, masonry walling and disposal of spoil.

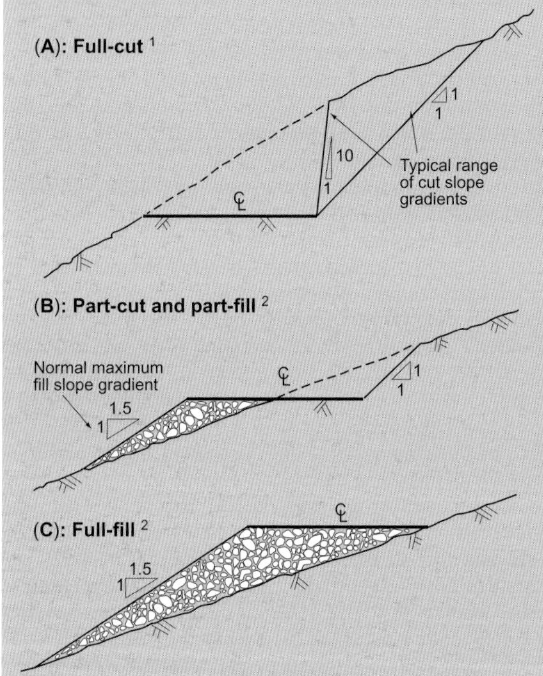

[1] May require cut slope protection or stabilisation above the road.
[2] Will require below-road retaining wall on slopes steeper than 25° unless rock fill or reinforced fill is used.

Cost comparisons for cross-section construction (combines earthworks, spoil disposal and fill slope retaining walls)

Natural slope angle (degrees)	*Cost per metre run of road expressed in US$ (2008 prices)*		
	Full-cut	*Half-cut, half-fill*	*Full-fill*
10	15	5	12
20	43	16	39
30	77	170	437
40	204	319	559
50	305	505	n/a
60	388	n/a	n/a
70	521	n/a	n/a

NB high full-fill costs for natural slopes ≥30° are due to retaining wall requirements.
n/a, not applicable.

jointing and faulting in rock-dominated slopes and the presence of taluvium/colluvium or other inherently weak material, such as highly weathered clays, in soil-dominated slopes);

- high or perched groundwater levels and seepages; and
- excessive surface erosion and gullying.

Essentially, there are three approaches to the design of a cut slope. They are not mutually exclusive and a combination of all three will most likely yield the optimum outcome. The approaches are:

- knowledge-based;
- empirical; and
- analytical.

C2.2.1 Knowledge-based approach

A detailed ground investigation (Section B4) will help to identify and classify slope materials for the design of cutting angles in advance of excavation. In the absence of a detailed ground investigation, cut slope design for new road construction is usually based on commonly adopted angles

Table C2.1. *Comparison of road sections*

Type of section	*Advantages*	*Disadvantages*
Full-cut	• Road formation requires minimum compaction because it is formed entirely in natural ground.* • No requirement for fill slope placement or compaction. • Potential source of fill material for use elsewhere along the road. • Potential source of rock, if present, for masonry, aggregate and drainage backfill. • Usually the only practical solution if existing ground slope >50°.	• Greater height of cut may lead to greater instability and/or erosion. • May result in large volumes of spoil requiring safe disposal.
Part-cut and part-fill	• Volume of spoil minimized if balanced cut/fill can be obtained. • Minimum impact on landscape.	• Requirement for fill placement and compaction. • May require below-road retaining wall or reinforced fill to avoid excessive area of fill if existing ground slope >25°.
Full-fill (including wall-retained fill)	• Usually only practical solution when traversing re-entrants or water courses. • Could be the only practical solution (with fill retaining structure) on steep rock slopes if jointing is adverse to stability (see Fig. C2.3).	• Requirement for significant fill import, ground preparation (including benching), placement and compaction. • Will require below-road retaining wall or reinforced fill to avoid excessive fill area if existing ground slope >25°. • Impracticable if existing ground slope >40°.

*Transported soils and some low-density residual soils (Section A3.1) exposed in the subgrade of some full-cut sections will require compaction and possibly replacement.

Fig. C2.1. Use of retaining structures in steep mountain terrain.

Fig. C2.2. Bridges can be used where full-cut and retaining walls are impracticable due to excessively steep topography.

for the various types of materials anticipated or exposed (Sections B2 & B3). This relies on the knowledge of how the materials are likely to behave, based on published data and on experience of similar ground conditions elsewhere.

Table C2.3 shows some typical cut slope angles in materials where there is no significant adverse structural control on stability. The table is based on data derived from a range of road construction and improvement projects in Southeast Asia and uses the grade of weathering (Section A3.1) as the basis for assigning cut slope angles. Actual achievable angles will vary with the height of cut and the ground conditions existing at each location. TRL (1997),

Table C2.2. *Suggested cross-sections for unstable ground*

Instability type	*Alignment location*	*Preferred section*	*Notes*
Existing unstable ground, taluvium/ colluvium	Close to ridge crest	Full cut	Will reduce destabilising forces but locally may still require below-road retaining walls founded beneath any failure surfaces.
	Middle slope	Balanced cut and fill	If existing ground cannot be stabilised economically, then preference is for least disturbance, with reduced road width and flexible retaining structures* on either side of road.
	Foot of slope	Full fill	Will increase stabilising forces, but may require frequent and sizeable culverts and larger roadside drains. Scour protection required in riverside locations.
Adversely orientated discontinuities in rock	Applies to all cases	Full fill	Avoids excavation and undercutting of rock strata. Below-road retaining walls will need to be keyed and dowelled into a benched rock surface.

*Flexible in as much as some movement can be tolerated without structural failure.

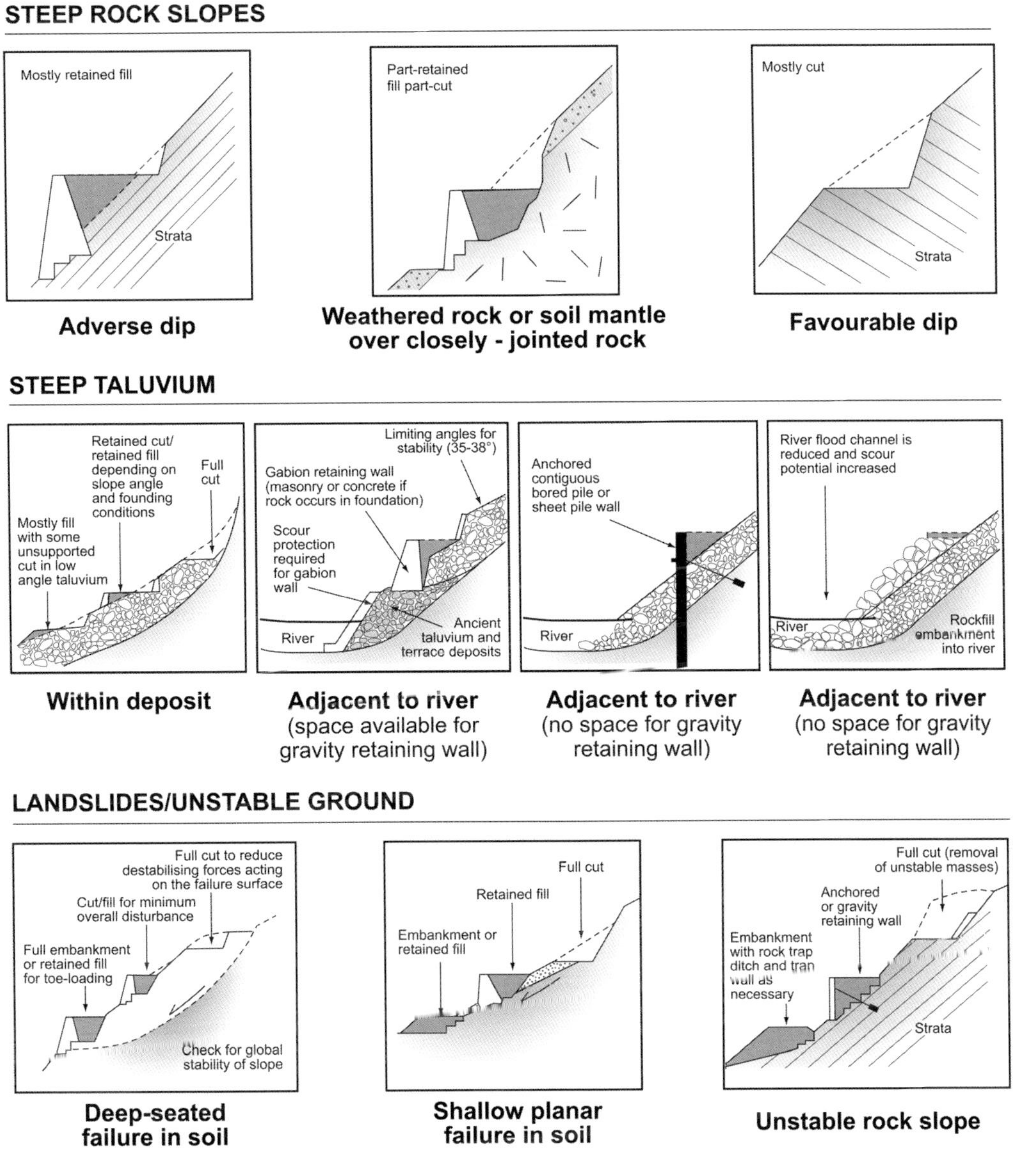

Fig. C2.3. Examples of schematic section in difficult and potentially unstable ground. Modified from TRL 1997.

for example, shows observed cut slope angles for rock and soil excavations in different height ranges in Nepal. Residual soil cuts are shown as being steeper than 1:1, due principally to the additional strength imparted by residual rock fabric, the temporary effects of negative pore pressures (or suctions) on slope stability (Section C3) and case-hardening of the cut slope surface (Section A3.1).

Table C2.3 is very generalized and must be used with caution. For instance, in locations where there are weaker bands or layers of material, water seepages or adverse structures exposed in the cut face, cut slope angles will usually need to be reduced or specific mitigation measures (Sections C3–C7) introduced in order to maintain long-term stability. In contrast, slopes along many mountain roads have been cut to 2:1 or steeper in weathering grade V and even VI soils and have remained stable for several decades. The cut slope shown in Figure C2.4 for example is between 30 and 40 years old. Despite its apparent stability over this period, it can be seen that the material forming the base of the cut slope is now softening and is being eroded as a result of

Table C2.3. *Typical cut slope angles*

Material type	*Weathering grade*	*Slope face angle (degrees)*	*Slope face gradient (vertical:horizontal)*
Competent rock*	I–II	80–85	6:1 to 10:1
Weathered rock*	III–IV†	60–75	2:1 to 4:1
Coarse-grained residual soil	V–VI	45	1:1
Fine-grained residual soil	V–VI	35‡	1:1.5 to 1:2‡
Coarse-grained taluvium and river terrace deposits	n/a	40	1:1.2
Fine-grained colluvium (varies)	n/a	35‡	1:1.5 to 1:3‡

*See Table C2.5 for further data on rock slope cut angles.
†Weathering grade IV is borderline soil/rock.
‡Depends on clay content and water table.
n/a, not applicable.

Fig. C2.4. Slope cut to 3:1 in weathering grade V material remains largely intact 40 years after construction.

seepage at this level. If allowed to go unchecked this will ultimately cause the cut slope to fail.

During construction, final cut slope angles will depend on an assessment of the materials and discontinuities exposed upon excavation. This means that:

- the exposed cut faces must be regularly inspected to check that the design assumptions remain valid; and
- individual slopes may need to be significantly re-profiled or stabilized through the use of mitigation measures. A stability analysis may be required to verify the revised design (Section C3).

Mitigation measures could include:

- retaining structures (Section C5);
- rock slope support and protection (Section C4);
- slope drainage (Section C6) to reduce pore-water pressures and to prevent seepage flows from causing softening and erosion on the slope face; and
- revetments and protective surfacing (Section C7) to minimize shallow failure and prevent weathering and erosion of exposed materials.

Cut slope profiles also need to reflect the strength of the weaker transported or *in situ* weathered material that usually occupies the upper part of the excavated slope. A shallower slope angle of between 30° and 45° is often applied to the top 1–2 m or so for this reason. The depth and angle of this reduced slope will depend on the strength of the materials exposed and can be determined in advance by reference to the performance of slopes along existing roads in the area.

C2.2.2 Empirical approach

In this approach Table C2.3 is further refined to reflect local observed data. The existing cut and/or natural slopes in the vicinity of a proposed road alignment are catalogued taking into account apparent stability, slope height and angle, type of soil and its geological origin (e.g. *in situ* weathered or transported), exposed bedrock geology and structure, probable groundwater regime (such as the absence/presence of seepages) and any other factors that might be significant (e.g. wet season rainfall). The conditions under which existing landslides and cut slope failures have occurred will also provide information on maximum achievable cutting angles, though each site must be evaluated on its own merits.

In this way it may be possible to develop a table or graphical plot giving limiting slope angles under a range of material and groundwater conditions (e.g. Ayalew *et al.* 2009). What will not be known is the factor of safety of the slopes included in the dataset unless they are observed to be failing (Section C3.2.3). Table C2.4 gives an example of this approach for natural slopes in Nepal. However, even under dry conditions the limiting natural slope angles shown in this table are slightly less than

Table C2.4. *Maximum observed angles (in degrees) for natural slopes (based on data from Nepal)*

Soil type	In situ *weathered soil*		*Transported soil*		*Surface seepage*‡
	*Dry**	*Wet*†	*Dry*	*Wet*	*Wet*
Clayey silt	33–36	16/17	28–31	14/15	11
Silt	33–36	16/17	28–31	16/17	12
Sandy silt	33–36	16/17	31–34	16/17	14
Silty sand	36–39	19/20	31–34	16/17	17
Silt & boulders	36–39	28/29	31–34	23/24	19
>50% boulders	36–39	31/32	33–36	23/24	19

*Dry soils: groundwater considered to be low all year round.
†Wet soils: groundwater considered to be seasonally high.
‡Surface seepage: groundwater seen to be seeping at the slope surface.

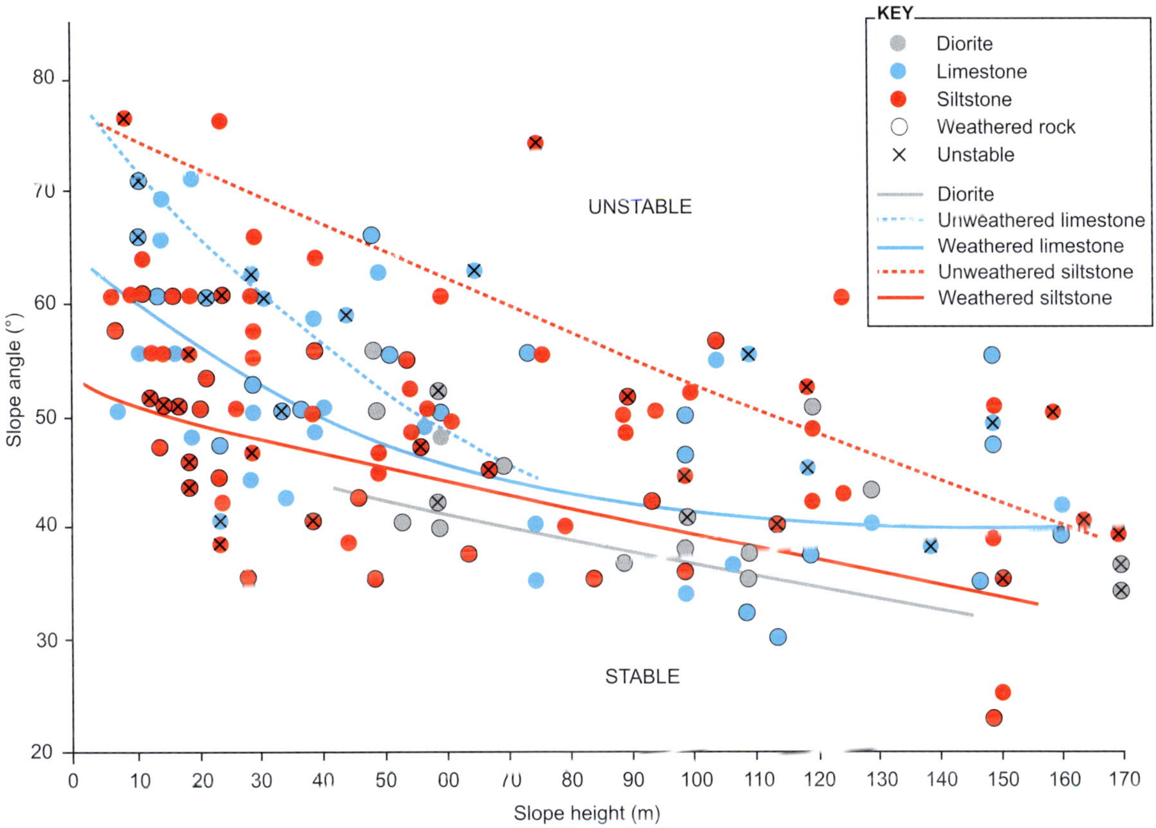

Fig. C2.5. Slope height–angle–stability relationships for rock slopes, Papua New Guinea.

those given in Table C2.3 for the corresponding materials; there is a significant further reduction under wet or saturated conditions. The limiting natural slope angles in Table C2.4 are based on slopes that will have been formed in response to groundwater fluctuations over considerable periods of time, unlike cut slopes that are expected to remain stable over the much shorter design life of a road. The table demonstrates the significant influence that groundwater condition has on the long-term stability of natural slopes.

Figure C2.5 shows *envelope* curves derived from observations of cut slopes and natural slopes in the Star Mountains of Papua New Guinea. The curves have been drawn by eye and divide predominantly stable slopes above from unstable slopes below. The siltstone was hornfelsed while the diorite was highly fractured and faulted and this was considered to be the main reason for the difference in slope angles observed (Hearn 1995). The scatter in the points illustrates the difficulty in making simple comparisons such as these. Discontinuity patterns and depths of weathering, for example, exert critical controls on stability (Sections A3.1–A3.3) and are not differentiated in Figure C2.5.

C2.2.3 Analytical approach

An analytical approach is a useful and necessary supplement to both the knowledge-based and empirical approaches:

- in developing a better understanding of the factors affecting stability;
- for assessing factors of safety; and
- for making rational decisions.

In the analytical approach the strength parameters of the underlying soils are determined or estimated, the design groundwater regime is observed or anticipated and a slope stability analysis is carried out (Bond & Harris 2008; Section C3.2.3). Ideally the analytical approach should be based on a detailed ground investigation with borehole sampling and derivation of shear strength parameters from the testing of undisturbed soil samples. However, for low-cost roads such investigations are usually confined to major cuts and high-risk locations (such as where occupied buildings or high-investment infrastructure are in close proximity) or where repair in the event of failure would be very costly or difficult. The soil/rock relationship in hilly and mountainous terrain is invariably complex, and does not lend itself to giving a reliable picture to even the best designed and conducted ground investigation. Simplifying assumptions will have to be made in respect of the material layer boundaries and the groundwater regime and this may render the analysis speculative. Confidence levels can be improved by a re-analysis once the slope has been excavated and the soil boundaries and groundwater assumptions are adjusted to match the observed conditions.

For rock slopes, an analytical approach is usually only applied to deep cuts (perhaps 10 m or more) or where potential slope failure poses a significant risk. The approach adopted usually combines knowledge-based and empirical approaches and involves the assessment of rock mass characteristics and the orientation and angle of joints and other exposed discontinuities (Section C4.2). Prior to excavation some relevant information can be obtained from nearby natural exposures and excavations, but it is usually not until a slope is excavated that a full assessment can be made.

Where the stability of the rock mass is controlled by the strength of the rock continuum and its small-scale jointing, indices such as the Geological Strength Index (GSI) are normally used (Section C4.2). Table C2.5 shows the range of slope angles derived for cut slopes of up to 10 m in height using the GSI for weathering grade II–III rock masses that display no significant structural control on stability. The cut slope angles are derived from analytical methods that do not reflect the entirety of ground conditions found on site and should be used as a guide only. The angles shown are for dry slope conditions and a factor of safety of 1.0, and are therefore considered to represent absolute maximum values. They would need to be adjusted to provide a suitable margin of safety.

Where the stability of a rock mass is controlled by persistent joints and other discontinuities, kinematic analysis using a stereonet provides the main means of assessment (Section C4.2).

The presence of any infill material along discontinuities usually acts to decrease friction angles (Section A3.2), leading to lower factors of safety and therefore lower permissible cut slope angles. Furthermore, the method of excavation (notably the blasting techniques employed; Section C2.3.1) can leave slopes in a damaged condition, conducive to rockfall (Fig. C2.6) and this may be a significant factor in the design of stable cut slope profiles.

C2.3 Choice of cut slope profile

Soil or rock cut slopes can be excavated with benched profiles or with continuous or compound slopes. The term 'berm' is sometimes used. In this book the term bench refers to a step in a cut slope profile and a berm refers to a step in a fill slope profile.

C2.3.1 Benched profiles

The various advantages and disadvantages associated with benched profiles are indicated in Table C2.6. Benched profiles are usually applied to slopes excavated in rock or weathered rock. Benching of cut slopes above a certain height is standard practice in many countries. A benched profile is beneficial where an excavation requires blasting, since access for drilling machinery is made easier. Pre-split (pre-shear) blasting should be used to minimize overbreak, thus reducing the potential for blasting-related failure of the final cut face. The technique comprises a line of closely-spaced blast holes drilled along the designed cut line. Simultaneous blasts are used to create a linear discontinuity in the rock mass between the line of blast holes. The rock in front of this discontinuity is then bulk blasted after a few milliseconds delay for excavation. With accurate and experienced drilling, pre-split blasting is usually only effective at drilling depths of up to 10 m, and therefore progressive drill and blast utilizing benches is more practicable on deeper cut slopes. For pre-splitting to be effective, the blast energy must be contained to generate the requisite rock shearing stresses. If the side burden is less than about

Table C2.5. *Indicative cut slope angles in rock*

Rock mass description	*Discontinuity roughness*			
	Rough, fresh unweathered surfaces	*Rough, slightly weathered, iron- stained surfaces*	*Smooth, moderately weathered and altered surfaces*	*Polished or slickensided, highly weathered surfaces with compact coatings or joint infilling*
Intact or massive Massive *in situ* rock with very few and widely spaced discontinuities	Discontinuity-controlled failures (see analytical method 2, Section C4.2)			
Blocky–interlocked Undisturbed to partially disturbed rock mass with multifaceted angular blocks formed by three orthogonal discontinuity sets	G1 & G2 60–65° G3 & G4 50–60°	G1 & G2 60–65° G3 & G4 50–60°	G1 & G2 50–60° G3 & G4 45–50°	G1 & G2 45–50° G3 & G4 40–50°
Blocky–disturbed Folded and/or faulted with angular blocks formed by many intersecting discontinuity sets	G1 & G2 45–50° G3 & G4 40 50°	G1 & G2 45–50° G3 & G4 40–50°	G1 & G2 40–45° G3 & G4 35–40°	G1 & G2 40–45° G3 & G4 35–40°
Disintegrated Poorly interlocked, heavily broken rock mass with a mixture of angular and subrounded rock pieces	G3 & G4 35–40°	G3 & G4 35–40°	G3 & G4 35–40°	G3 & G4 35–40°

G1 & G2, stronger rock types.
G3 & G4, weaker rock types.

7 m, the blast energy displaces the rock mass laterally. This leads to ineffective shearing of rock and diffusion of gases into the adjacent rock mass, which tends to tear (and not split) the rock along the proposed cut face. By definition, a box cut provides ample side burden.

Benching is often specified at vertical intervals of between 7 and 10 m, and Fookes & Sweeney (1976) suggest an absolute maximum of 12 m. The bench width varies according to the design norms from country to country, and Fookes & Sweeney (1976) suggest a minimum of 5 m to allow for maintenance access. Nevertheless, in many countries standard bench widths are much less than this and an absolute minimum of 1.5 m is recommended for safety reasons during maintenance. Taking into consideration the overbreak that inevitably occurs at the outer edge of the bench, it normally requires a 2.5 m design bench to achieve the 1.5 m minimum width. This is especially the case in rock cuts where some tearing behind the line-drilled face has occurred in the top 1.5 m of the drillhole. This can be stemmed by filling the drillhole with soil to contain the charge and to minimize air blast.

Inclined drill line holes are used to create a battered cut face in order to avoid instability problems associated with daylighting joints. However, for inclinations flatter than approximately 3V:1H the process can become impracticable and expensive due to bit wear and rod breakage (especially in hard, jointed rock). This is one reason why bench riser slopes are kept steep and bench width is adjusted to maintain the overall design slope. Further details on rock blasting techniques can be found, for example, in Wyllie & Mah (2004).

An inward crossfall to the bench facilitates collection of runoff and drainage longitudinally along each bench to a safe discharge location. In the case of benches cut in strong rock without open joints, this drainage can be formed by excavating an unlined channel into the surface of the bench. In open-jointed or softer materials potentially prone to seepage or erosion, masonry or concrete-lined bench drains are usually required to convey surface runoff safely along each bench before discharging it either to a turnout at the end of the cut or a relief cascade.

Benches are sometimes specified with a crossfall out of the slope without the benefit of any bench drains. This is intended to create an evenly distributed discharge of runoff over the cut face below. Whether this is achieved in practice is debatable, however, and the benefits of controlling runoff to prevent erosion in weathered rock and soil would appear to be negated. Outward-sloping benches are therefore not generally recommended.

Fig. C2.6. Blast-damaged rock.

In summary, a benched profile is recommended only for cut slopes greater than 7–10 m in rock (weathering grade I–III and possibly IV). For cut slopes in *in situ* weathered soil (weathering grade IV–VI), benched profiles are not recommended unless they are provided with adequate drainage systems that are properly maintained. Benched profiles are not recommended for slopes formed in taluvium or colluvium.

C2.3.2 Continuous slope profiles

The advantage of a continuous slope is that drainage is distributed uniformly, and therefore approximates to the natural condition. It avoids concentrated flows and, for soil slopes of up to 40°, plant growth will occur more readily on a uniform slope than on a benched slope to the same overall angle. The steeper cut faces of a benched slope make it more difficult for topsoiling and planting schemes to take hold, unless the overall angle of the excavation is reduced to compensate.

C2.3.3 Compound slope profiles

Where cut slopes are designed and constructed to a continuous slope, the implication is that the strength of materials exposed remains constant with depth. While this may apply to cuttings formed entirely in homogeneous rock (not common), colluvium or residual soil, it is usual to find multiple soil and rock layers of varying strength. This layering often comprises soil overlying weathered rock and weathered rock overlying fresh rock, but it may also relate to the varying strengths and competencies of different rock types with depth (as might be the case in a sedimentary or volcanic sequence). In these circumstances, a compound slope profile may be more appropriate if the depth of these

Table C2.6. *Advantages and disadvantages of benched cut slopes*

Advantages	*Disadvantages*
• Benches slow down the rate of surface runoff, and therefore reduce surface erosion. • Benches permit the construction of mid-slope longitudinal drains much more easily, and these can form part of an overall slope drainage system. • Where excavation is to be undertaken in softer materials, such as weathered rock, benching can help prevent long erosion furrows from developing by interrupting and controlling the flow of surface runoff. • Shallow failures are usually limited to one bench at a time. • Shallow failures are usually contained on the bench below, and are thus often prevented from reaching the road. • Benches offer advantages in terms of access for drilling equipment and excavation plant. • Benches permit access to the slope face for maintenance purposes.	• Benches are nearly always inadequately maintained on low-cost roads as they are not easily accessible to road maintenance crews. This can quickly lead to drainage failures. • The cut faces in a benched slope profile will be steeper than a continuous slope cut to the same overall angle. This may encourage localised failures to occur in soft materials and may create conditions of instability in adversely-jointed rock that might otherwise not occur (Text box C2.2). • Conversely, if the risers of a benched cut slope are cut to the same slope as a continuous cut the overall height and volume of cut will usually be greater. • Vegetation is less easy to establish on a benched slope profile to the same overall angle (i.e. where steeper risers are required between benches). • Defective bench drainage systems due to erosion or blockage can lead to uncontrolled rainfall runoff and concentrated erosion that ultimately leads to slope failure.

Text box C2.2. Discontinuity-control on the stability of earthworks

An example from Azerbaijan illustrates the case where the benching of a cut slope results in the creation of sections of slope that become unstable due to bedding plane orientation. The slope is underlain by folded Pliocene-age sediments comprising beds of silty and fine sandy clay. Partings occur within the sequence at approximately 200 mm intervals and the parting surfaces have a thin coating of plastic clay. The strata dip at an angle of 21–22° directly out of the slope and the overall slope of the benched profile is 20°. Individual cut faces have been excavated to 23–24°, thus enabling failure to take place along the clay partings within the sequence of bedding. These failures are confined to individual cut faces while the overall 20° benched cut profile shows no signs of movement.

Text box C2.3. Variable soil and rock sequences encountered in Laos and Ethiopia

For a road project in Laos, where no ground investigation had been carried out, the designer decided to overcome the problem of determining where the soil/weathered rock interface was by requiring all slopes to be cut at an angle of 45° in their entirety, no matter what material was to be excavated. Unfortunately this generally meant that the cut angle was too steep in the weaker soils and too shallow in the stronger materials. The result was that the upper portion of some of the cut slopes became unstable. Furthermore, earthwork quantities were significantly greater and the cut slopes generally higher than they needed to be. This exposed a greater area of excavated slope to erosion. The construction cost was unnecessarily increased, as was the potential for future slope maintenance expenditure. Spoil disposal also became a significant problem.

The construction of the Gundewein to Mekane Selam road across the Blue Nile gorge in Ethiopia involved similar considerations, although they were made more complex by differing rock types with depth. Drilling was carried out in some of the deeper cuts to investigate these materials before excavation, but there were funds available for only a limited number of holes and the poor quality of drilling meant that core recovery was generally very low. The reference condition mapping carried out for this project (Section B3.2) identified materials at outcrop. It also showed the anticipated near-surface sequence of volcanic and sedimentary rock with depth where this could be ascertained from exposures on neighbouring slopes. Each soil and rock type had a prescribed cutting angle for various cutting height ranges, but the difficulty was in knowing what proportion of each cut would be occupied by the various reference condition units. In the shallower cuts this was not important but in the deeper cuts, often in excess of 20 m, a decision had to be made on the profile to be adopted in advance of excavation. Due to limited subsurface information this was usually based on an assumed average rock condition from the anticipated sequencing of rock types. This enabled the cut line to be set out and excavation to commence. If stronger materials were encountered closer to the ground surface than anticipated, the cut line was taken further forward and a steeper cut slope was formed for the remainder of the excavation (thus reducing cut quantities). As there had been a degree of conservatism built into the scheduling of cutting angles within each reference condition unit, this usually provided sufficient flexibility when the depth to stronger materials proved greater than anticipated. In some of the excavations through basalt and tuff sequences it was found that the materials exposed, while correctly predicted geologically, were in fact stronger than had been expected; parameters for the reference condition were adjusted accordingly during construction.

strength boundaries can be reliably predicted or observed in a ground investigation. Text box C2.3 illustrated this issue of variable material strength with depth by reference to two contrasting projects in Laos and Ethiopia.

By far the most common form of compound slope profile relates to the case where provision is made for soil overlying weathered rock. Text box C2.4 illustrates the importance of an accurate assessment of the depth of this interface when scheduling a compound cut slope profile to accommodate the strengths of the two materials. This can also be used to determine the adequacy of the width of the right of way and whether or not additional land acquisition is necessary. In reality, there will be occasions when the depth of soil overlying rock is significantly underestimated during design; it is therefore recommended to take a conservative view since this gives greater flexibility during construction to yield a safer final cut slope profile. However, there will be circumstances in which the cost and practicality of land acquisition prevents such a conservative approach from being adopted and measures will need to be taken to allow steeper cuts to be formed (Sections C3, C4 and C7).

If an existing cut slope is to be re-formed, for reasons of stability or road widening, then it is likely that the original weathering profile will already be exposed and the profile of the new cut slope can be defined more accurately.

Standard computer programs used for generating cross-sections and quantities from topographic ground models and design slopes can accommodate the use of benched, continuous and compound slope profiles.

C2.4 Fill slopes

Fill slope failures are most commonly the result of:

- inadequate compaction, creating a material with a reduced shear strength, resulting in a fill that is more susceptible to 'wetting up' during heavy rainfall. This causes softening, surface erosion or shallow failure, often along the fill/natural ground interface;
- inadequate removal of existing vegetation and topsoil from the underlying slope and failure to bench the fill into sloping ground in order to provide a suitable shear key. These conditions may otherwise create preferential slip planes and impede drainage during periods of heavy rainfall;
- uncontrolled rainfall runoff, mainly from the road surface, leading to either wetting of the fill or erosion of the fill slope. This runoff can originate from blocked roadside drains or from side drain turnouts and culverts;

Text box C2.4. Importance of the accurate assessment of the soil/rock interface

Figures A and B below show the cross-section of a natural slope and the location of the actual soil/rock interface. A road is to be constructed and the road alignment dictates that the base of the cut slope must be at point X.

In Figure A the assumed interface is lower than the actual interface. With a design cut slope angle of 45° in residual soil and 70° in weathered rock, the anticipated cut slope profile is ADX. The cut line commences in soil at point A and when the actual soil/weathered rock interface is reached at B the slope can either be cut along the interface to point C to enable the final cut in weathered rock to reach point X, or the weathered rock can be cut along BX at a shallower angle than envisaged in the original design. In either case this will result in an overall height and quantity of cut that is greater than is necessary, a corresponding increase in cost and the need to dispose of, or use as fill elsewhere, the additional cut material.

In Figure B the assumed interface is higher than the actual interface, and the anticipated cut slope profile is ABX. The cut line commences at point A but since the soil/weathered rock interface is not reached until C, the contractor is forced either to increase the cut slope angle in the weathered rock to about 85° (which would be the easiest option) so that the cut slope profile is ACX, or to cut the soil slope back to A′ C′ (which could be difficult) so that the final cut slope profile is A′C′X. In the former case, the increased slope angle in the weathered rock is likely to lead to increased instability but results in a reduction in cut quantity and therefore initial cost. In the latter case, the contractor is obliged to cut back the entire soil slope face at significant additional cost.

In Figure B, if point A is located at the edge of the right of way defined on the basis of the original design, then either the additional land between A and A′ has to be acquired (which is often a time-consuming process) or the contractor/owner is forced to accept a cut slope profile ACX or ABX or some intermediate combination that will result in a slope more susceptible to instability.

(A)

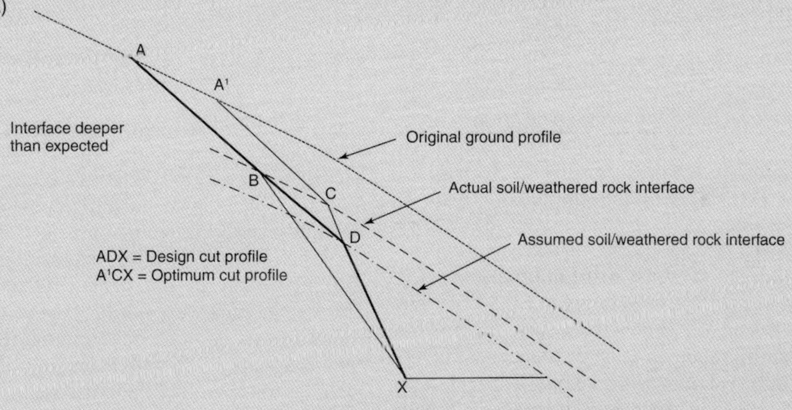

(B)

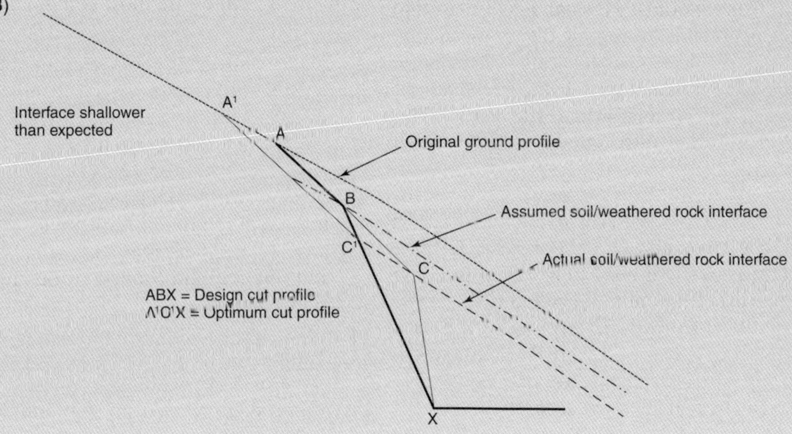

Fig. C2.7. Failure of a fill slope constructed on a landslide.

- instability of the underlying ground, either due to a previous failure of the natural hillside (Fig. C2.7) or failure induced by the loading effects of the fill slope itself.

Most fill slope failures are therefore the result of poor construction practices rather than inadequate design.

Where the natural hillside slopes are inclined at an angle of less than 20°, the existing ground should be cleared of vegetation and scarified. At natural slopes greater than 20°, steps should be cut into the hillside, preferably a minimum of 3 m wide to allow for machine access and to provide a shear key. In both cases, properly compacted granular fill is normally constructed to a slope angle of 1V:1.5H (34°) on low-cost roads. Where the fill is predominantly fine-grained, a shallower slope angle (maximum 1V:2H or 26°) will be more appropriate. In exceptional circumstances where rockfill is used, the fill slope angle may be increased to 1V:1.25H (38°). If fill slope angles greater than this are required, then reinforced fill should be considered. More commonly, however, fill retaining structures are constructed (Section C5). For traffic safety reasons, the fill slope angle is often decreased to 1V:3H (18°) for the uppermost 1–2 m of the fill slope. A less steep section of fill slope adjacent to the road is more easily negotiable if vehicles stray from the carriageway.

For high fill slopes, greater than say 10 m, the introduction of an intermediate berm may be appropriate. The berm width should be at least 2 m and may be 4–5 m on high fills. This can have two main advantages:

- it can act in the same way as a bench on a cut slope and help control surface runoff and therefore reduce surface erosion; and
- it can improve the stability of the fill slope if the overall slope angle of the fill slope is thereby reduced.

As noted earlier, fill slope failures can also arise from the failure of the underlying natural ground. Under these circumstances it may be necessary to:

- remove all the failed material above the failure surface and key the new fill into the stable ground below; or
- construct a fill retaining wall or reinforced fill founded in stable ground beneath the failure surface.

Unless construction can be carried out within a single dry season, it may be necessary to undertake temporary measures to minimize potential surface runoff from entering the area to be filled. These measures might include the construction of earth bunds at the crest of the slope or the temporary diversion of roadside drains.

Fill is normally compacted in horizontal layers not exceeding 150 mm final thickness, although this may be varied following field compaction trials. Compliance with the specified degree of compaction should be checked by *in situ* density testing. Tests are usually carried out at the rate of one per 1500 m^3 of placed fill or change in material type. In order to ensure that the correct degree of compaction is achieved to the edge of the slope, the slope should be overfilled and compacted with the remaining overfilling material removed later, usually by backactor. Further information on guidance for fill slope construction can be found, for example, in Department of Transport (1994), Monahan (1994), Trenter (2001) and Wesley (2008) (for residual soils especially).

C2.5 Earthworks balance

Once the initial cut and fill cross-sections for a new road construction have been designed, a mass haul diagram needs to be prepared. This is a plot of cumulative volume of cut v. fill against distance along the road, taking due account of bulking factors arising from materials handling (bulking occurs because compaction is usually unable to replicate the *in situ* density of the original materials). This can be done very easily if the alignment has been generated from a computerized ground model. A typical mass haul diagram is shown in Figure C2.8 (Department of Main Roads WA 2005) and is based on the assumption that all excavated material is suitable for use as fill. The quantities would have to be adjusted where this were not the case.

On the mass haul diagram a rising curve (1) indicates an increasing volume of cut and a falling curve (2) indicates an increasing volume of fill. The maximum earthworks point (3) occurs at the end of an excavation and the minimum earthworks point (4) occurs at the end of an embankment. The vertical distance between a maximum and the next minimum point indicates the volume of embankment; the vertical distance between a minimum and the next maximum point indicates the volume of excavation. The horizontal distance between (3) and (4) indicates the length of road on embankment. Where the curve cuts the baseline the volume of excavation equals the volume of fill. Points *a* and *c* on Figure C2.8, for example, show that the earthworks are balanced between A and C (i.e. the material excavated from AB would form the embankment up to point C), assuming it was all suitable for

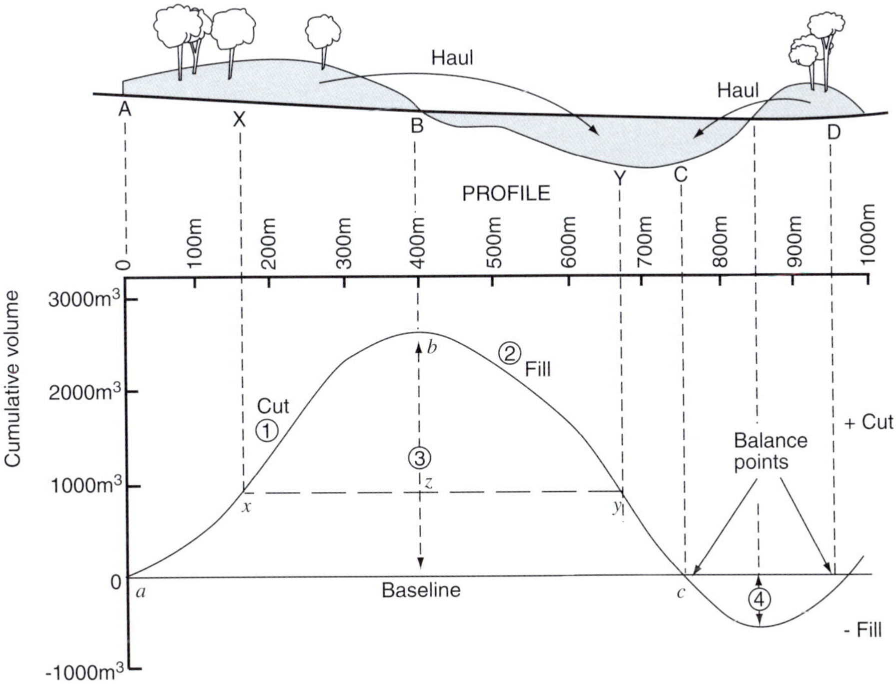

Fig. C2.8. Mass haul diagram.

use as fill. Any horizontal line intersecting the mass curve similarly indicates lengths over which cutting and filling are balanced. Thus, *xy* is the balancing line. The cut from X to B just balances the fill from B to Y, with the volume moved being represented by *bz*.

When the mass curve lies above the balancing line, the excavated material must be hauled forward; when below, the direction of haul is reversed. The length of the balancing line intercepted by the mass curve represents the maximum haul distance in that section. Taking the base line as the balancing line, the greatest haul distance involved in the disposal of excavation AB is AC, so that no material should be hauled beyond C.

In the design of a new road in mountainous terrain, the production of a mass haul diagram can be of considerable use in refining the horizontal and vertical alignment to minimize excessive volumes of surplus cut or fill and/or excessive haul distances. Where surplus material is generated, safe disposal sites should be identified (see below) well in advance of construction.

C2.6 Spoil disposal

The disposal of surplus or unusable excavated material and landslide debris is a major consideration in the design, construction and maintenance of mountain roads.

C2.6.1 New roads and road improvement schemes

Spoil material relates primarily to surplus excavated material arising from the construction of a road. The preferred approach is to design the road so that the excavated material can be fully utilized in the construction of embankments and fill slopes within a short distance of the source location, but this is often not feasible and may be uneconomic. Safe disposal of spoil is therefore required.

Uncontrolled or inappropriate spoil disposal can give rise to a variety of problems, including:

- end tipping of spoil onto formed embankments and fill slopes, often resulting in significant settlement, slope failure and erosion (Fig. C2.9);
- end tipping of spoil onto natural ground, smothering vegetation and often resulting in significant erosion and instability, some of which may extend into the underlying natural slope material (Fig. C2.10);
- blockage and disruption to slope drainage and siltation of drainage courses;
- inundation and disruption to adjoining agricultural areas; and
- destruction of existing habitats in both the surface vegetation and in stream courses.

In view of the importance of safe spoil disposal, it is recommended that the identification of disposal areas is carried

Fig. C2.9. Failure of end-tipped spoil into watercourse.

Fig. C2.10. Failure of natural ground beneath a spoil dump due, at least in part, to loading.

out as part of the design and that consultations with landowners and farmers are undertaken during the environmental assessment phases of a project. Payment to the construction contractor for spoil management should be provided through separate pay items in the bills of quantities, rather than included as a contractor's overhead to be covered in general earthworks items. The latter is likely to encourage the contractor to minimize the associated costs, thus avoiding his responsibilities with regard to haulage and safe disposal. If there are separate pay items, the contractor will have greater incentive to comply with the specifications in relation to spoil disposal.

The following guidelines are given on the location and management of safe spoil disposal sites.

- Preference should be given to large level areas, such as river terraces (away from active river channels) or flat areas formed by structural control in the underlying rock. Spoil should be spread and compacted sufficiently by tracked vehicle to minimize any potential erosion or instability.
- In the absence of large level areas, a contractor will often resort to side-casting of spoil. Where there is no choice but to side-cast material, the designated locations should be such that slope stability and environmental impacts will be minimal. In unpopulated areas, these locations might include the steeper slopes protected by resistant rock. In populated areas, side casting will have to be prohibited and the contractor must haul to a safe disposal site.
- It is normally preferable to utilize a large number of small areas rather than a small number of larger areas in order to minimize the risk of slope overloading and limit the potential failure volume at any one location.
- Side-casting can have significant negative environmental impacts and so would normally need to be balanced by appropriate offset measures.
- Spoil check structures should be obligatory, in the form of gabion or timber structures behind which small volumes of spoil are dumped. Timber is less preferable to gabion as a local forest source is usually required

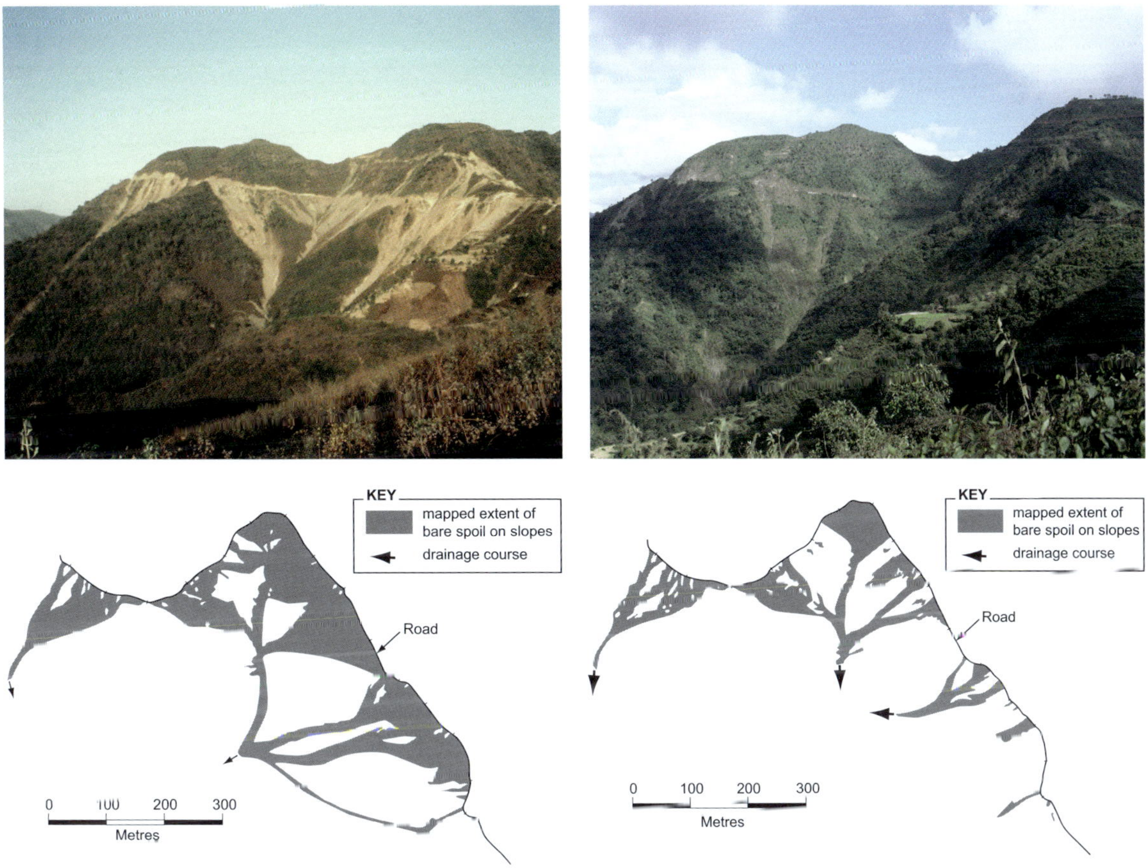

Fig. C2.11. Rapid vegetation regrowth within twelve months on side-cast material.

and the timber will rot quickly in warm, moist climates if not properly treated.
- Wherever possible, the natural vegetation cover should be maintained and spoil surfaces should be planted with appropriate grasses and shrubs to minimize erosion (Section C7.2.4).

Locations that should be avoided include:

- areas where spoil could impinge on springs, streams or river channels, resulting in impeded drainage, erosion, increased sediment load and downstream environmental impacts;
- slopes where road alignments, adjacent housing or farmland might be at risk; and
- areas where past, active or potential future slope instability or erosion could be reactivated, exacerbated or initiated, respectively.

Observations of spoil disposal on road construction projects in Nepal, for example, lead to the following conclusions.

- In areas of unstable ground and where slopes are underlain by weak and landslide-prone materials (such as mudstone, siltstone and phyllite), it is worthwhile investing in a spoil management plan that allows for the safe disposal of spoil outside these areas. This might include haulage over several kilometres.
- Attempts to create large spoil dumps on locally flat areas within otherwise sloping ground can lead to failure of the underlying slope, if site suitability is not confirmed by an engineering geologist. In the worst cases, this can lead to regressive ground movement and the failure of the road formation itself.
- The side-casting of small volumes of spoil onto steep slopes usually causes stripping of the vegetation and topsoil. However, in the humid tropics and subtropics, recovery rates can be remarkably quick (Fig. C2.11) unless:
  - there is an underlying geological weakness;
  - there is a persistent source of surface runoff, such as a road turnout or culvert outlet; or
  - the majority of the spoil material is rock debris of weathering grade I–III.
- The vegetation that colonizes spoil is rarely an ecological or economic replacement for what has been destroyed; a

Fig. C2.12. Use of bio-engineering (principally grass and shrub planting) to protect a spoil slope from erosion (see Section C7.2.4).

different range of pioneer species, perhaps alien to the area, will not necessarily protect the spoil against erosion. Selective planting schemes should therefore be put in place (Fig. C2.12; Section C7.2.4).

C2.6.2 Maintenance of existing roads

Spoil disposal along existing roads mainly involves landslide and rockfall debris and the material arising from the construction of remedial works (e.g. excavations for retaining walls) and the clearing of sediment from drainage structures.

Usually, once a landslide blockage has occurred, the requirement to reinstate access in the shortest possible time is paramount. This frequently results in landslide debris being dumped downslope in the immediate vicinity of the landslide, often creating further instability and erosion on slopes and in stream channels below. The occurrence of a landslide could be indicative of weak and landslide-susceptible materials underlying the slope as a whole. Such ground conditions should be avoided by selecting a stable disposal site as close to the landslide as possible, ensuring that it is acceptable as far as environmental and land use considerations are concerned.

The following recommendations should be observed.

- Wherever possible, the first choice should always be to remove debris by truck to a safe disposal area, or to stockpile it temporarily along one side of the road (in such a manner that it does not overload existing earthworks) until it can be removed after the road has been opened and the emergency resolved.
- Landslide debris and other spoil material should not be placed downslope in an area where the failure extends underneath the road (e.g. areas where tension cracks or a vertical displacement of the road have occurred). At the very least, it should be disposed of at a location outside the boundary of the failure.
- Utilize a number of suitable disposal sites rather than a single location to reduce the risk of slope overload and potential failure volumes.
- Avoid disrupting springs and natural water courses since this may result in significant erosion and stream channel blockage, creating further instability problems.
- Avoid dumping material over retaining walls as this may cause slope erosion, slope failure and impeded drainage in front of the wall foundation.
- Avoid damage to roadside drains or the road surface during debris clearance operations.
- Ensure that newly-formed spoil slopes are compacted to specification and provided with surface protection measures (Fig. C2.12 and Section C7.2.4).

References

Ayalew, L., Moeller, D. & Reik, G. 2009. Geotechnical aspects and stability of road cuts in the Blue Nile Basin, Ethiopia. *Geotechnical and Geological Engineering*, **27**, 713–728.

Bond, A. & Harris, A. 2008. *Decoding Eurocode 7*. Taylor & Francis Group, London.

DEPARTMENT OF MAIN ROADS, WA 2005. *Optimisation – Mass Haul, Chapter 8, Geometric Design, Roadways Technical Guidelines*. Department of Main Roads, Government of Western Australia, **67-08-62H1.**

DEPARTMENT OF TRANSPORT 1994. *Design Methods for the Reinforcement of Highway Slopes by Reinforced Soil and Soil Nailing Techniques*. Design Manual for Roads and Bridges, **HA68/94**, Part 4.

Fookes, P. G. & Sweeney, M. 1976. Stabilization and control of local rock falls and degrading rock slopes. *Quarterly Journal of Engineering Geology*, **9**, 37–55.

Hearn, G. J. 1995. Landslide and erosion hazard mapping at OK Tedi copper mine, Papua New Guinea. *Quarterly Journal of Engineering Geology*, **28**, 47–60.

Monahan, E. J. 1994. *Construction of Fills*. 2nd edn. Wiley, USA.

Trenter, N. A. 2001. *Earthworks – a Guide*. Thomas Telford, London.

TRL 1997. *Principles of Low Cost Road Engineering in Mountainous Regions*. Overseas Road Note 16, Transport Research Laboratory, Crowthorne.

Wesley, L. D. 2008. A simple compaction control method for slope construction. *In*: Chen, Z., Zhang, J.-M., Ho, K., Wu, F.-Q. & Li, Z.-K. (eds) *Landslides and Engineered Slopes. From the Past to the Future. Proceedings of the 10th International Symposium on Landslides and Engineered Slopes, 30 June – 4 July, 2008*, Xi'an, China, Taylor & Francis Group, London, 1005–1010.

Wyllie, D. C. & Mah, C. W. (eds) 2004. *Rock Slope Engineering*. 4th edn, Taylor and Francis Group, Suffolk.

C3 Soil slope stabilization

G. J. Hearn*, T. Hunt & S. d'Agostino

URS Scott Wilson Ltd, Scott House, Alençon Link, Basingstoke, Hampshire RG21 7PP, UK
*Corresponding author (e-mail: gareth.hearn@scottwilson.com; garethhearn@talktalk.net)

C3.1 Distinction between soil and rock

This section focuses on techniques of soil slope stability analysis and stabilization and provides selected case studies to illustrate some of the approaches commonly adopted.

The first distinction to make is whether the materials forming the slope under consideration behave primarily as soil or rock. A soil slope will fail through its granular mass, whereas a rock slope (Section C4) usually fails principally along discontinuities (bedding, foliation or other joints). Relict joints may be present in soil weathering profiles, associated with the original rock structure (Section A3.1). These normally have a weakening effect and, if adversely orientated, may become preferred planes of failure.

In practice, a clear distinction between soil slopes and rock slopes cannot always be made. As described in Section A3, rock structure, weathering profiles and taluvium/colluvium deposits on mountain slopes can be highly variable; it is not uncommon to find that a given slope will fail partly through rock and partly through soil. In *in situ* weathered soil profiles it is often the weathering grade boundaries, for example between weathering grade III and IV and IV and V, that tend to form the basal slip surfaces of shallow rainfall-induced landslides in the humid tropics and subtropics. However, in weathering grade I–III rock, the presence of jointing, folding and faulting often results in the development of complex weathering and strength profiles to the extent that zones of 'rock' with strengths not much greater than that of soil may be present at considerable depth (as illustrated in Sections B4 & B5). This will often exert significant control on the depth and configuration of existing or potential failure surfaces.

C3.2 Soil slope stability assessment and analysis

C3.2.1 Ground model

Before a slope stability assessment or analysis is made it is first necessary to establish a ground model. This is a normal progression from the field mapping and ground investigation activities described in Sections B3 and B4 whereby soil and weathered rock profiles, relict rock jointing patterns, groundwater conditions and the depth and configuration of existing or potential failure surfaces are assessed and represented in one or more cross-sections. The ground model may simply reflect the presence of one soil type if the failure is entirely in colluvium, for example. However, more commonly on mountain slopes it will be necessary to introduce more than one material type, such as colluvium or taluvium, overlying residual soil or weathered rock.

Table C3.1 summarizes the information required to establish a ground model for soil slope stability assessment and stability analysis in the context of low-cost roads. The table differentiates between information considered to be essential, desirable and optional. An illustration of a ground model is provided in Figure C3.1 with the methodology used to derive it.

Although there are important distinctions to be made between slope materials and failure mechanisms in the development of the ground model, it is the role of groundwater (pore pressure) during or immediately after intense or prolonged rainfall that is usually the key factor that initiates the failure of a slope through any of the following:

- the dissipation of soil pore suctions (negative pressures between soil particles); or
- a lowering of effective stress and hence a reduction in shear strength of the slope materials, either
  - as a groundwater table rise within a slope that comprises a single soil mass, or
  - as 'perched' water within a slope that comprises soil layers of varying permeability or in a soil layer above an impermeable rock head surface.

Slopes with deep weathering profiles often occupy prominent or 'high points' in the landscape, such as rounded ridges or spurs. The relative stability of these locations allows weathering profiles to slowly develop unhindered by erosion or slope instability. These profiles often comprise soils of moderate permeability and are frequently associated with low water tables due to their elevation. Pore suctions can therefore develop, and these will enhance the stability of cut slopes. On more sloping ground, pore suctions can quickly dissipate during heavy rain and positive pore

From: Hearn, G. J. (ed.) *Slope Engineering for Mountain Roads*. Geological Society, London, Engineering Geology Special Publications, **24**, 165–188.
DOI: 10.1144/EGSP24.12 0267-9914/11/$15.00

Table C3.1. *Level of importance of information for soil slope stability assessment and analysis*

Information	*Qualitative stability assessment*			*Numerical stability analysis*		
	Essential	*Desirable*	*Optional*	*Essential*	*Desirable*	*Optional*
Engineering geological map		✓			✓	
Slope profile	✓			✓		
Groundwater	✓			✓		
Soil/rock boundaries	✓			✓		
Soil layer geometry		✓		✓		
Soil strength parameters from published data		✓			✓	
Soil strength parameters from field assessment		✓		✓		
Soil strength parameters from lab testing			✓		✓	
Slip surface configuration (back analysis)	✓			✓		

pressures can develop if water infiltrating the soil profile cannot drain away quickly enough. A raised or perched water table can develop along or above permeability boundaries, for example between soils of different weathering grades, and contribute to slope instability.

C3.2.2 Slope stability assessment

Slope stability assessment is largely descriptive and judgement-based, and takes the following into consideration:

- the identification of ground movement indicators such as tension cracks, compression ridges or bulges, hummocky ground and disturbed vegetation;
- the apparent or known shear strength of the soil;
- the evidence for groundwater levels, including springs and seepages;
- the depth of soil over observed or anticipated *in situ* rock (the deeper the soil the more opportunity there is for a large slope failure to develop);
- the orientation of the rockhead surface as this could form a low-strength surface along which overlying soil (*in situ* weathered soil or taluvium/colluvium) could fail; and
- the presence of low-strength discontinuities within the soil such as relict joints and existing failure surfaces, as these could form preferred planes of movement.

When carrying out slope stability assessment it is important to record the observations, assumptions and judgements that are made. The vast majority of slope assessments for alignment selection, earthworks design and slope management adopt those qualitative methods that allow decisions to be made as to whether a ground investigation and a slope stability analysis are required.

C3.2.3 Slope stability analysis

For low-cost roads, *slope stability analysis* is usually used:

- when carrying out standard designs for cut slopes according to the range of material types and groundwater conditions observed or anticipated; and
- where landslides and potential slope failures pose significant risk to road operations and adjacent land uses.

Analysis usually takes the form of one of the following.

- Analysis of an unfailed slope where a factor of safety needs to be determined for a potential first-time failure. A typical example of this might be the analyses carried out for different cut slope heights and angles for a range of soil types. Alternatively, it could be in association with a particular cut slope, fill slope or retaining wall where there is a concern for stability, either in connection with a new road or the maintenance of an existing road.
- Back analysis of a failed slope for remedial design purposes. In these circumstances a factor of safety of approximately 1.0 is assumed to have applied at the moment of failure.

C3.2.3.1 Basic concepts

The basic concepts associated with soil slope stability analysis can be found in most relevant civil engineering text books (e.g. Lamb & Whitman 1979; Bromhead 1986; Abramson

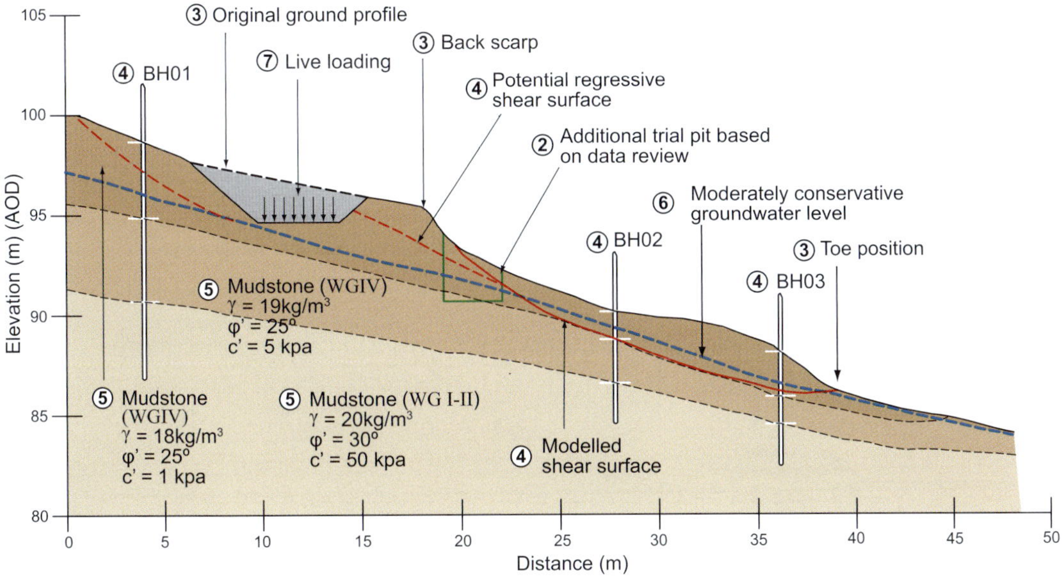

STEP	DESCRIPTION
①	**Data collection** Collect all available data **a) General:** Published geological maps, topographical maps, interpretation of remote sensing, field mapping and observations **b) Site specific:** Natural and construction exposures of soils and rocks, ground investigation (GI) data
②	**Data review** Assess data coverage Collate laboratory test data and *in situ* test data Identify shortfalls in dataset and potentially carry out further investigative work
③	**Slope form** Generate "critical" slope profile through landslide, earthworks and road, as appropriate (i.e. at maximum slope height and angle) Identify potential back scarp and toe positions based on field observations
④	**Stratum designation** Classify stratum horizons as engineering geological units from field observations and GI Plot GI points onto slope profile Interpolate horizons between plotted points on profile Plot location of shear surface if identifiable in field exposures, from GI and inclinometer readings
⑤	**Parametric designation** Applied to each stratum horizon Based on laboratory test data, cross-referenced with *in situ* test data Where insufficient site-specific data exist, parameter designation can be undertaken by literature review and engineering judgement
⑥	**Groundwater level determination** Observations of springs and seepages during field mapping Data from piezometers, preferably over period of > 1 year, to assess seasonal fluctuations If piezometer data are unavailable, information on water strikes during GI can be used according to its timing in relation to the wet season Use of groundwater seepage modelling software if sufficient data exist
⑦	**Load designation** Surcharge due to short term or live loading (i.e. traffic loading or construction plant) Surcharge due to long term loading (i.e. by structures) Selsmic loading based on a design earthquake event

Fig. C3.1. Ground model methodology used to analyse slope failure in weathered mudstone.

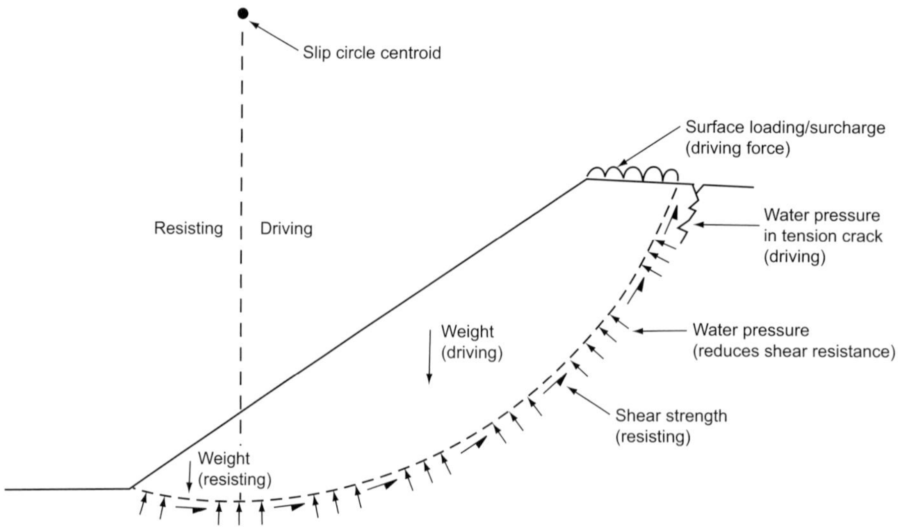

Fig. C3.2. Driving and resisting forces acting on an existing or potential failure surface.

et al. 2002; Duncan & Wright 2005). Very briefly, stability analysis involves balancing forces and moments for the sliding soil above a particular slip surface. The forces acting can be broadly divided into two groups as illustrated in Figure C3.2 and described below:

- the destabilizing (driving) forces, that is, the weight of the soil or weathered rock above an existing or potential slip surface, the water pressure force acting in a tension crack and surcharge loads (e.g. traffic loads) located at or near the top of the slip;
- the stabilizing (resisting) forces, that is, the mobilized shear strength of soil or weathered rock acting along the existing or potential slip surface and, depending on the shape of the slip surface, the weight of soil at or near the toe of the slope.

If the ratio of the resisting forces divided by the driving forces (i.e. the factor of safety against sliding) is >1.0, then the slope is considered to be stable; if the ratio is <1.0 the slope is considered to have failed or to be in the process of failing.

Slope stability analysis requires the following input data:

- the topography of the slope. This needs to be sufficiently detailed to enable cross-sections to be drawn at a usual scale of 1:100;
- the known or assumed depth and orientation of soil and weathered rock layers;
- material parameters, including density and shear strength, normally expressed in terms of effective cohesion (c') and friction angle (φ');
- the known or assumed shape and depth of the failure surface (in the case of a back analysis);
- the known or assumed critical groundwater condition (at the time of failure in the case of a back analysis). This may be represented as
  - a groundwater table, below which all slope material is deemed to be saturated, or
  - an r_u value (where r_u is the pore pressure ratio and defined as the water pressure divided by the total overburden pressure at any particular point along the critical slip surface);
- external loads, for example generated from traffic, retaining walls and spoil heaps. Earthquake loads will also need to be considered in seismic areas.

In clay soils, the soil strength is modified by changes in pore pressure that result from recent excavation or filling. After a change in load, the soil cohesion has an undrained shear strength (c_u) and a friction angle of zero ($\varphi_u = 0$). The period of ongoing strength modification can last for years after the load change due to very low permeabilities. However, once the load-induced pore pressures have dissipated, the clay strength will again be governed by c' and φ'. While this strength modification should be borne in mind when considering the stability of newly cut clay soil slopes (where suctions can enhance short-term stability) or recently constructed embankments (where increased pore pressures reduce short-term stability), it is rarely an issue when analysing natural slopes if the topography has remained largely unchanged for significantly long periods of time.

Circular slip surfaces are more likely in homogeneous, fine-grained soils slope (Fig. C3.3). Non-circular slips occur in granular soils and heterogeneous soils, for example where high-strength soils overlie low-strength soils or vice versa. Translational (planar) failures occur most frequently in cohesionless soils close to the slope surface, or where

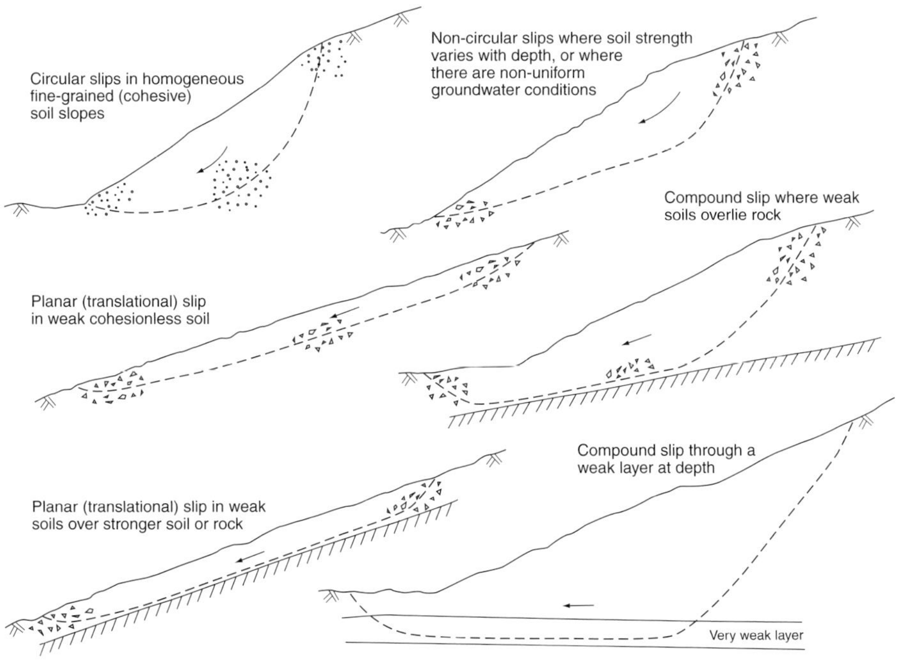

Fig. C3.3. Common soil slope failure mechanisms (modified from Craig 2004 with permission of Taylor & Francis).

weak soils overlie a rock head surface at relatively shallow depth. Compound slip surfaces in soils involve more than one failure mechanism, for example where the basal surface is a planar failure governed by a deeper underlying stronger horizon or rock head surface and where the upper part of the failure breaks through the soil to the slope surface, either as a circular or linear shear in fine-grained and coarse-grained soils, respectively (Fig. C3.3).

Before the advent of computers, circular slip surfaces were assessed by hand using the method of slices. The most widely accepted of these methods were those of Taylor (1937), Bishop (1955) and Janbu (1957). Later, these hand methods were speeded up by the introduction of stability coefficients and design charts by Bishop & Morgenstern (1960) and Hoek & Bray (1981). More recently Cavaleiro *et al.* (2010), for example, describe the use of charts to gain rapid assessments of slope stability in granite residual soils. However, design charts assume homogeneous soils and horizontal ground above the slope in question, and neither of these conditions occurs in the majority of slopes in hilly or mountainous areas. Computer programs have automated these methods, and allow actual slope topography to be represented and numerous trial slip surfaces in complex soil profiles to be rapidly evaluated. Nevertheless, the original methods are still useful for cross-checking computer outputs and for preliminary assessments. In this respect, the design engineer therefore needs to take care that the ground properties (inputs) and predictions (outputs) are realistic before embarking on more detailed analyses.

The most appropriate analyses are likely to be based on limit equilibrium, using non-circular or circular methods, depending on the observed or anticipated slope failure mechanism. The analysis of planar failures by infinite slope analysis and compound and non-circular failures by two- or three-part wedge analysis is described in Morgenstern & Price (1965), Hoek & Bray (1981) and Craig (2004), for example. Sloan (2012) highlights the importance of selecting the analytical technique to suit the failure mechanism by demonstrating how the factor of safety can be significantly over-estimated if non-circular failures are analysed using circular geometry.

The finite element method (FEM) has become increasingly used in the analysis of slopes (e.g. Duncan 1996; Griffiths & Lane 1999). The FEM does not require a shear surface to be pre-determined and treats the soil mass as a continuum rather than a number of discrete slices, as is the case with the limit equilibrium approach. The method is able to model progressive failure up to and including ultimate shear by treating the slope as an elastoplastic medium. Sloan (2012) describes how Finite Element Limit Analysis (FELA) is more robust in accounting for heterogeneity, anisotropy, fissures, soil structure interaction and variable pore pressures.

The FEM and FELA might be considered for large and complex landslides at high risk sites, where soil and pore pressure parameters can be determined accurately enough for modelling purposes. Limit equilibrium methods will probably suffice for most low-cost road applications.

C3.2.3.2 Selection of parameters

Strength parameters (c' and φ') are usually assessed by combining and corroborating as many of the following data sources as possible:

- published data;
- observations of natural slopes in the area to identify maximum angles for different soil types (Section C2.2.2). On most mountain slopes it is appropriate to assume that coarse-grained soils can be characterized as $c' = 0$ soils, and therefore marginally stable slopes or unstable slopes approximate in angle to φ' providing the water table is not close to the ground surface (i.e. the slope is 'dry');
- field descriptions of exposed soils, either at the surface or through trial pitting and other ground investigation techniques (Section B4);
- laboratory testing on undisturbed samples to determine shear strength parameters (Section B4);
- laboratory testing on disturbed samples to determine grading curves and Atterberg limits (where appropriate) and enable strength parameters to be evaluated (Section B4) and extrapolated; and
- back analysis of failed slopes with an assumed critical groundwater table to determine strength parameters (see below).

C3.2.3.3 First-time failure analysis

In this type of analysis the computer program is required to calculate the minimum factor of safety and the location of the critical failure surface. First-time failure analysis for existing slopes along low-cost roads is infrequently carried out unless the cut slope is particularly deep or the risk associated with failure is high, as might be the case where there are high-value land uses on the slopes above, such as housing or service installations. In first-time failure analysis, the degree of uncertainty in the selection of strength parameters and the choice of groundwater table can be high. To compensate, it is usual to adopt either:

- a pessimistic ground model (i.e. low strength parameters, high water table);
- higher partial factors (Section C3.2.5); or
- a higher factor of safety for design (Section C3.2.4).

C3.2.3.4 Back analysis

As noted earlier, in cases where a slope has already failed or is in the process of failing (as shown by tension cracks, bulges and other telltale features), it is often appropriate to carry out a back analysis to confirm soil strength parameters and groundwater conditions on the assumption that the factor of safety at the moment of failure was or is 1.0. For an effective back analysis, the depth and configuration of the slip surface also need to be known. If pre-failure ground mapping is unavailable the shape of the original topography can usually be assessed from the morphology of the surrounding slopes and combined with field survey to establish a likely profile of the pre-failure slope.

The major unknowns are the shear strength parameters along the failure surface and the groundwater condition at the time of failure. In the case of the former, and in the absence of any soil sampling and testing, it is usually possible to carry out preliminary field-based assessments of shear strength parameters for the failed material. In the case of the latter, it may be possible to estimate the location of the groundwater table from observations on site (e.g. surface seepages). In order to obtain a ground model that matches site conditions at the time of slope failure, sensitivity analyses can be carried out by varying:

- values of c' and φ' (usually averaged over the full length of the failure surface);
- groundwater levels; and
- depth and configuration of the failure surface, if this is not known from ground investigation.

The analytical ground model can then be tested on other nearby failed and unfailed slopes in similar materials to determine whether it yields a sensible range of factors of safety. As mentioned earlier it is usual to assume that c' is equal to zero when carrying out back analyses of slope failures along mountain roads, as most hillsides are mantled in non-cohesive soils and the analysis is particularly sensitive even to small changes in c'.

As with first-time failures, it is often the case that without intensive ground investigation and laboratory testing there will remain significant uncertainties over the depth and configuration of the failure surface, the groundwater table at the time of failure and the strength parameters that apply. Furthermore, if slope failure results in significant topographic change and involves multiple failure surfaces and regressive movements, it can be extremely difficult to replicate in a back analysis.

C3.2.3.5 Forward analysis

Once the analytical ground model is established from the back analysis it is then possible to assess the future stability of a slope taking into consideration:

- the difference between peak and residual shear strengths;
- earthquake ground accelerations (if appropriate);
- rainfall-induced rises in groundwater; and
- the influence of any remedial works that are planned or designed.

When a slope failure occurs, the strength of the soil along the slip surface will reduce to a residual shear strength (due to effects such as particle realignment caused by large movements). During the course of failure, the friction angle at the failure surface may reduce by as much as one-third to its residual value. In carrying out a forward analysis there are three conditions to be borne in mind:

- condition at failure, that is, first-time failure (φ_{peak});

- condition after failure, that is, reactivation potential ($\varphi_{residual}$); and
- that c′ is usually assumed to be zero along the failure surface even in cohesive soils, despite the possibility that it may have been greater than zero when failure originally occurred.

These reductions in strength, and the changes to topography and groundwater that occur due to failure itself, need to be taken into account in any forward analysis, particularly if much of the slipped mass is to be left in place. In some cases, there will be insufficient information to be confident about pre- and post-failure soil strength and groundwater conditions even after the back analysis is undertaken. It is therefore difficult to be completely confident in the calculated factors of safety for forward analysis. In low-risk solutions it may prove more practicable therefore to design remedial works on the basis of a percentage improvement rather than on an absolute factor of safety.

C3.2.4 Lumped factors of safety

Typically, higher factors of safety are required for designs where uncertain ground conditions coincide with a higher risk potential and a long low-maintenance design life requirement.

For cut slopes of up to 5 m in height on low-cost roads, it is usual to accept a minimum lumped factor of safety of 1.2 or even 1.1 in the knowledge that, depending upon the ground conditions exposed, some slopes will have lower factors of safety while others will have higher. When the risk associated with cut slope failure is low, as would be the case for temporary blockage to the side drain and the adjacent carriageway on a low-volume road for example, then these relatively low factors of safety are usually regarded as an acceptable outcome. Deeper cuttings may require a minimum factor of safety of 1.3 due to the greater risk they pose should they fail.

For the design of stabilization works for a failed slope, a minimum lumped factor of safety of 1.2 is usually an acceptable goal for low-cost roads although even this might not be achievable (a value closer to 1.1 may have to be accepted if conservative parameters have been used). Given the usual uncertainties in the analysis, a decision will have to be made as to whether or not to invest in remedial works that yield such low levels of improvement in factors of safety. A decision to invest in road maintenance and protection, rather than stabilization, may be the most practicable approach (Section A4.3).

Table C3.2 provides indicative factors of safety for negligible, low and high risk to life situations for new roads utilizing conservative (1 in 10 year rainfall) groundwater levels in the stability analyses. In the context of low-cost roads the negligible and low risk categories are probably most appropriate, with the high risk category only relevant to urban or populated areas, or where traffic volumes are high. Although similar standards should be sought for existing roads this is not always the case (Table C3.3) as there may be less opportunity to engineer the preferred solution due to space and working constraints. However, these factors of safety should be used with caution and, particularly in high risk situations, the analyses should be based on conservative groundwater levels where rigorous geological and geotechnical studies have been carried out, and where the modified slope conditions remain substantially unchanged from the existing slope. Where national design standards and codes of practice are available, these should be used.

Eurocode 7 (BSI EN 2004) provides European member states with options and guidelines for the calculation of factor of safety for slope design (Section 11 of that document). According to the Code, all limit states should be considered, including:

- loss of overall stability of the ground and associated structures;
- excessive movements in the ground due to shear deformations, settlement, vibration or heave; and
- damage or loss of serviceability in neighbouring structures, roads or services due to ground movements.

The effects of the following circumstances should be taken into account:

- construction processes;
- new slopes or structures on or near the particular site;

Table C3.2. *Suggested lumped factors of safety for the analysis of new slopes with reasonably certain ground conditions*

Potential economic loss should failure occur	*Recommended minimum factor of safety against risk to life*		
	Negligible Risk	*Low Risk*	*High Risk*
Negligible	>1.0	1.2	1.4
Low	1.2	1.2	1.4
High	1.4	1.4	1.4

Modified from GCO (1984).

Table C3.3. *Suggested absolute minimum lumped factors of safety for the analysis of existing slopes and for remedial or preventative works*

	Recommended minimum factor of safety against risk to life		
	Negligible Risk	*Low Risk*	*High Risk*
Factor of safety	>1.0	1.1	1.2

Modified from GCO (1984).

- previous or continuing ground movements from different sources;
- vibrations;
- climatic variations, including temperature change, drought or heavy rain;
- vegetation or its removal;
- human or animal activities;
- variations in water content or pore-water pressure; and
- wave action.

Most geotechnical engineers would consider these guidelines as being part of standard procedure regardless of European jurisdiction. In seismic areas the effects of earthquake loading would need to be added to the above list.

C3.2.5 Partial factors of safety

Table C3.4 shows the recommended partial factors to be employed in the computation of factor of safety for the United Kingdom, for example, which has adopted Approach 1 Combination 2 of Eurocode 7 whereby partial factors are applied to unfavourable variable actions and material properties (Bond & Harris 2008). The variable actions, such as transient external loading and water table fluctuations, are multiplied by the partial factor, while the material strength properties are divided by the partial factor, in order to account for variability and uncertainty in ground conditions. After application of these partial factors the ratio of stabilizing to destabilizing forces should then be greater than 1. The British Standard Code of Practice for Earthworks (BSI 2009) notes that partial factors may not be adequate where the risk of a slope failure is very high, and may be too high for situations where residual strength is adopted (as would be the case for landslides and failed slopes).

As far as future practice on low-cost roads is concerned, Eurocode 7 simply formalizes the approach previously adopted whereby the implications of variability and uncertainty in input parameters are taken into account when deciding on an acceptable factor of safety. The cost and risk outcomes for adopting average, conservative and worst-credible parameters should still form important considerations. Eurocode 7 requires good quality ground information for the Table C3.4 partial factors to apply, and therefore it is probably inappropriate for use on most slope designs for low-cost mountain roads.

The option of including a probabilistic approach to supplement more routine deterministic methods should also be considered. For example, Table C3.5 shows the accepted annual probabilities of failure adopted by the Norwegian national guidelines for building and civil engineering works (Nilsen 2000). However, this approach requires good historical records and knowledge of the probability of landslide-triggering events such as rainstorms of a recorded intensity (see also Jaiswal & van Westen 2009; Wu & Chen 2009) and earthquakes (Owen *et al.* 2008); this information is unlikely to be available in many low-cost road cases.

Table C3.4. *Recommended Partial Factors for use with Approach 1 Combination 2 of Eurocode 7 (BSI EN 2004)*

Factors	*Values*
Action factors	*Multiply By*:
Permanent	1.0
Variable	1.3
Material factors	*Divide By*:
Effective angle of shearing resistance tan φ'_{peak}	1.25
Effective cohesion c'	1.25
Undrained shear strength c_u	1.4
Unconfined compressive strength q_u	1.4
Weight/density γ	1.0

Table C3.5. *Acceptable annual probability of failure according to consequence in Norway*

Safety class	*Consequence of failure*	*Maximum annual probability of failure*
1	Minor	10^{-2}
2	Medium	10^{-3}
3	Major	$<10^{-4}$

C3.3 Soil slope stabilization

As described in Section A4 and illustrated in Figure C3.4, landslides can impact a mountain road:

- as failures in the cut slope (type 1);
- as failures in the fill slope (type 2);
- as failures of the natural hillside above the road (type 3), either due to the presence of slope instability that existed prior to construction or, more commonly, due to the removal of toe support by cut slope excavation (in this case the failure surface daylights either at, or close to, road level);
- as failures of the slope below the road that regress and remove support for part or all of the road (type 4); or
- as failures of the entire slope upon which the road is constructed (type 5). These landslides usually pre-date road construction though occasionally they can be triggered as first-time failures during construction and operation as a result of river scour, heavy rainfall or seismicity (they often develop as extensions to type 4 failures).

Table C3.6 summarizes the methods of engineering management commonly applied to each failure type. With

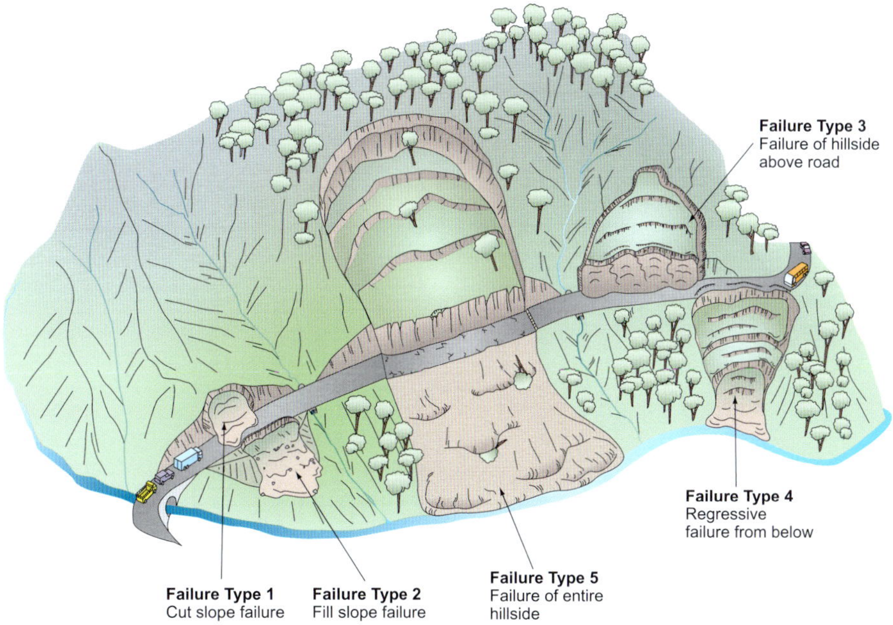

Fig. C3.4. Five common configurations of soil slope failure along mountain roads.

respect to stabilization, there are principally three ways of improving the stability of a slope:

- reducing the destabilizing (driving) forces – regrading, that is, removal and flattening;
- increasing the stabilizing (resisting) forces – external support, for example, toe weighting; and
- increasing soil strength – remove and replace with stronger materials, soil reinforcement and drainage.

There are numerous textbooks that deal with techniques of slope stabilization, including those by Turner & Schuster (1996), Ortigao & Sayao (2004), Bromhead (2005) and Cornforth (2005). Table C3.7 lists those measures commonly adopted in the stabilization of soil slopes and identifies what their limitations might be for low-cost roads. The remainder of this chapter provides further discussion and case studies of soil slope stabilization for each of the failure types depicted in Figure C3.4.

C3.4 Slope failure type 1: cut slopes

This type of slope failure usually inflicts the least damage and, in most cases, is dealt with by routine clearance of the failed material only. If the failed material originates from soil and weathered rock towards the top of the cutting, the process of failure itself often leads to the immediate removal of unstable material and a reduction in slope angle, thus improving stability. Nevertheless, the potential for erosion of the exposed failure plane will need to be considered. Sometimes revegetation can be left to natural processes; at other times, additional planting is required (Section C7.2.4). In most cases the former will apply.

In the case where a cut slope failure involves the entire height of the cutting, a back scarp will be created and the failed mass will usually remain on the lower portion of the slope and on the adjacent road and side drain. The back scarp will usually be concave in cross-section, with an upper steepened part that may not be stable in the longer term. This may need to be cut back to reduce overall slope angles in the weaker materials and to remove overhangs. Removal of the failed mass from the road and the side drain may still leave failed debris on the slope above, posing a potential future hazard. This can either be removed in its entirety, or stabilized. For small failures, perhaps up to 500 m^3 in volume, it may prove most effective to remove the entire slipped mass, thus exposing the majority (if not all) of the original failure plane. This can then be protected from erosion by planting measures. For larger failures, stabilizing the landslide debris may be a better option rather than removing it, especially if significant excavation at the toe of the slope could cause regressive ground movements further up the hillside.

Where a decision is made to stabilize a failure, the usual practice is to clear the landslide material from the road and side drain and to construct a wall to retain the remaining debris. The retaining walls most frequently used in this

Table C3.6. *Typical engineering management options for the failure types shown in Figure C3.4*

Failure type	*Engineering management*				
	Avoid	*Remove*	*Stabilize*	*Protect*	*Accept*
1	These failures are often triggered as a result of slope excavation and therefore avoidance during route selection is usually not an option.	Removal of slipped debris is an option in the case of the smaller failures if the remaining slope is stable and can be protected against erosion.	Can be achieved usually through earthworks, drainage and retaining structures (see Table C3.7).	Catchwalls or fences may be provided to protect road from falling debris (see also Section C4).	Small failures onto the road can be accepted if source removal, stabilization or protection is too difficult or costly compared to the damage inflicted.
2	Shift road into hillside to avoid unstable fill slope below. However, this may initiate type 1 and 3 failures.	Not usually practicable where large fill slopes are involved. Also either partial or complete removal will result in loss of road width.	This is usually achieved through excavation and recompaction, improved drainage or by the use of retaining structures founded beneath failure surfaces.	Construction of road edge and road fill retaining walls founded beneath failure surfaces may isolate the road from the failing fill slope.	Ongoing movements that cause progressive loss of all or part of the road cannot usually be accepted without loss of road function.
3	These failures are often caused by slope excavation. Therefore avoidance during route selection is usually not an option. In the worst cases, where landslides frequently cause road blockage, realignment might become cost-effective in the longer term, if a suitable alternative exists.	Not usually practicable given large volumes, access difficulties and uncertainties over the stability of the remaining slope.	May not be practicable or economically feasible to achieve stabilization in the case of the larger slope failures, though improvements can be achieved through earthworks, drainage and retaining structures (Table C3.7).	Catchwalls or fences may be provided to protect road from rockfall debris (Section C4), but these are unlikely to be appropriate for the larger soil slope failures.	It is usually only feasible to accept ongoing movements if these are slow and can be accommodated by maintenance.
4	Avoid through alignment selection (new roads) or realignment (existing roads) if a suitable alternative exists. Roads are often shifted into the hillside to avoid developing problems below. However, this may initiate type 1 and 3 failures.		If the slope failure is local to the road, then stabilization by retaining structures and drainage may be possible, though unlikely at low-cost.	Construction of road edge and road fill retaining walls founded beneath failure surfaces may isolate the road from the slope failure below.	Ongoing movements that result in the progressive loss of all or part of the road cannot usually be accepted without loss of road function.
5	Avoid through alignment selection (new roads) or realignment (existing roads) if a suitable alternative exists.		Stabilization of large landslides is usually beyond the scope of low-cost roads.	Road cannot usually be protected against ground movements.	It is usually only feasible to accept ongoing movements if these are slow and can be accommodated by road surface and drainage repairs. A gravel road surface should be considered.

Table C3.7. *Options for soil slope stabilization*

Requirement	*Technique*	*Where?*	*Limitations*
Reduce driving forces	Regrade slope to reduce angle (Section C2)	On any slope where reduction in cut slope angle is feasible	Unlikely to be feasible in steep terrain, regraded surface will need erosion protection
	Drain surface (Section C6)	Anywhere where surface runoff is apparent or water table/perched water table is at or close to the slope surface	Will only reduce surface infiltration, therefore combine with other techniques
	Drain subsurface (Section C6)	Anywhere where the water table can rise above the slip surface	Depends upon depth to which drains can be constructed in relation to depth of water table beneath slope surface
Increase resisting forces by application of an external force	Construct retaining wall (Section C5)	Anywhere where space and foundations allow	Moderate cost; must be founded below slip surface; may need to be combined with other techniques
	Construct toe berm	Anywhere where space allows	Usually requires significant space at toe and may not be feasible in steep terrain
Increase resisting forces by increasing internal strength	Drain subsurface (Section C6)	Anywhere if water table is above slip surface	Depends upon depth to which drains can be constructed in relation to depth of water table beneath slope surface
	Install soil nailing (Section C5)	Usually used to steepen cut slope angle e.g. for road widening	High cost; specialist installation equipment needed. Applicable mostly to unfailed slopes only.
	Use bio-engineering to enable roots to bind soil together (Section C7)	Anywhere where slip surface is very shallow (less than 1m deep)	Not suitable for deep-seated failures. Planting mix must include deep and strong-rooted shrubs.
	Use reinforced fill (Section C5)	Anywhere where space is limited for conventional fill	High cost; requires fill slope reconstruction

situation are masonry and gabion due to their ease of construction and relative low-cost. Because of their stiffness and strength compared to gabion, masonry walls are usually preferred where:

- bearing capacities are sufficient, there are no soft spots in the foundation and differential settlement is not anticipated (rock, weathered rock and possibly residual soil foundations);
- seepages and groundwater levels within the failed mass are generally considered to be low or controllable through drainage; and
- masonry stone is available locally.

A gabion wall may be more appropriate if any of these conditions are not met and where ground movements are expected to continue after wall construction.

C3.4.1 Case study

Figure C3.5 shows a cut slope failure that took place following heavy rain along the Hirna to Kalubi road in the Eastern Highlands of Ethiopia. By the time the photograph was taken, the failed mass had been removed from the road and a temporary cut slope had been formed. The underlying geology comprises basalt overlying tuff, and the failure took place at the approximate boundary between the two.

Fig. C3.5. Type 1 slope before and after implementation of remedial measures.

Black cotton soil (Section A3.1) was present on the majority of the hillside above the cut slope.

Two inspection trenches were excavated, one through the tension crack that formed the upslope extent of the slope failure (Fig. C3.6) and the other in the central portion of the slide, to assess the inclination of the slip plane and the nature of the materials involved. The materials exposed in both trenches comprised highly fractured basalt (loose, angular cobbles and boulders of basalt in a clay matrix) overlying completely weathered tuff (soft, light brown, silty clay). A slip plane was present in both the inspection trenches extending along the boundary between the basalt and the tuff. The slip surface was at a depth and orientation that was consistent between the trenches and where it daylighted on the face of the temporary cut slope. The material forming the slip plane was soft clay with a high moisture content and high plasticity. The slip plane itself was inclined at 11° to 17°, dipping out of the slope in the direction of landslide movement. This also corresponded to the dip and dip direction of the bedding within the tuff. Seepage was noted from a zone extending from the slip plane to 0.5 m above it, both in the trial trenches and where it daylighted in the cut slope.

The remedial design (Fig. C3.6) comprised a gabion retaining wall founded just beneath the slip plane and a masonry revetment constructed from the base of the gabion wall to side drain level in order to protect the lower unfailed slope. The gabion wall was scheduled because it would be able to accommodate some residual ground movements through its flexibility; the masonry revetment wall was chosen because its foundation was anticipated to be within *in situ* weathered material (hence a stiff, brittle structure was acceptable). The slope behind the gabion wall was backfilled with granular material and a system of herringbone drains (Section C6.2.2.2) installed to facilitate the lowering of the water table. The computed factor of safety against failure of the reconstructed slope was 1.29. Figure C3.5 shows the slope after the implementation of these measures.

In this case study, drainage was a key component of the remedial design. Drainage measures should be implemented where:

- there is an obvious source of water above the landslide back scarp that continues to channel water into the slide mass;
- there are springs evident in the back scarp;
- there is suspected seepage from the landslide back scarp into the landslide debris in front of it (this may not be evident from surface inspection, but might be inferred from the drainage condition of the debris if there are no other apparent sources of water); and
- there is a requirement to increase soil strength through a reduction in pore water pressure, and hence an increase in effective stress.

C3.5 Slope failure type 2: fill slopes

The construction and stability of fill slopes are discussed in Section C2.4. Fill slope failures are a common occurrence along mountain roads and are usually associated with:

- movements in the underlying natural ground;
- movement along the natural slope/fill boundary where this has not been adequately benched to form a shear key;
- undercutting and loss of support at the toe of the fill slope due to erosion in adjacent streams and below culvert outlets;
- saturation of the fill by uncontrolled rainfall runoff from the road due to either
  - a blocked roadside drain at the inside (hillside) edge of the road, causing the flow to be diverted across the road and onto the slopes below, or
  - the presence of an adverse road camber (i.e. the road surface sloping outwards towards the fill slope) that causes road runoff to flow to the outside edge;

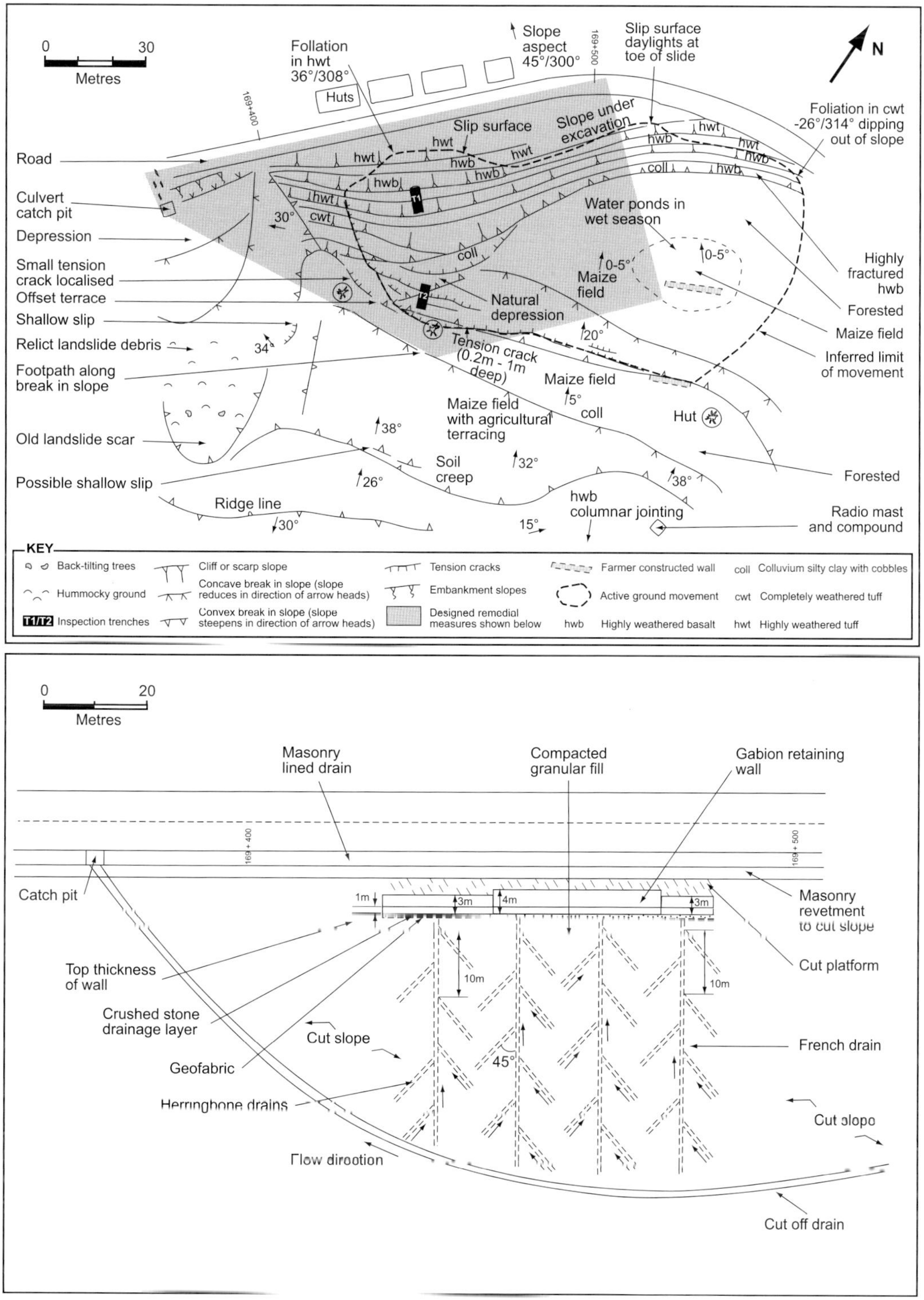

Fig. C3.6. Geomorphological map and plan of remedial measures.

- inadequate compaction;
- fill constructed to too steep an angle.

There are essentially two options for remedial works:

- to excavate and, if necessary, replace the existing fill, ensuring adequate compaction and benching into the underlying hillside (the common problem with this approach is the need for machine access to the base of the fill slope and for temporary storage of the excavated material prior to re-use); or
- to construct a below-road retaining wall founded beneath the slip plane in original ground (the problem with this approach is the usual need to utilize at least half the road width to excavate for the retaining wall foundation and the potential cost, compared to replacement or recompaction of the original fill).

Whichever option is selected, it is important to identify the cause of the failure and to take the necessary steps to ensure that it is remedied. If road runoff is the principal cause then provision of the following should be considered:

- an upstand to the hillside drain so that future slipped debris is retained rather than allowed to block the drain; and
- a roadside drain or an upstand along the outer edge of the road, so that surface runoff is channelled along the road edge and discharged at a safe location.

A third option (which is not a remedial measure *per se*) is to realign the road into the hillside thus placing it wholly in cut and not dependent upon the stability of fill. This option is commonly employed on low-cost roads as a means of reinstating access as quickly as possible following a type 2 failure. However, it can result in the creation or aggravation of types 1 and 3 failures above.

Fig. C3.7. Type 2 fill slope failure.

Fig. C3.8. Below-road masonry wall constructed at crest of fill slope in Figure C3.7 with grass planting on eroded slope face.

C3.5.1 Case study

Figure C3.7 shows the result of uncontrolled runoff from a roadside drain that became blocked by a very minor cut slope failure along a road in Laos. This caused runoff across the road and into the adjacent fill slope, causing it to fail. Due to access difficulties, a decision was made to construct a below-road masonry retaining wall founded on original ground beneath the fill/original ground interface (Fig. C3.8) rather than reconstruct the fill slope. The failed material below the wall was planted with grass to reduce any potential for further erosion.

C3.6 Slope failure type 3: Above-road slopes

The principal difference between this case and the type 1 failure is the practicality in achieving an acceptable factor of safety against further movement of large landslides if stabilization is attempted with a limited budget. If slope movements into the road are both significant and frequent, this will represent a recurrent hazard to traffic and an ongoing maintenance cost. There will be little option, therefore, other than to implement measures that lead to an acceptable level of improved stability, even if total stabilization is not practicable. The three principal stabilization methods of slope regrading, drainage and toe support by retaining structures (Table C3.7) are likely to apply, although regrading may be problematic due to access difficulties that are usually encountered on steeper slopes. Regrading also becomes impracticable in steep terrain when the regraded face would otherwise extend a considerable distance upslope, thus exposing a large new area of slope to potential erosion.

Text box C3.1. Stabilization of a type 3 landslide in Bhutan

This type 3 landslide in Bhutan originally took place in intact phyllite, creating a deposit of taluvium estimated to be up to 5 m in depth. This illustration is probably not unlike many of the larger slope failures encountered along mountain roads. By combining drainage measures with retaining structures, a design FoS of 1.3 was achieved against failure of the remaining landslide debris.

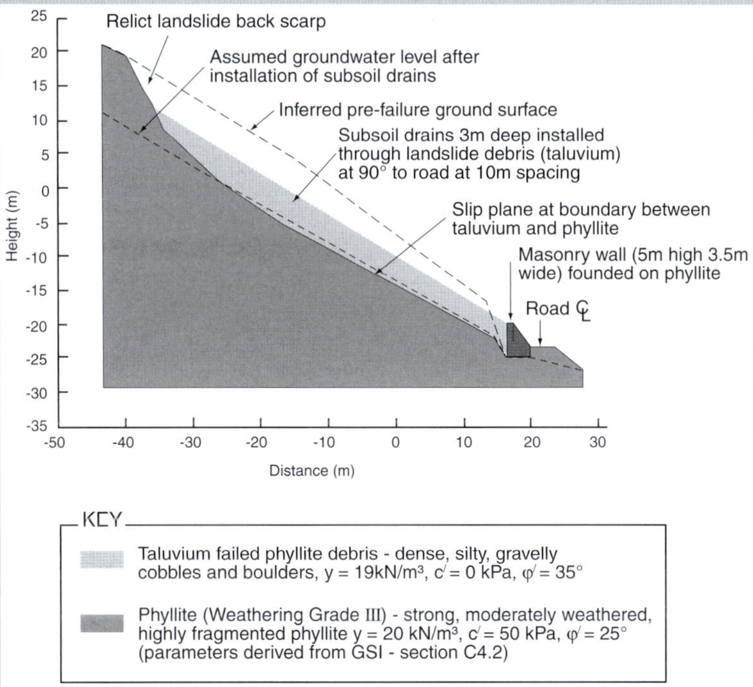

A combination of slope drainage and retaining structures is most commonly used, as is illustrated in Text box C3.1. Whether or not designed drainage measures can be constructed to function as intended will depend on access for excavation plant, the drainage pattern on the slope, soil permeability, and any obstacles to drain construction such as large boulders and rock outcrops.

In higher risk situations, for example on more highly trafficked roads or where residential property is potentially at risk, it may be necessary to consider more sophisticated techniques either individually or in combination. Figure C3.9 illustrates this from Hong Kong where a slope has been stabilized by the use of benching, soil nails, drainage and geofabric slope protection.

Fig. C3.9. Type 3 slope stabilization in Hong Kong.

C3.6.1 Case study

The Mekane Selam to Gundewein road in Ethiopia crosses very steep terrain associated with the gorge of the Blue

Fig. C3.10. Taluvium exposed in access track excavation.

Nile and its tributaries. A section of the alignment (Fig. C3.10) was required to cross the upper portion of a 35–40° slope formed in weathered rockfall material (taluvium) located beneath limestone cliffs (as described in Section C1.4; Fig. C1.5). Although this is not a type 3 failure as such, it is a relevant illustration given the fact that the cut slope was to be formed in failed material. The alignment is at maximum allowable gradient from the top of the plateau to the river below and thus the vertical alignment across the taluvium slope was essentially fixed. The material comprised predominantly medium-dense to dense angular cobbles and boulders in a silt matrix to a depth of more than 20 m (bedrock was not described in any of the boreholes that were terminated at this depth). Most of the slope was considered to be close to limiting angles for stability, as it is uncommon to encounter this material at a slope of much more than 38°. Nevertheless, a full-cut design was chosen because:

- it would allow load to be removed from the top of the overall slope, thus improving general slope stability;
- of uncertainties over adequate bearing capacities for fill retaining walls (see the *within deposit* case for *steep taluvium* in Fig. C2.3), the slopes below the road being too steep for unretained fill.

The hazard posed by potential activation of type 1 and type 3 failures therefore needed to be considered and the design reviewed the following options:

- option 1: unsupported and unprotected slopes cut to 2:1 (V:H);
- option 2: unsupported and unprotected slopes cut to 1:1;
- option 3: unsupported slopes cut to 1:1 and protected against shallow failure through the use of soil nails (Section C5) and netting;
- option 4: unsupported slopes cut to 1:1 and protected against shallow instability using gabion or masonry revetments (Section C7);
- option 5: cut slopes supported by reinforced concrete retaining walls (Section C5);
- option 6: cut slopes supported by gabion retaining walls (Section C5); or
- option 7: cut slopes and adjacent hillside supported by piled retaining structures (Section C5).

Option 1 was discounted on the grounds that, while temporary excavations in these slope materials were standing at angles of 2:1 (shown in Fig. C3.10) or even greater, in the longer term these slopes would fail as a result of both shallow and deeper-seated movement brought about by groundwater rise or more shallow, sub-surface drainage. The temporary cut slopes at 2:1 were up to 5 m in height; the final designed height of a 2:1 slope to accommodate the required road formation width would need to be 20 m in height and therefore much more vulnerable to failure. Cut slopes of 1:1 (option 2) were considered the steepest achievable in the longer term and would require a total cut height of 40 m. However, due to expected 'soft spots' in the material it was anticipated that failures would occur; an unsupported and unprotected design (option 2) was therefore rejected.

Option 3 was discounted because soil nails require specialist drilling and installation and the heterogeneous nature of the taluvium is such that drilling would encounter variable soil and boulder material, and possibly some voids. The use of netting, nailed into the slope above the cutting and draped over the cut face, would help contain rock and soil falls originating in the cut slope, but would have little influence on the potential for deeper instability.

Option 4 comprised a 1:1 cut protected locally against shallow slope failures in the weaker materials exposed by prescribing revetments (Section C7) during excavations, but deeper failures in the longer term could not be ruled out.

Options 5 and 6 were analysed according to the cross-sections shown in Figure C3.11.

For option 5, a 10 m high cantilevered retaining wall was designed to support the cutting. This was considered to be the maximum practicable height of retaining wall that could be constructed. Sufficient factors of safety against sliding, overturning and bearing capacity failure were obtained when the wall section was examined in relation to the cut slope. However, the sloping ground above would exert additional loads onto the wall, and the wall could not be designed to withstand these. Calculations showed that, if the slope behind the wall could be cut back horizontally by a distance of 15 m, then the wall could perform as required. However, this would create a steeper slope on the hillside above which would be prone to instability. If this steeper slope were to fail significantly then the retaining wall below would be surcharged and would fail.

The same conclusion for option 5 applied to option 6. A gabion structure offered the advantages of being free-draining and able to withstand some ground movements, should small volumes of debris surcharge the wall by failure from above. Nevertheless, the wall would not be able to withstand even moderate surcharges, and therefore

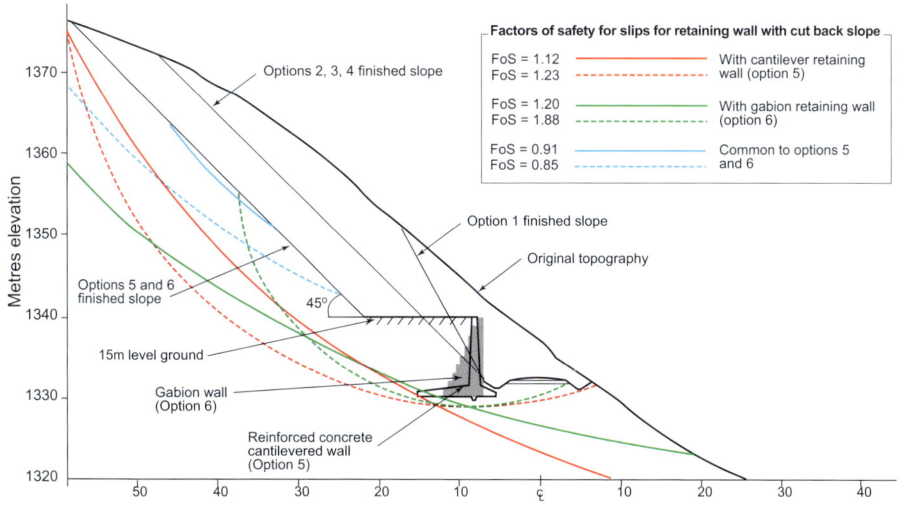

Fig. C3.11. Cross-section showing options and analysed slopes with reinforced concrete and gabion retaining walls.

would not be sustainable without a long-term commitment to maintaining the slope above.

In the case of option 7, only a high-investment, deeply founded piled structural design would provide a total solution against slope instability. This was considered to be beyond the financial scope of the project.

The recommendation was therefore to proceed with option 4 and to accept a degree of hazard from slope failures from the cut slopes extending into the hillside above. This illustrates a common situation that occurs whereby only partial solutions are both practicable and affordable within a low-cost framework.

C3.7 Slope failure type 4: below-road slopes

This is a common occurrence along many mountain roads constructed in side-long ground (Text box C3.2), especially where landslides and potentially unstable ground occur on the steeper slopes adjacent to eroding rivers and streams. These landslides often regress upslope through headward extension until an equilibrium slope angle is reached. The situation is frequently aggravated by uncontrolled road runoff and the side tipping of construction spoil that adds extra load to the slopes below (Section C2.6).

The options available in this case are essentially threefold.

- Construct a wall along the outside edge of the road shoulder to support the road formation with a foundation beneath the failure surface. There is usually a maximum practical depth to which this can be achieved, with 5–10 m being the usual range on a low-cost road; bored pile walls would offer a deeper, but much more expensive solution.
- Attempt to stabilize the landslide below. This is often difficult to achieve due to access difficulties and the steepness of slope, and can usually only be a practicable and affordable option where the ground movements below the road are shallow and localized.
- Realign the road into the hillside. However, this will usually have to be repeated if ground movements continue to regress towards the realigned road. If the realignment results in an increased height of cut, as is usually the case, then this may precipitate slope failures from above that will also have to be dealt with.

The first option is usually the most appropriate in the longer term provided an adequate foundation can be located beneath the failure surface, or the anticipated future upslope extension of it. Knowledge of the depth and extent of slope failure is therefore required in advance of the design. A ground investigation can assist in determining the required depth while geomorphological mapping will identify the length of road at potential risk from landslide regression, and hence the length of wall required.

C3.7.1 Case study

Figures C3.12 and C3.13 show a section of road where a landslide on the slopes below regressed upslope to cause loss of road formation over a distance of *c*. 50 m. The situation was exacerbated by the dumping of spoil over the landslide immediately adjacent to the road during construction. The underlying geology comprises phyllite rock, but this is highly disturbed and weakened by tectonism and weathering. During field mapping, an exposure of silty clay was observed towards the toe of the slope and this was considered to form part of the up thrusted landslide failure surface.

A ground investigation was undertaken comprising boreholes and trial pits. Rock was encountered at 6 m depth close

Text box C3.2. Type 4 regressive landsliding leading to significant loss of road formation

The Halsema Highway in the Philippines illustrates how regressive landslides can cause serious damage and loss. The road is located close to a ridge line through the Cordillera of Luzon where the slopes are impacted by typhoons and seismicity. The underlying igneous rocks are highly tectonized and weathered, and erosion and landsliding are major hazards. During earthquake and typhoon damage reinstatement works (Hart *et al.* 2002), the emphasis was placed on the construction of below-road retaining walls designed to support the road bench but with foundations beneath the failure surface of regressive landslides. While some drilling investigations were carried out (Text box B5.3), the majority of the investigation required to develop the design was based on engineering geological mapping and trial pit investigation.

Fig. C3.12. Road located across the head of a type 4 regressive landslide (realignment in progress into hillside towards right).

to the outside edge of the road, and the cross-section shown in Figure C3.14 was prepared accordingly. The design at this location comprised a road edge retaining wall, founded on bedrock beneath the failure surface. However, the decision was made by the road authority to realign the road into the hillside (as shown on the cross-section in Fig. C3.14).

The outside edge of the realigned road is *c.* 15 m from the upslope extent of ground movement. Unfortunately, the excavated material derived from the realignment was dumped over the landslide, creating further instability below the original road alignment. The slope materials exposed in the excavation for the realignment were mostly weathering grade IV phyllite and probably strong enough to remain stable in the cut face. The proximity of the ridge line above means that any type 1 failures created in the realigned cut slope will not have the opportunity to develop into significant slope instability.

The cost of excavation alone for the road realignment was almost twice the estimate for the road edge wall originally

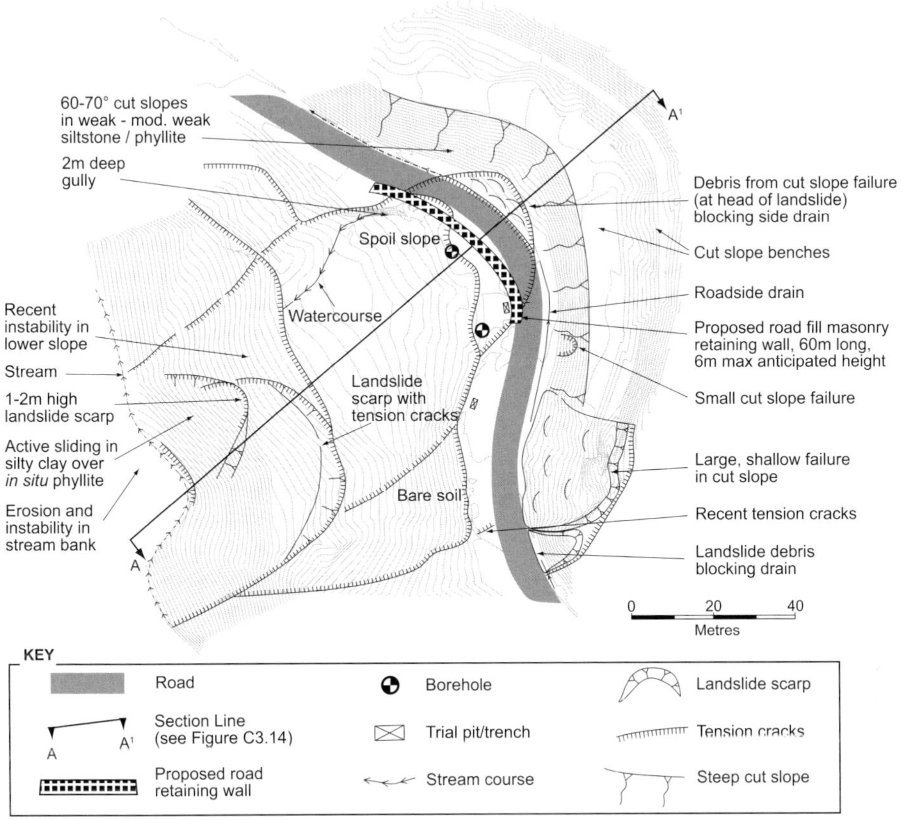

Fig. C3.13. Plan showing road across landslide head.

proposed. Furthermore, the realigned road remains potentially at risk from any future landslide regression. This illustrates that retaining walls in certain circumstances can be considerably cheaper and more effective than earthworks solutions: a fact that is not always appreciated by road authorities.

C3.8 Slope failure type 5: failure of the entire slope

Where a low-cost road is required to cross, or is already located on, a large deep-seated landslide the options are often very limited. The cost of stabilizing these landslides may be up to five times or more the per-kilometre cost of road construction and is usually well in excess of road improvement and maintenance budgets. A study of slope failures along the Laos road network (Hearn *et al.* 2008) identified that these large deep-seated landslides constituted only 3% of all recorded slope failures, and this figure probably represents a good indication of their incidence along most mountain roads as a whole. These slope failures (Fig. C3.15) tend to be slow moving and the rates of movement usually remain manageable in terms of low-volume road access. The recommended course of action is to:

- confirm that an alternative alignment or local realignment is not feasible (e.g. as illustrated in Fig. C3.16);
- consider modifying the vertical or horizontal alignment to reduce loads acting on the failure surface where applicable and to lessen the impact of ground movements on the road;
- establish a slope monitoring system in order to determine and review ongoing rates of movement (Section B5);
- carry out measures designed to slow down rates of movement (these may comprise improved slope drainage and local regrading in order to reduce driving forces);
- carry out measures to protect the road as much as possible from ongoing movements (these might include above-road and below-road retaining walls constructed in gabion that are able to withstand movements without total failure); and
- construct the road with a gravel rather than a sealed surface (the former being easier to maintain in moving ground).

All of the above options will require careful engineering geological assessment prior to selecting a course of action.

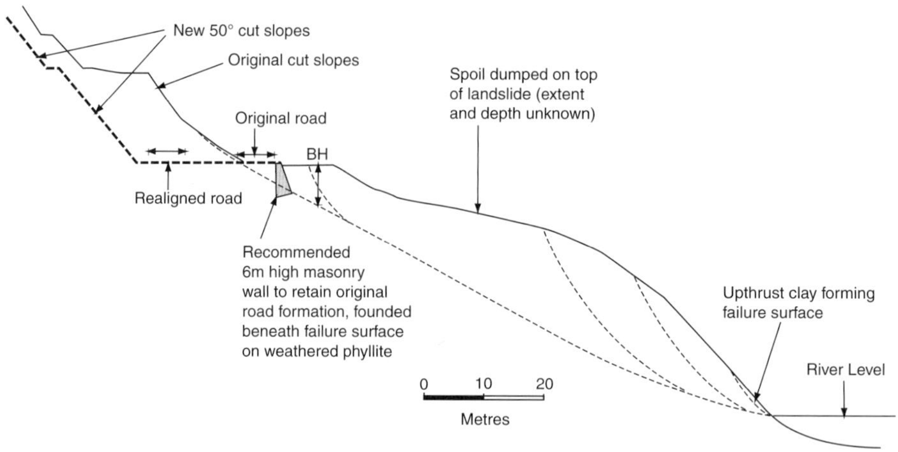

Fig. C3.14. Cross-section through landslide at A–A[1].

C3.8.1 Case study

Heavy rainfall in July 1993 with a return period of between 50 and 130 years (Dhital *et al.* 1993) caused widespread damage in central Nepal and extreme flooding in the Trisuli River. Significant damage occurred to the Naubise to Mugling road, a large proportion of which is located on the valley side slopes adjacent to the river. Road improvement works (alignment improvement, widening and pavement reconstruction) had been completed just prior to the flood. At one location, a combination of river scour on a meander bend and a rise in groundwater table caused a large deep-seated landslide and the loss of the road edge and deformation to the road surface over a length of *c.* 110 m. This initial failure was sufficient to trigger movement of the slope above the road. The inferred slip surfaces used in the back analysis are shown in Figure C3.17. Heavy seepage was noted at the base of the road cut slope and this supported the interpretation of there being two separate failures: one extending into the road carriageway from below and the other involving the entire slope above but 'breaking out' in the base of the cut slope. Boreholes were put down to 20 m and encountered limestone and conglomerate boulders in a silt/clay matrix. Bedrock was not found in any of the boreholes. The *in situ* valley side slope above the landslide comprised limestone and conglomerate overlying phyllite.

Fig. C3.15. Road access maintained across a type 5 landslide.

The results of the slope analyses carried out are shown in Table C3.8. Back analysis yielded a factor of safety of 1.01 for the two slip surfaces (A below and B above the road) shown on Figure C3.17. The calculated factor of safety for a continuous failure surface underlying the entire slope was 1.13 (not shown on Fig. C3.17) and was therefore not considered to apply.

Two stabilization options were considered:

- option 1: a gabion toe retaining wall founded at river bed level with a sloping backfill wedge and a system of

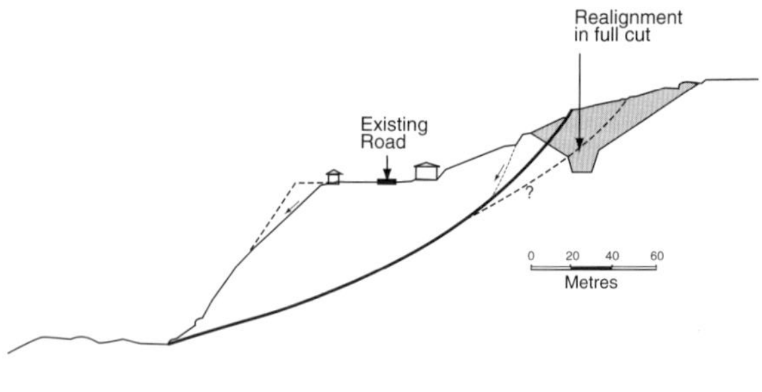

Fig. C3.16. Road realignment to avoid a deep-seated landslide (however, in this instance the landslide essentially becomes a type 4 regressive failure).

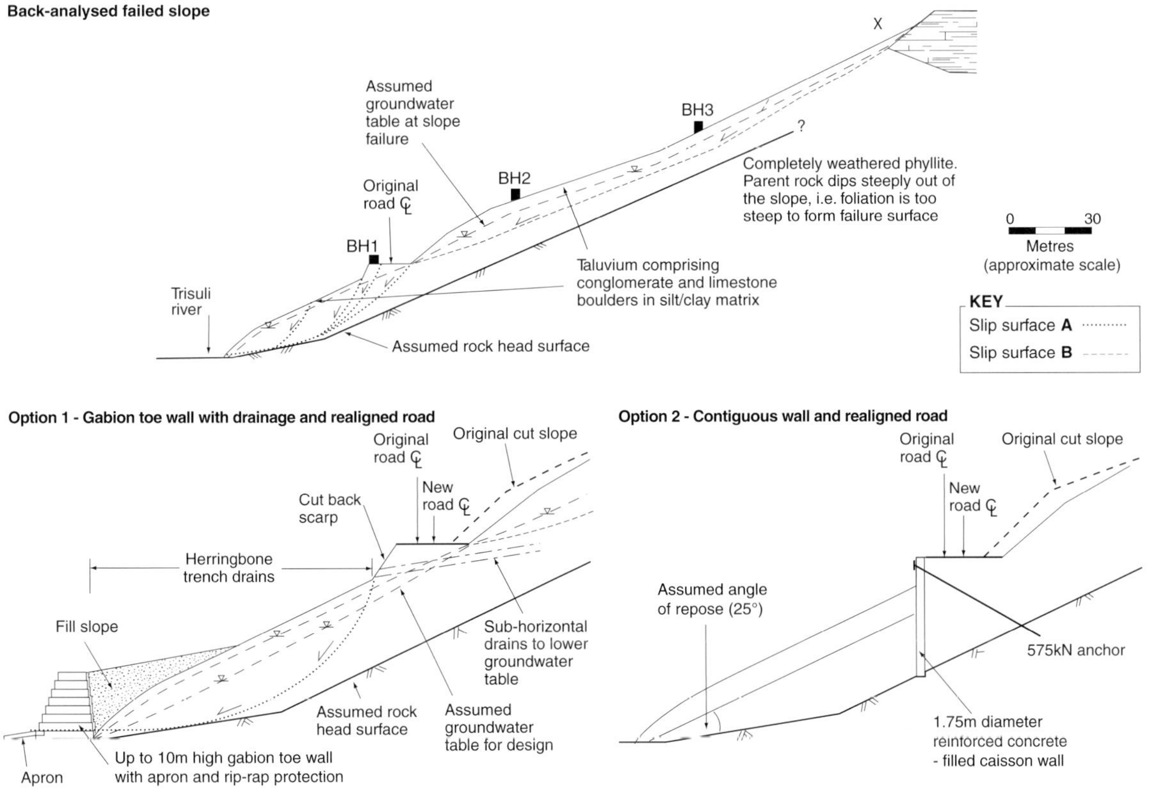

Fig. C3.17. Analysed sections.

surface drains below the road and sub-horizontal drains (Section C6.2.2.4) drilled into the landslide back scarp beneath the road to lower the groundwater table; and

- option 2: a reinforced concrete-filled anchored contiguous caisson wall designed to support the road and with a foundation into rock beneath the failure surface.

Cut slope revetments and surface drainage improvements to the slope above the road formed part of both options.

The outline concept of both options is shown in Figure C3.17 and the calculated factors of safety associated with each are given in Table C3.8. In the case of the gabion toe retaining wall with backfill and slope drainage, the lowest computed factor of safety was 1.26. The factor of safety against further movement of the slope above the road (slip surface B) is shown in Table C3.8 as remaining at 1.01 because the drainage measures implemented could not be relied upon to give a guaranteed reduction in groundwater table for the majority of the upper slope. The anchored contiguous caisson wall option resulted in a factor of safety of 1.5, although the slopes below the wall and above the road would (in theory at least) continue to fail.

On the grounds of cost and construction practicalities, the gabion retaining wall and surface/subsurface drainage option was selected. Slope stability analyses were carried out on five surveyed cross-sections and these were used to develop the design. In order to achieve the required factors of safety, a sizeable free-draining backfill wedge had to be designed. This required constructing the gabion toe wall into

Table C3.8. *Calculated factors of safety*

	Existing condition	*Design options*	
		1. Toe wall with drainage	*2. Contiguous wall*
Slip below road extending to inside edge of road (slip surface A)	1.01	1.26	1.5*
Slip above road with toe in cut slope (slip surface B)	1.01	1.01	1.01

*Factor of safety controlled by stability of wall as wall foundation is in *in situ* rock and wall is unaffected if slope in front continues to fail (wall is only designed to support the road fill).

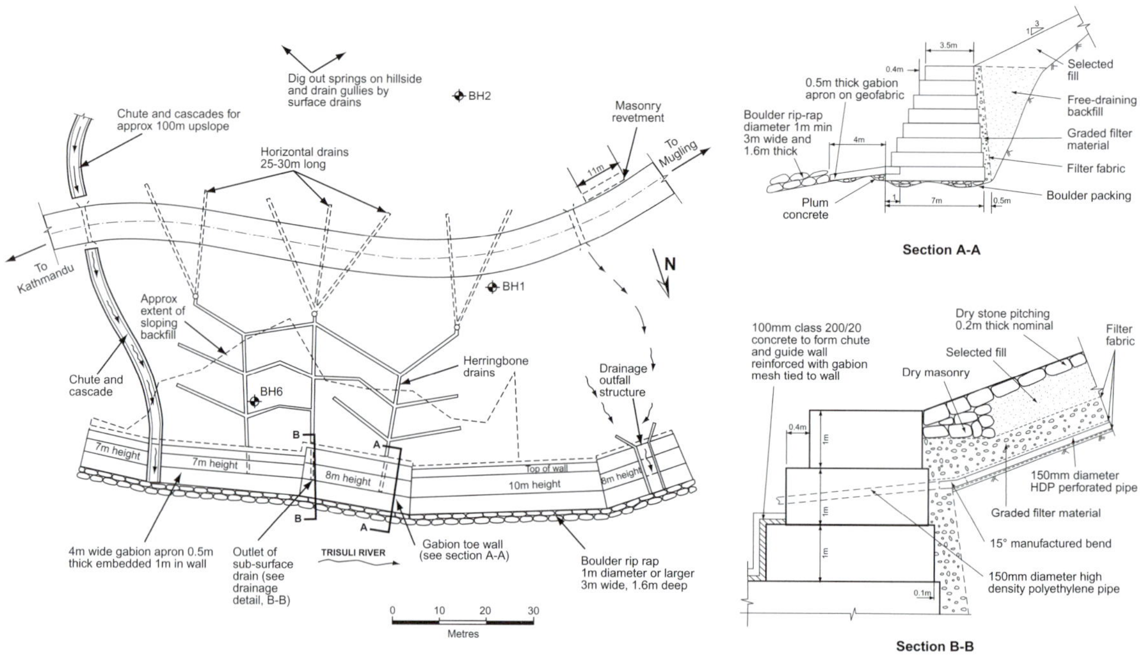

Fig. C3.18. General layout and design details for the selected option.

Fig. C3.19. Gabion toe wall and fill slope shortly after construction.

Fig. C3.20. Previous type 5 landslide slope with gabion toe wall and full growth of planting to the lower slope (September 2009).

the channel of the river, thus placing it at additional risk from river scour. Figure C3.18 shows the plan layout and some of the section details of the stabilization measures constructed. Construction of the gabion wall took place in the dry season, with localized temporary river diversion to allow foundations to be dug as deeply as possible. The lowermost 3 m of wall was assembled using double-mesh gabion on its front face in order to provide additional protection against river scour. A mass concrete foundation, together with a 4 m wide gabion apron and boulder rip rap, provided further protection.

Figure C3.19 shows the gabion toe wall and the fill slope supporting the road shortly after construction. The photograph in Figure C3.20 was taken 15 years later. The gabion structure and the road itself remain intact and the drainage is still functioning. Shrub and tree planting on the slopes below and above the road has helped to stabilize surface soils and improve drainage. The total cost of road reinstatement for this 100 m section of road amounted to *c.* US$ 300 000 at 1994 prices. The average cost of the road improvement works had been *c.* US$ 42 000 per 100 m and so the cost of landslide stabilization and road reinstatement at this location amounted to seven times that of the original road improvement project per unit length.

References

ABRAMSON, L. W., LEE, T. S., SHARMA, S. & BOYCE, G. M. 2002. *Slope Stability and Stabilisation Methods*. John Wiley and Sons, New York.

BISHOP, A. W. 1955. The use of the slip circle in the stability analysis of slopes. *Géotechnique*, **5**, 7–17.

BISHOP, A. W. & MORGENSTERN, N. 1960. Stability coefficients for earth slopes. *Géotechnique*, **10**, 129–150.

BOND, A. & HARRIS, A. 2008. *Decoding Eurocode 7*. Taylor & Francis Group, London.

BROMHEAD, E. N. 1986. *The Stability of Slopes*. Surrey University Press, London.

BROMHEAD, E. N. 2005. Geotechnical structures for landslide risk reduction. *In*: GLADE, T., ANDERSON, M. & CROZIER, M. J. (eds) *Landslide Hazard and Risk*. J Wiley & Sons Ltd, Chichester, England, 549–594.

BSI (EN) 2004. Eurocode 7: geotechnical design (part 1 general rules). British Standards Institution EN ISO BSI 1997-1: 2004.

BSI 2009. Code of practice for earthworks. 09/30093255 DC. British Standards Institution, United Kingdom.

CAVALEIRO, V. M. P., RISCADO, J. F. M. C. & PAIS, L. J. A. 2010. Determination of the safety factor of slopes for the granite residual soil in the region of Covilha, using charts. *In*: WILLIAMS, A. L., PINCHES, G. M., CHIN, C. Y., MCMORRAN, T. J. & MASSEY, C. I. (eds) *Geologically Active. Proceedings of the 11th IAEG Congress, Auckland, New Zealand, 5–10 September 2010*, Taylor & Francis Group, London, 71–76.

CORNFORTH, D. H. 2005. *Landslides in Practice. Investigation, Analysis, and Remedial/Preventative Options in Soils*. John Wiley and Sons Inc., New York.

CRAIG, R. F. 2004. *Soil Mechanics*. 7th edn. Taylor & Francis Group, London.

DHITAL, M. R., KHANAL, N. & THAPA, K. B. 1993. The role of extreme weather events, mass movements, and land use changes in increasing natural hazards. International Centre for Integrated Mountain Development (ICIMOD), Kathmandu, Nepal.

Duncan, J. M. 1996. State of the art: limit equilibrium and finite-element analysis of slopes. *Journal of Geotechnical and Geoenvironmental Engineering, American Society of Civil Engineers*, **122**, 577–596.

Duncan, J. M. & Wright, S. G. 2005. *Soil Strength and Slope Stability*. John Wiley and Sons, New Jersey.

GCO 1984. Geotechnical Manual for Slopes. 2nd edn. *Geoguide 5*. Geotechnical Control Office, Hong Kong.

Griffiths, D. V. & Lane, P. E. 1999. Slope stability analysis by finite elements. *Géotechnique*, **49**, 387–403.

Hart, J., Hearn, G. & Chant, C. 2002. Engineering on the precipice: mountain road rehabilitation in the Philippines. *Quarterly Journal of Engineering Geology and Hydrogeology*, **35**, 223–231.

Hearn, G. J., Hunt, T., Aubert, J. & Howell, J. H. 2008. Landslide impacts on the road network of Lao PDR and the feasibility of implementing a slope management programme. *International Conference on Management of Landslide Hazard in the Asia-Pacific Region*, Sendai, Japan.

Hoek, E. & Bray, J. 1981. *Rock Slope Engineering*. Revised 3rd edn. Institution of Mining and Metallurgy, London.

Jaiswal, P. & van Westen, C. J. 2009. Estimating temporal probability for landslide initiation along transportation routes based on rainfall threshold. *Geomorphology*, **12**, 96–105.

Janbu, N. 1957. Earth pressures and bearing capacity calculations by generalised procedure of slices. *Proc 4th International Conference on Soil Mechanics and Foundation Engineering*, London, **2**, 207–112.

Lamb, T. W. & Whitman, R. V. 1979. *Soil Mechanics*. John Wiley and Sons, New York.

Morgenstern, N. R. & Price, V. E. 1965. The analysis of the stability of general slip surfaces. *Géotechnique*, **15**, 79–93.

Nilsen, B. 2000. New trends in rock slope stability analysis. *Bulletin of Engineering Geology and the Environment*, **58**, 173–178.

Ortigao, J. A. R. & Sayao, A. S. F. J. 2004. *Handbook of Slope Stabilisation*. Springer, Rio de Janeiro.

Owen, L. A., Kamp, U., Khattak, G. A., Harp, E. L., Keefer, D. K. & Bauer, M. A. 2008. Landslides triggered by the 8 October 2005 Kashmir earthquake. *Geomorphology*, **94**, 1–9.

Sloan, S. W. 2012. Geotechnical stability analysis. The 51st Rankine Lecture. *Géotechnique*, **62**, in press.

Taylor, D. W. 1937. Stability of earth slopes. *Journal of the Boston Society of Civil Engineers*, **24**. Reprinted in: Contributions to Soil Mechanics 1925 to 1940, Boston Society of Civil Engineers, 337–386.

Turner, A. K. & Schuster, R. L. 1996. Landslides: investigation and mitigation. Special Report, United States National Research Council, Transportation Research Board, Washington DC.

Wu, C.-H. & Chen, S.-C. 2009. Determining landslide susceptibility in Central Taiwan from rainfall and six site factors using the analytical hierarchy approach. *Geomorphology*, **112**, 190–204.

C4 Rock slope stabilization

G. J. Hearn

URS Scott Wilson Ltd, Scott House, Alençon Link, Basingstoke, Hampshire RG21 7PP, UK
(e-mail: gareth.hearn@scottwilson.com; garethhearn@talktalk.net)

C4.1 Controls on rock slope stability

The assessment of rock slope stability is dependent on whether stability is controlled by:

- the strength of the intact rock (this relates predominantly to rock masses that are continuous, homogenous and isotropic); or
- failure on discontinuities.

C4.1.1 Failure controlled by the intact rock strength

For continuous rock masses, where discontinuities play no role in controlling stability, a given slope might be cut vertically for rock with high intact strength (Fig. C4.1) or treated as a soil if weak and cut to a shallow angle. Rock strength is a function of the mineralogy, the interaction (physical/chemical) between the grains and the processes which have affected the rock after its formation (e.g. Anon 1977).

Hawkins (1998) compares various methods in place (e.g. Deere & Miller 1966, IAEG 1981, ISRM 1981, BSI 1999) for classifying rock by unconfined compressive strength (UCS) in terms of MPa. Field methods for estimating UCS are described, for example, in BSI EN (2003).

C4.1.2 Failure controlled by discontinuities

The most frequently occurring rock slope failures are due to either displacement along dominant persistent discontinuities or multiple closely-spaced discontinuities (as illustrated in Fig. C4.2). The strength of the intact rock is usually ignored when rock slope stability is discontinuity-controlled.

C4.2 Assessing rock slope stability

C4.2.1 Discontinuity-controlled stability (assessment method 1)

Usually, rock failures occur predominantly along a single discontinuity or a combination of two or three. The geometrical relationship between these discontinuities and the angle and orientation of the hillside or cut slope is usually the most important factor in determining rock slope stability. Discontinuity-controlled rock slope failure mechanisms are described and illustrated in Part A and include planar, wedge and toppling failures (Fig. C4.3 and Hoek & Bray 1981; Wyllie & Mah 2004). Along mountain roads, true wedge failures and topples are usually far less frequent than planar failures. If the rock mass is highly jointed then failure could occur along a number of discontinuities, combining a range of failure mechanisms.

Planar failures usually occur when all of the following conditions are satisfied:

- where discontinuities are orientated within $\pm 20^\circ$ of the slope face direction (although consideration should be given to the potential for wedge failure along joints marginally outside of this requirement);
- where discontinuities daylight out of (intersect) the slope face;
- where the dip angle of discontinuities generally exceeds their effective angle of friction, which includes a joint roughness component; and
- where lateral release surfaces (which provide negligible resistance to sliding and have very low to negligible tensile strength) are present in the rock mass (these define the lateral boundaries of the potential sliding block).

With respect to the second condition listed above, movement can take place along joints that dip more steeply than the angle of the slope face if basal shear is able to propagate onto the slope through failure along other joint sets or through the intact rock itself, or both.

In order to assess the stability of a given rock slope, there are three principal approaches that are utilized.

- The first concerns the kinematic feasibility of slope failure, as defined by the orientation of the principal discontinuities in relation to the slope topography. Stereonets are the principal means by which failure kinematics are assessed and the procedure is illustrated in Text box C4.1.
- The second involves the use of limit equilibrium techniques to calculate factors of safety for both planar and wedge failures based on a fairly simple resolution of forces. The relevant equations are given in Hoek & Bray (1981) and Wyllie & Mah (2004) for example.

From: Hearn, G. J. (ed.) *Slope Engineering for Mountain Roads*. Geological Society, London, Engineering Geology Special Publications, **24**, 189–208.
DOI: 10.1144/EGSP24.13 0267-9914/11/$15.00

Three-dimensional applications of limit equilibrium techniques for slope stability analysis are described, for example, in Hungr *et al.* (1989).

Fig. C4.1. Sandstone cut vertically where discontinuities play no role in stability.

- The third involves the numerical modelling of rock slopes taking into consideration the internal deformations that occur within rock masses and the relationships between material anisotropy, non-linear behaviour, *in situ* stresses and the effects of interdependent parameters such as pore (*cleft* in rock discontinuities) water pressure and seismic loading. Numerical analyses divide a rock mass into zones, each with its own material properties and stress-strain relationships.

Numerical modelling methods can be broadly subdivided into the following (e.g. Lorig & Varona 2004):

- discontinuum methods
- continuum methods

Discontinuum methods treat the rock mass as an assemblage of discrete blocks, or elements, whereby stability is controlled by block deformation and the movement of blocks relative to each other. Discontinuities in a discontinuum model are represented explicitly in terms of location and orientation. The method is capable of simulating large displacements due to ground movement or the opening up of the rock mass along joint sets.

Fig. C4.2. Discontinuity-controlled rock slope stability (**a**) persistent; (**b**) multiple, closely-spaced.

Continuum methods assume continuous material with discontinuities treated as interfaces between continuum bodies. These approaches cannot model multiple intersecting joints easily, unless homogenization methods are used. They include, for example, finite element and finite difference analysis and allow for complex failure mechanisms, creep deformation, dynamic loading, variability in strength parameters and three-dimensional analysis. Finite element limit analysis methods, on the other hand, are naturally discontinuous and optimize the failure mechanism using velocity discontinuities within the rock mass (Sloan 2012). Continuum approaches are further subdivided into

Text box C4.1. Rock slope stability analysis for slope design

The valley side slopes shown in the photograph are composed of fractured dolomitic limestone, phyllite, marble and intrusive dykes of dolerite. The design studies carried out at critical sections in advance of road construction across these slopes comprised the following:

- engineering geological mapping;
- mapping of adjacent slopes to ascertain broader pattern of rock structure and persistence of discontinuities;
- derivation of principal discontinuity sets (dip/direction);
- assessment of rock strength characteristics from field testing (friction angle and unconfined compressive strength);
- assessment of the intact rock using the GSI rock mass rating scheme (see text);
- use of stereonets to display the discontinuity geometries in stereographic projection;
- identification of slope angles and orientations that were potentially unstable according to plane and wedge failure modes;
- use of a proprietary software program to analyse the factors of safety for those slopes found to be kinematically unstable (plane or wedge failures).

The stereonet analysis confirmed that there was wedge failure potential along the intersection of joint sets J1 and J2. Joint set J1 fell within the window of planar instability and therefore acted predominantly as the sliding plane, with joint set J2 forming the lateral release plane. Bedding dipped steeply into the slope and delineated the upper release plane of the wedge. Back analysis was applied using limit equilibrium techniques, and it was assumed that under current conditions the factor of safety of the slope was *c*. 1.0. Consequently, stabilization options were proposed, including rock reinforcement and cutting back.

Stereographic projection uses stereonets to project the orientation and angle of slope faces and rock discontinuity planes measured in three dimensions onto a two-dimensional grid for analysis of rock slope stability (e.g. Hoek & Bray 1981; Goodman & Shi 1985; Wyllie & Mah 2004; Pedrazzini *et al.* 2011).

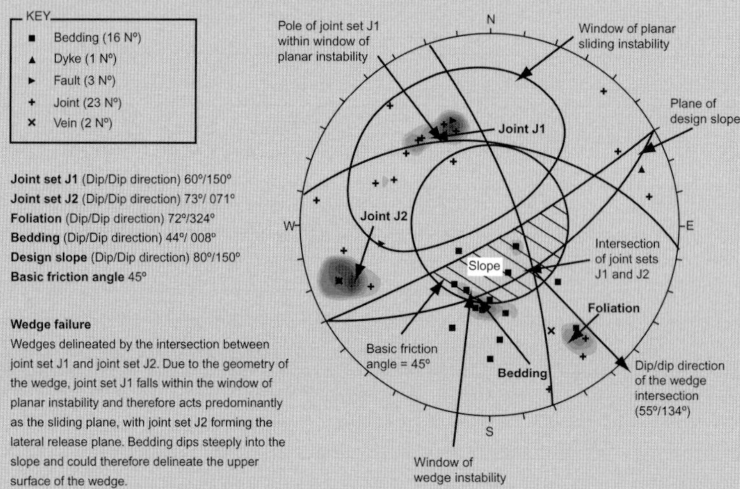

Material type	*Description*	*Friction angle basic (°)*	*UCS (MPa)*
Phyllite	Weak to moderately strong, dark grey moderately to highly weathered phyllite	30	0–25
Limestone	Strong to very strong, dark grey to grey, slightly to moderately weathered limestone	45	100

Discontinuity type	*Dip (°)*	*Dip direction (°)*	*Joint compressive strength (MPa)*	*Joint roughness coefficient*	*Friction angle (instantaneous)*	*Cohesion (instantaneous) (MPa)*
Joint Set 1	60	150	50	4	53	0.01
Joint Set 2	73	071	50	6	60	0.01
Foliation (related to JS 1)	72	324	30	4	55	0.003
Bedding	44	008	30	8	58	0.07

two-dimensional and three-dimensional modelling techniques and an example of the former is provided in Text box C4.2.

Due to the need for intensive and high quality input data and the rigour of the required analyses, these numerical techniques are usually only applied at high risk locations. It is

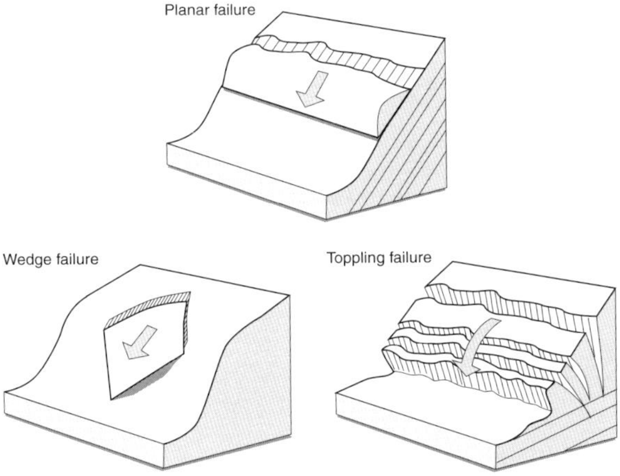

Fig. C4.3. Planar, wedge and toppling discontinuity failure mechanisms.

important to note that the output from these methods does not provide a slope factor of safety, as would be determined by a limit equilibrium analysis. Instead, factors of safety can be determined indirectly by reducing the rock mass and joint shear strengths in the analysis until failure occurs, then calculating the ratio between shear strength at failure and actual shear strength.

Hybrid techniques include the incorporation of limit equilibrium and finite element analysis, such as the Finite Element Limit Analysis (FELA) of Sloan (2012) for example.

C4.2.2 Multiple, closely-spaced discontinuity-controlled stability (assessment method 2)

This type of failure occurs in rock masses dominated by multiple (typically more than four) closely spaced discontinuity sets that include thinly laminated or foliated, tectonically sheared weak rocks. When considering these types of rock masses it is assumed that there are a sufficient number of closely spaced discontinuities such that failure along multiple and interlinking discontinuities can take place (Hoek & Brown 1997).

Rock mass classification schemes are used principally as part of assessment method 2. These systems were developed empirically by establishing the most important parameters, and giving each a numerical value and weighting. The most useful scheme for classifying rock masses for slope design is the Geological Strength Index (GSI) (Hoek 1994; Hoek & Brown 1997; Hoek 1999; Tsiambaos & Saroglou 2010). The GSI provides an estimate of the strength of jointed rock masses based upon an assessment of the intact strength of the rock, the number and persistence of discontinuities and the roughness condition (and infilling) of the surfaces along these discontinuities (see also Marinos *et al.* 2005). Hoek *et al.* (1998) describe an extension to the GSI to include reference to foliated and sheared rock structure. Pantelidis (2010) describes the use of a rock mass rating scheme to assess rock slope failure hazards. Other schemes widely used include those of Bieniawski (1989) and the Q-System (e.g. Barton *et al.* 1974). However, these tend to be used more for mining and tunnelling applications.

For low-cost roads it is usual to classify rock slopes by their constituent rock type(s) and whether their discontinuity structure is generally adverse or favourable to stability. During the design of a new road, published geological maps and field mapping will allow exposed rock types and geological structure to be recorded; some prediction can therefore be made of the jointing pattern, degree of weathering and strength of the rock that will be encountered during excavations. However, considerable uncertainty will remain especially concerning the composition of the deeper rock cuts and where the structural geology is complex. Drillhole investigations for the deeper excavations will advance the stability assessment, but these will not usually provide any information on discontinuity orientations at depth. Normal practice, therefore, is to develop an interim assessment of rock characteristics and structure based on surface exposures and to record rock type, structure and weathering profiles during excavation, adjusting the design as required. In the case of existing roads this information will already be available in road cuts; removal of the weathered mantle and soil cover from cut faces may be all that is necessary to ascertain rock strength and structure.

Table C4.1 summarizes the information required to be able to establish a ground model for rock slopes in the context of low-cost roads. The table differentiates between information considered to be essential, desirable and optional for rock slope stability assessment (qualitatively based) and rock slope stability analysis (numerically based).

Where discontinuity and digital elevation data are available, GIS-based methods can be used to identify slopes and areas where mapped discontinuity planes and topography combine to create conditions adverse to stability for plane and wedge failure mechanisms (e.g. Irigaray *et al.* 2010; Hearn *et al.* 2012). However this can be an extremely time-consuming exercise, especially if the requisite discontinuity data are required to be collected from fieldwork, and would probably not be cost-effective for low-cost road studies.

C4.3 Managing rock slope stability along mountain roads

The management of rock slope stability along mountain roads requires an assessment of the depth and mechanism of failure and the potential risk it poses to road operation and the stability of adjacent structures and land uses. The following modes of failure commonly apply:

- deep seated failures in low-strength and/or highly fractured rock masses;
- major structural instability in the form of sliding, wedge and toppling failures in hard rock masses;

Text box C4.2. Use of two-dimensional discrete element analysis to assess rock slope stability at a major bridge site

The proposed Konkan railway line between Katra and Qazigund crosses the river Chenab in the Indian state of Jammu and Kashmir. The Chenab Bridge, once constructed, would be the highest bridge in the world at a height of 359 m above river bed level. The 1315 m long bridge consists of a 467 m span steel arch over the river, supported by a total of 18 piers. The loads exerted by these piers could affect the stability of the slopes upon which they are to be constructed. These slopes are underlain by closely-jointed dolomitic limestone, chert and quartzite breccia with frequent fault and other tectonic shear structures. Two-dimensional discrete element analysis was undertaken in order to assess slope stability under static and dynamic (earthquake loading) conditions. Rock mass and discontinuity data were collected using a combination of field mapping, trial pits, drillholes and exploration adits, and the latter were used to conduct *in situ* shear and plate load tests. The locations and magnitudes of displacements along discontinuities, and the ratio of available strength to applied stress, were determined to identify the potential failure modes and displacement magnitudes within the rock mass.

The analysis indicated that the hillside was stable under static loading conditions, with no credible failure modes predicted and with displacements at pier loading locations not exceeding 10 mm. However, under design seismic loadings of 0.31 g, severe pier displacements of up to approximately 400 mm were predicted as a result of failure of the supporting slopes along discontinuities to a depth of 10–20 m. Tensioned cable anchors were recommended in order to maintain the integrity of the rock mass.

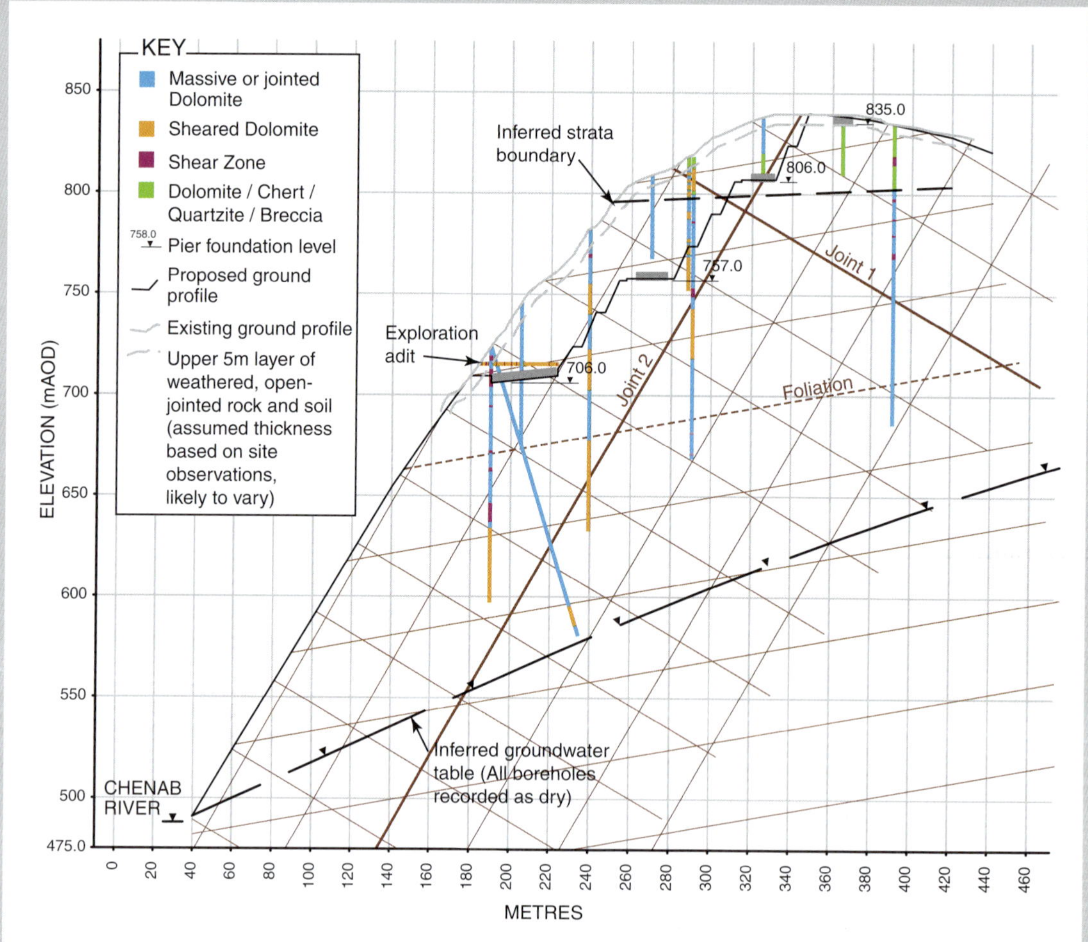

Cont.

Table C4.1. *Level of importance of information for rock slope stability assessment and analysis*

Information	*Qualitative stability assessment*			*Numerical stability analysis*		
	Essential	*Desirable*	*Optional*	*Essential*	*Desirable*	*Optional*
Engineering geological map	✓			✓		
Rock mass description	✓			✓		
Discontinuity orientations	✓			✓		
Slope profile		✓		✓		
Groundwater		✓		✓		
Rock material strength		✓		✓		
Rock mass classification (e.g. GSI)		✓		✓		

- minor structural instability, also described as face insecurity, in the form of sliding, wedge and toppling failures in cut slopes; and
- ravelling and erosion in low strength and/or highly fractured rock masses.

Deep-seated and major structural instabilities normally require stabilization measures, while ravelling and localized rockfall usually necessitate surface treatments and protection measures (Fookes & Sweeney 1976). Protection, in this context, refers to the protection of the road and other structures from rock failure impact. Figure C4.4 summarizes typical rock slope failure mechanisms and the prescriptive measures that might be employed to stabilize or mitigate them. Table C4.2 provides a more detailed review of the measures summarized in Figure C4.4 for low-cost road applications. Figure C4.5 illustrates how these stabilization and protection measures might be applied along a typical mountain road constructed in rocky terrain. Sections C4.4 and C4.5 provide further discussion on methods of rock slope stabilization and protection, respectively.

C4.4 Stabilization measures

Stabilization measures are described in relation to:

- reinforcement and support;
- drainage; and
- removal.

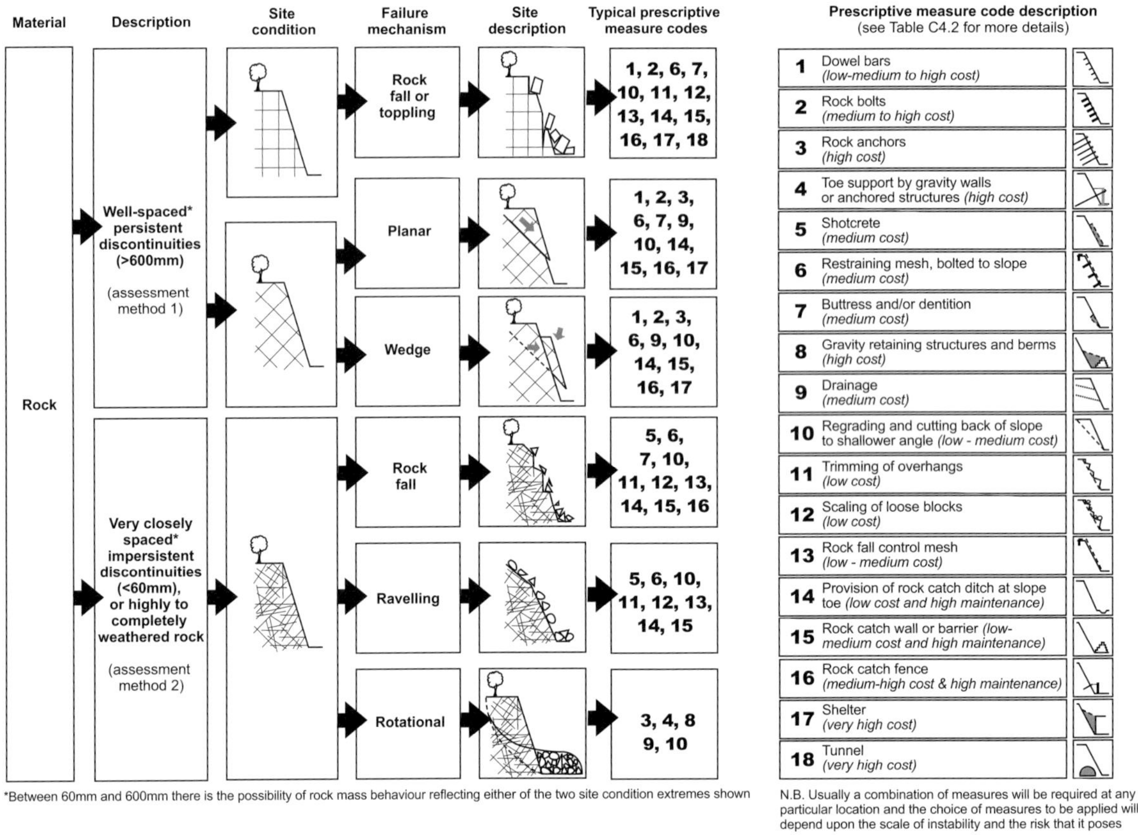

Fig. C4.4. Simple classification of rock slope materials, failure mechanisms (excluding avalanches) and outline prescriptive remedial measures.

C4.4.1 Reinforcement and support

As noted in Table C4.2, this can include:

- dowels;
- bolts;
- anchors;
- tied-back walls;
- shotcrete (Section C7.2.3);
- restraining mesh;
- underpinning buttresses; and
- toe support to rock slopes undergoing deep-seated failure.

Typical details for dowels, bolts and anchors are given in Figure C4.6.

C4.4.1.1 Dowels

Dowels are normally installed as shear pins through individual rock blocks and across potentially adverse sliding or release joints. They typically comprise a reinforcing bar inserted into a pre-drilled hole and grouted into place along the full length of the bar. The bar is not normally less than 16 mm diameter and can range up to 32 mm diameter. The pre-drilled hole is grouted immediately prior to bar insertion to minimize the potential for voids. Bars must be provided with sufficient length beyond the joint to prevent sliding or release. A maximum length of 3 m is usual.

Dowels are designed on the basis of tensile restraint with a relatively limited tensile capacity. They are suited, therefore, to a maximum block size of about 1–2 m^3. Where dowels are supporting larger blocks, they may be provided with a cross bar welded at the exposed end to create a 'T-piece' to provide added restraint. The exposed crossbar is then encased by a nominal-sized concrete cover for corrosion protection. Stability calculations may be performed for the larger individual blocks, although it may be more effective to undertake a generic set of calculations to cover a wide range of block sizes and orientations that may then be used as a prescriptive dowel solution for a particular geometry and block size.

Installation of multiple dowels in a single rock block is not unusual and these should be equally spread around the

Table C4.2. *Options for rock slope stabilization and protection*

Requirement		*Technique*	*Where?*	*Cost and Practicality*
Stabilization	Slope Stabilization – Reinforcement and Support	1. Dowels	Any potentially unstable block that can be kept in place by dowels	Low - medium cost; use usually restricted to blocks 1-2m thick, careful design through face mapping required
		2. Bolts	Any potentially unstable block that can be bolted back to stable material	Medium - high cost; installation using specialist equipment; long term corrosion/creep problems. Careful design through face mapping required
		3. Pre-stressed ground anchors	Any potentially unstable block or rock mass that can be anchored back to stable material	High cost; installation using specialist equipment; long term corrosion/creep problems, long term monitoring required. Careful design through face mapping required
		4. Tied-back walls and similar surface structures	Where areas of rock surface require reinforcement and face support	High cost; same as for rock bolts and anchors
		5. Shotcrete	Closely fractured or degradable rock face	Medium cost; specialist equipment required, though hand-applied concrete or chunam might be considered (Section C7.2.3)
		6. Restraining mesh	Mesh can have a face reinforcing effect if dowelled tightly against the rock face	Medium cost; potential access problems; good anchorage required throughout
		7. Underpinning buttresses	Cavity on rock face, overhang	Medium cost; potential access problems
		8. Retaining structures, berms	Deep-seated rock failures	High cost; may be impracticable due to space restrictions and size of failure
	Slope Stabilization – Drainage	9. Drainage	Any rock face where water pressures in fissures create instability	Medium cost; drilling equipment necessary for drain holes. Drain holes may not function well in fractured rock
	Slope Stabilization – Removal	10. Regrading	Instability at crest of rock face Over-steep slopes	Low – medium cost; potential access problems; difficult in very steep terrain; pre-split blasting may be required
		11. Trimming	Overhangs	Low – medium cost; pre-split blasting may be required
		12. Scaling	Loose rock on surface	Low cost; labour-intensive; potential access and safety problems
Protection	Protection to Road	13. Rock fall control mesh	Loose/weak rock on surface	Low - medium cost; will not retain major blocks; good anchorage required at top of face
		14. Catch ditch	Base of slope where space permits	Low cost; shape of ditch dependent on height and slope of rock face
		15. Catch wall or barrier	Base of slope where space permits	Low - medium cost; wall dimensions dependent on height and slope of rock face
		16. Catch fence	Mid-slope or base of slope	Medium - high cost depending on impact capacity; cost increases if special anchorage and founding conditions are required (eg in soil/debris)
		17. Shelter	At base of high unstable face where other measures are not feasible	Very high cost
		18. Tunnel	If relocation is the only solution	Very high cost

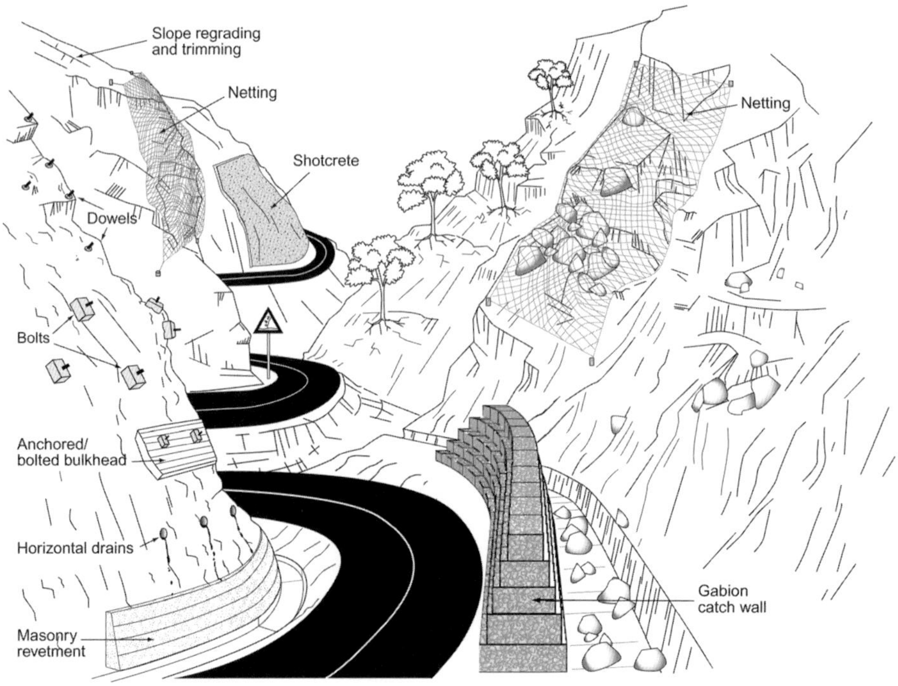

Fig. C4.5. Rock stabilization and protection measures commonly applied to mountain roads.

centre of gravity of the block. They may also be installed in a regular grid spacing (pattern dowelling) on the more blocky slope faces, sometimes in conjunction with a slope facing (e.g. shotcrete; Section C7.2.3). The arrangement of dowels on a rock slope is usually designed on the basis of the structural geometry of the rock mass using face mapping (e.g. Fig. B3.9). Pattern dowels are usually specified where the rock mass is closely jointed in order to lock up the slope face on a mass basis rather than stabilizing individual blocks. As joint spacing decreases, dowel length usually increases in order to counter the effect of the rock mass acting more like a soil and potentially failing by deeper shear through the mass.

C4.4.1.2 Bolts

Rock bolting involves the insertion of a high-yield reinforcing bar into a drilled hole with a defined bond length along which the bar is tensioned at the slope face via a plate and threaded nut (torque head) assembly. The tensioning of the bar creates compression in the grouted drill hole and surrounding rock mass and keeps joints tightly closed. Bolts are usually larger than dowels, depending on the depth to the sliding plane or joint surface.

Rock bolts act most efficiently when they are normal to the sliding plane and therefore some consideration needs to be given to the orientation of individual rock bolts. However, a few degrees inclination below the angle perpendicular to the sliding plane is normally specified in order to ensure that a nominal component of the loading is applied in the upslope direction.

C4.4.1.3 Anchors

An anchor is a pre-stressed rod or cable with an anchorage or bond length within the slope and a reaction plate at the slope face. The pre-stressed element is then tensioned to the design loading following installation, and relied upon to maintain the pre-stress load in the long-term. Anchors can be up to 30 m or more in length, depending on the depth of movement or jointing. Anchors are usually used, therefore, to stabilize or prevent large rock slope failures and other ground movements at depth. Published specifications (e.g. GEO 1997; FHWA 1999) and Wyllie & Mah (2004) provide guidance on the design and installation of pre-stressed ground anchors and anchored systems for slope works. However, it is recommended that the manufacturer of any system adopted is consulted prior to finalizing a design.

Given the large scale of potential ground movement to which these measures are applied, the higher levels of pre-stress and the incorporation of multiple corrosion protection measures, the following actions should be regarded as mandatory:

- a regular programme of maintenance; and
- confirmation of torque or prestress through lift-off tests.

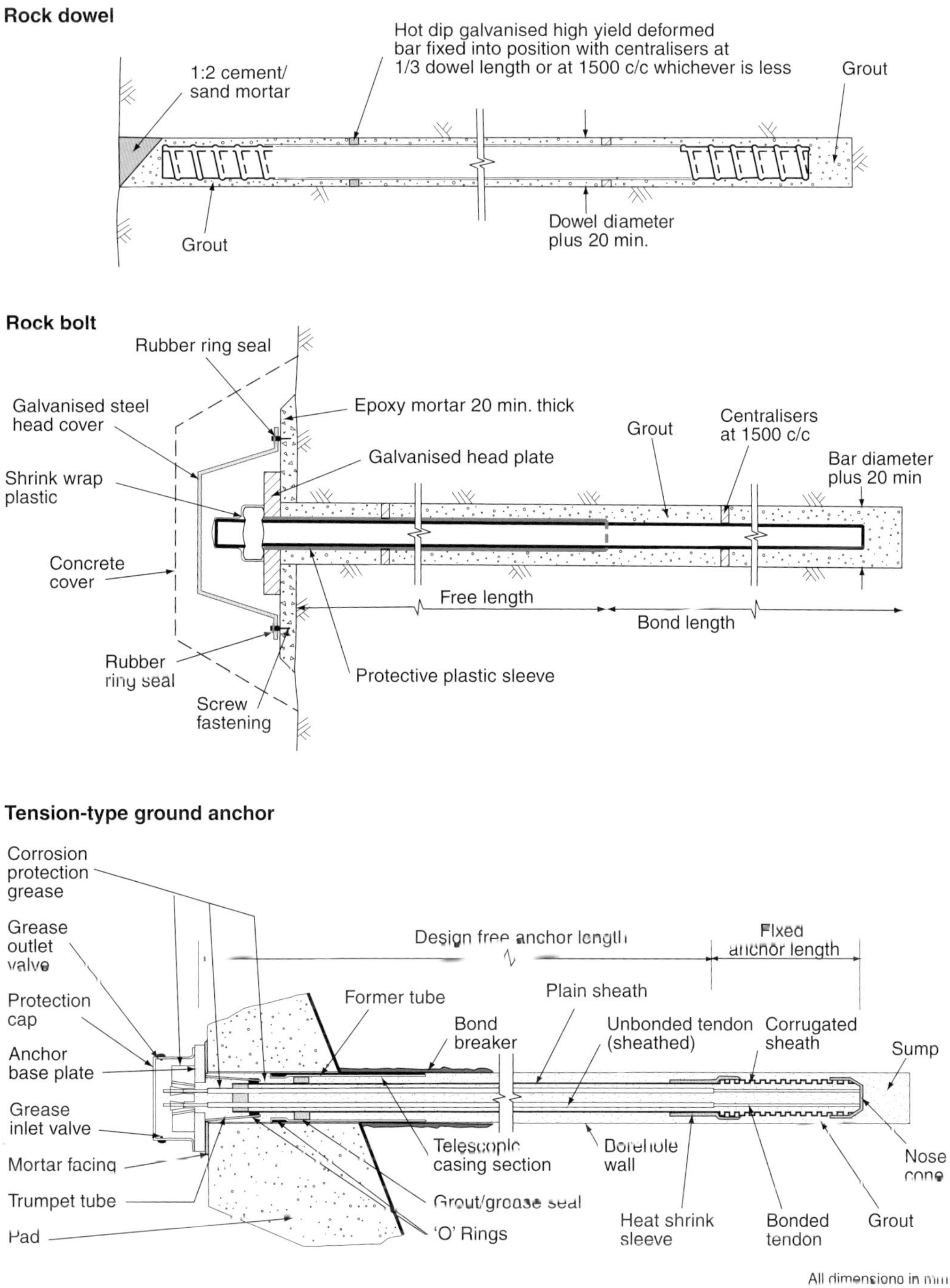

Fig. C4.6. Typical details for dowels, bolts and anchors (modified from GEO 1997, with permission from the Head of the Geotechnical Engineering Office of the Director of Civil Engineering and Development, the Government of Hong Kong Special Administrative Region).

In most cases these actions will probably form a prerequisite for maintaining a manufacturer's performance guarantee. As with tensioned rock bolts, careful consideration should be given to the specification of rock anchors as compared to other possible stabilization and protection measures, due to the long-term monitoring and maintenance commitment required.

C4.4.1.4 Tied-back walls and similar structures

These structures combine anchors or tensioned bolts with reinforced concrete or steel surface structures, such as walls and steel grids or grillages, in order to reinforce and retain failed or failing rock material. The example given in Figure C4.7 shows the construction of a tied-back structure on a slope composed of loose rock debris. The slope is irregular and is not to a uniform angle, and large boulders on the surface are allowed to dictate the detailed layout of the structure. Its integrity is reduced as a result.

C4.4.1.5 Shotcrete

Shotcrete acts principally as a form of surface protection against erosion, ravelling and spalling of weathered and jointed rock slopes, and is discussed in Section C7.2.3. Nevertheless, a degree of slope reinforcement is also provided by the dowels or bolts that are used to fix shotcreted mesh to a slope surface. Weep holes to provide through-drainage are an essential component of all shotcrete applications.

C4.4.1.6 Restraining mesh

The primary function of mesh is as a form of protection (see later) to prevent rockfall debris from falling, rolling or bouncing onto an adjacent road or building. However, mesh that is secured tightly to a rock face (Figs C4.8 & C4.9) through the use of dowels or bolts can provide a degree of restraint and reinforcement of the slope surface in the same way as shotcrete. Mesh is usually designed to restrain in-place blocks of up to 1.5 m^3 in size.

C4.4.1.7 Underpinning buttresses and dentition

Buttresses are used as compression elements providing support to blocks that form an overhang. Buttresses are typically designed as mass concrete (Fig. C4.10), but are provided with sufficient reinforcement to limit cracking on the exposed faces and are tied into the face with a series of steel dowels grouted into pre-drilled holes. A key aspect of buttress design is in ensuring the effective load transfer of the weight of the overhanging block through the buttress to a competent founding layer below.

Dentition is the filling of the space left by a detached block, or as a result of differential erosion or local failure of the material exposed in the slope face, in order to maintain support to the surrounding area and prevent enlargement. Dentition material normally comprises mass concrete or shotcrete that may be held into the slope face by short steel dowels. Masonry is sometimes used as dentition to fill cavities in weathered rock cut slopes along low-cost roads (Fig. C4.11).

For both buttressing and dentition it is important to ensure that there is adequate provision for through-drainage (Figs C4.10 & C4.11) to prevent build up of water pressure.

C4.4.1.8 Toe support

The provision of toe support to rock slopes undergoing deep-seated failure usually takes the form of retaining structures or earth bunds/berms. Due primarily to space and cost considerations, the retaining capacity of gravity structures (Section C5) on low-cost mountain roads is limited to the support of small slides of up to 250 m^3 or so in volume per metre run of slope. The practicality of using earth bunds in mountainous terrain to support rock slope failures is usually severely limited by lack of space.

C4.4.2 Drainage

As far as low-cost roads are concerned, drainage measures normally considered for rock slopes relate principally to the surface drainage measures discussed in Sections C6 and C2.3 for benched cut slope profiles. Sub-horizontal drains might be considered in order to relieve water pressures or to lower water tables in high-risk rock slopes, but their use on low-cost roads is uncommon (Section C6.2.2.4) due to cost considerations and the need for specialist installation equipment. These drains only function as required if they are able to intercept an aquifer or drainage path within the rock mass. The location, depth and orientation of these drainage paths can sometimes be assessed through face mapping and structural analysis assisted, where appropriate, by ground investigation. Where this information is not known, the use of prescriptively applied sub-horizontal drains will probably only be partially successful at best. If excavation exposes seepage zones in the cut face during construction that could significantly affect the stability of the slope, then sub-horizontal drains should be considered.

C4.4.3 Removal (scaling)

This includes the removal of one or more of the following:

- rock overhangs;
- loose blocks of rock;
- weathered rock at the top of a slope (principally through cutting back);
- load from the head of rock failures to increase the factor of safety on the sliding surface(s); and
- an entire failed mass.

The practicality and justification for adopting any of these measures is primarily dependent upon considerations of access and safety and particularly on the size and frequency of rockfall and the infrastructure at risk. For most rock slopes along low-cost roads it is usually only justifiable to carry out localized scaling, such as the removal of overhangs and

Fig. C4.7. A tied-back surface retention structure under construction on a type 3 slope (Fig. C3.4).

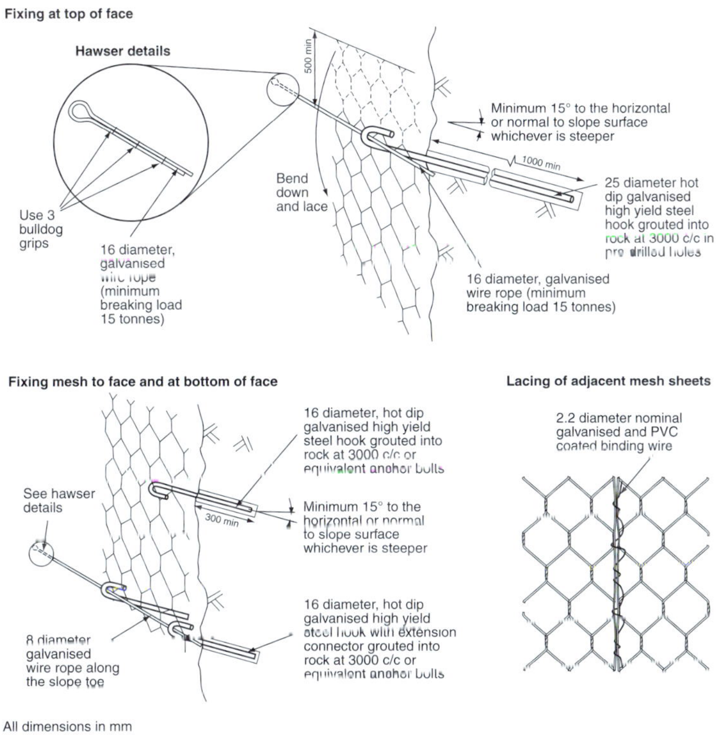

Fig. C4.8. Wire mesh dowelled to rock face (modified from GEO 2003 with permission of Civil Engineering and Development Department of the Government of Hong Kong Special Administrative Region).

Fig. C4.9. Mesh tightly bolted to a rock face to restrain against localized rock topples, wedge and plane failures.

loose or weathered material from the slope face. Scaling has to be done with great care as it can worsen the situation by creating fresh opportunities for rockfalls.

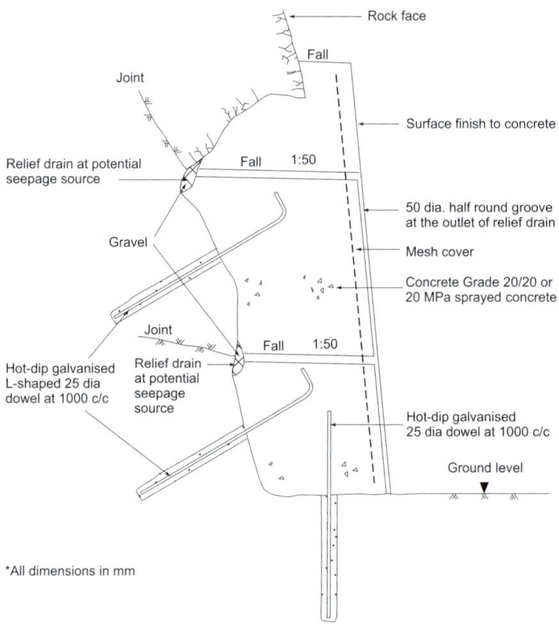

Fig. C4.10. Typical detail for a concrete rock buttress (modified from GEO 2003 with permission of Civil Engineering and Development Department of the Government of Hong Kong Special Administrative Region).

C4.5 Protection measures

Protection measures are designed to contain falling rock debris and prevent it from impacting a road or other structures and pedestrian areas. Analyses of rockfall trajectories for the design of protective barriers are described, for example, in Ritchie (1963), Paronuzzi (1989), Robotham *et al.* (1995), Wyllie & Mah (2004) and Salzmann *et al.* (2010). However, the bounce trajectory of falling rock on a benched cut slope will be different to that on a continuous slope of the same overall angle, and this should also be considered.

C4.5.1 Containment

C4.5.1.1 Hanging mesh nets

In their simplest form these comprise chain-wire mesh nets that are fixed to anchors installed along the rock slope crest or on benches and laid or hung down over the slope face (Fig. C4.12). The principle behind this measure (Bertolo *et al.* 2009) is the control of the travel of detached rock blocks such that they are deposited at the slope toe and do not encroach onto the adjacent road or pedestrian areas. The absence of fixings at the toe of the slope face allows the movement of detached blocks, and provides for the convenient removal of accumulated rockfall debris during maintenance.

In the case study described in Section C4.6, gabion netting was used as an alternative to chain wire mesh to protect the face of a large rockslide/rockfall zone where the main source of rockfall was towards the top of the slope. Loose blocks were removed manually from the face using crowbars

Fig. C4.11. Masonry dentition as cavity infill to cut slopes. Note the use of dry stone panels in the top left application to facilitate through-drainage in a seepage zone and the use of weepholes in the dentition in the lower areas.

before the gabion net was rolled over from the top of the scarp. The netting was secured by anchor bars driven into the rock mass several metres back from the top of the slope and beyond the zone of any surface cracking.

C4.5.1.2 Buffer zones and catch ditches

Where space is available, a buffer zone can be formed at the slope toe and separated from trafficked, inhabited or pedestrian areas by a barrier (including trees) to provide an effective runout area for rockfall. However, there is often inadequate space for buffer zones along mountain roads, and catch ditches and walls or barriers (below) may provide a suitable alternative for rockfall containment.

C4.5.1.3 Catch walls and rigid barriers

Catch walls are often constructed at the base of a rock slope to protect a road and passing traffic against falling rock and debris. They are also sometimes used where stream channels have been truncated by slope excavation leaving a 'hanging gully' at the top of a cut slope. The essential requirement of a catch wall is that there is sufficient space behind the wall for failed material to accumulate without overtopping, and that there is machine access to allow accumulated debris to be cleared. Ideally, the wall is best constructed in gabion (Fig. C4.13) or mesh-reinforced fill, since these structures are capable of absorbing some dynamic load without structural damage. However, reinforced concrete is sometimes used where space is very limited (Fig. C4.14). For obvious reasons it is preferable for any failed debris that accumulates behind the wall to be removed each dry season. In practice, however, this is infrequently undertaken on low-cost roads and the retention capacity of the structure should be considered in this event. Given the cost of these structures it is important to ensure that they are positioned correctly. In the example illustrated in Figure C4.14, the catch wall has been constructed with little apparent source of rock fall or soil fall debris on the slopes above.

C4.5.1.4 Catch fences

These comprise steel framed chain-wire fences, usually up to 3 m in height but occasionally 5–6 m. Rockfall trajectory analyses should be carried out using proprietary software and surveyed rock slope profiles when designing the location, height and impact absorption capacity of catch fences. The highest specification fences can absorb up to 5000 kJ of impact energy, but it is more common to find

Fig. C4.12. Wire mesh hung over a rock slope.

Fig. C4.13. Gabion catch wall with access for machine clearance.

Fig. C4.14. Reinforced concrete catch wall.

Fig. C4.15. Shelter to protect against rockfall. Note the upstand wall at the outer edge of the roof structure to retain a layer of debris as an impact absorber from future falls.

fences designed to absorb 100–250 kJ (equivalent to 1200 kg of rock travelling at 20 m/s). Fences may be provided with permanent foundations, but are often also designed with a pre-cast concrete base in order to facilitate their removal for periodic clearance of rockfall debris from the area behind. Steel cable ties, extending from the top of the fence to anchors on the rock slope, are also often included to reduce the likelihood of the fence being toppled by large-volume rockfalls and to economize on the foundation design.

C4.5.2 Shelters and tunnels

Rock shelters are considered in cases where:

- the scale of rock failure is so great that stabilization is either uneconomical or impracticable, or both; and
- the volumes, frequency and proximity of rockfall runout are considered to be too high to be manageable by the use of fences, buffer zones and catch walls.

Figure C4.15 illustrates a rock shelter where the steepness of slope above and the proximity of the river below meant that there was no other practicable option for keeping the road open. There are no standard details for the design of rock shelters; each situation requires its own site-specific design. However, important considerations to be borne in mind include:

- the stability of the slope upon which the shelter is to be constructed;
- the potential for lateral loads to act on the structure as a result of deeper ground movements within the adjacent *in situ* rock;
- the potential for river scour to undermine the structure from below (Fig. C4.15);
- the lateral loads acting on the back of the structure due to the wedge of accumulated rockfall debris;
- vertical loads acting on the roof of the structure as a result of accumulated rockfall debris (Fig. C4.15);
- impact loads from rock masses falling onto the roof of the structure (Fig. C4.15);
- the potential for earthquake loading;
- ventilation, lighting, traffic and pedestrian safety; and
- cost and cost-effectiveness.

Most of the above can be addressed through structural design. However, if the slope is undergoing deep-seated failure as well as rockfall from above, it may not be possible to design a structure capable of withstanding these loads. Although this situation is relatively rare, it will need to be considered when reviewing options.

C4.6 Case study

A study carried out in Nepal reviewed the options of stabilization and protection where a large rock failure (Fig. C4.16) impacted the main road from India to Kathmandu located alongside the Trisuli River. The rock failure comprised a combination of rockslide and rockfall mechanisms and led to regular road blockages. These often created traffic tailbacks of several kilometres. The following options were considered:

- option 1: do very little, other than reconstruction of the damaged road surface and regular clearance of failed material from the road;
- option 2: minor realignment of the road towards the river, together with a below-road retaining wall, to create additional space at the base of the rock slope. This option also included slope improvement works involving localized trimming, scaling, netting, buttressing, drainage control and catch wall protection (150 m long, 6 m high gabion wall at the slope toe);
- option 3: total slope stabilization;
- option 4: construction of a viaduct along the existing centreline but designed to elevate the road above rockfall and rockslide improvements;
- option 5: road realignment to the opposite side of the valley; and

Fig. C4.16. Large type 3 rock slope failure.

- option 6: construction of a two-lane 200 m long rock shelter.

Option 3 (total slope stabilization) was not considered feasible, and the viaduct and road realignment options (options 4 and 5) were judged to be too expensive. Outline designs and cost estimates were therefore prepared for options 1, 2 and 6 and the estimated costs associated with each of these at 1993 prices were:

- option 1: US$160 000;
- option 2 (assuming future road clearance and reinstatement would not be required): US$240 000;
- option 6: US$312 000.

Option 1, although the cheapest, was considered to leave an unacceptable residual risk to road users. The rock shelter was rejected on the basis of cost plus the fact that a stable foundation on rock could not be guaranteed for its entire length thus placing the structure at risk from potential river scour (unless piled at increased cost). Consequently, the slope treatment and protection option (option 2) was selected and implemented. Within a 5-year period the rock slope had begun to show signs of revegetation and has subsequently stabilized. Other major rock failures have occurred along the road since this work was undertaken and these have required treatment and protection by the road authority. Had either of the high cost options (including the rock shelter) been implemented at the original site, this might have served as a precedent for over-investment along the remainder of the road.

References

ANON 1977. The logging of rock cores for engineering purposes. *Quarterly Journal of Engineering Geology*, **10**, 45–52.

Barton, N. R., Lien, R. & Lunde, J. 1974. Engineering classification of rock masses for the design of tunnel support. *Rock Mechanics*, **6**, 189–239.

Bertolo, P., Oggeri, C. & Peila, D. 2009. Full-scale testing of draped nets for rock fall protection. *Canadian Geotechnical Journal*, **46**, 306–317.

Bieniawski, Z. T. 1989. *Engineering Rock Mass Classifications. Technology and Engineering*, John Wiley Ltd, New York.

BSI 1999. *Code of practice for site investigations*. BSI 5930, British Standards Institution, HMSO, London. First published in 1981 (Amendment 2, 2010).

BSI (EN) 2003. *Geotechnical investigation and testing. Identification and classification of rock. Part 1: Identification and description*. British Standards Institution. EN ISO BSI 14689-1:2003.

Deere, D. U. & Miller, R. P. 1966. *Engineering classification and index properties of intact rock*. Air Force Lab. Tech. Rep. AFNL-TR-65-116 Albuquerque, NM.

FHWA 1999. *Ground anchors and anchored systems*. Geotechnical Engineering Circular No 4. Report No FHWA-IF-**99-015**. Prepared by Sabatini, P. J., Pass, D. G., Bachus, R. C. Washington, DC.

Fookes, P. G. & Sweeney, M. 1976. Stabilization and control of local rock falls and degrading rock slopes. *Quarterly Journal of Engineering Geology*, **9**, 37–55.

GEO 1997. *Model Specification for Prestressed Ground Anchors*. 2nd edn. Geospec 1. Geotechnical Engineering Office, Hong Kong. (1989), reprinted 1997.

GEO 2003. *Guidelines on the Use of Prescriptive Measures for Rock Slopes*. GEO Report No **161**, Yu, Y. F., Siu, C. K., Pun, W. K. Geotechnical Engineering Office, Hong Kong.

Goodman, R. E. & Shi, G.-H. 1985. *Block theory and its application to rock engineering*. Prentice-Hall, Englewood Cliffs, New Jersey.

Hawkins, A. B. 1998. Aspects of rock strength. *Bulletin of Engineering Geology and the Environment.*, **57**, 17–30.

Hearn, G., Wise, D., Hart, A., Morgan, C. & O'Donnell, N. 2012. Assessing the potential for future first-time slope failures to impact the oil and gas pipeline corridor through the Makarov Mountains, Sakhalin Island, Russia. *Quarterly Journal of Engineering Geology and Hydrogeology*, **45**, in press.

Hoek, E. 1994. Strength of rock and rock masses. *International Society for Rock Mechanics, News Journal*, **2**, 4–6.

Hoek, E. 1999. Putting numbers to geology – an engineer's viewpoint. The second Glossop Lecture. *Quarterly Journal of Engineering Geology and Hydrogeology*, **32**, 1–19.

Hoek, E. & Bray, J. 1981. *Rock Slope Engineering*. Revised 3rd edn. Institution of Mining and Metallurgy, London.

Hoek, E. & Brown, E. T. 1997. Practical estimates of rock mass strength. *International Journal of Rock Mechanics and Mining. Science and Geomechanics Abstracts*, **34**, 1165–1186.

Hoek, E., Marinos, P. & Benissi, M. 1998. Applicability of the Geological Strength Index (GSI) classification for very weak and sheared rock masses. The case of the Athens Schist Formation. *Bulletin of Engineering Geology and the Environment*, **57**, 151–160.

Hungr, O., Salgado, F. M. & Byrne, P. M. 1989. Evaluation of a three-dimensional method of slope stability analysis. *Canadian Geotechnical Journal*, **26**, 679–686.

IAEG. 1981. Report on the commission on engineering geological mapping. *Bulletin of the International Association of Engineering Geology*, **19**, 364–371.

Irigaray, C., El Hamdouni, R., Jimenez-Peralvarez, J. D., Fernandez, P. & Chacon, J. 2010. GIS application to a kinematic analysis and slope-stability assessment of slope cuts on rock massifs along a national road in Spain. *In*: Williams, A. L., Pinches, G. M., Chin, C. Y., McMorran, T. J. & Massey, C. I. (eds) *Geologically Active. Proceedings of the 11th IAEG Congress, Auckland, New Zealand, 5–10 September 2010*, Taylor & Francis Group, London, 1949–1957.

ISRM. 1981. Brown, E. T. (ed.) *Suggested Methods of Rock Characterisation, Testing and Monitoring*. Pergamon Press, Oxford.

Lorig, L. & Varona, P. 2004. Numerical analysis. *In*: Wyllie, D. C. & Mah, C. W. (eds) *Rock Slope Engineering*. 4th edn, Taylor and Francis Group, Suffolk, 218–244.

Marinos, V., Marinos, P. & Hoek, E. 2005. The geological strength index: applications and limitations. *Bulletin of Engineering Geology and the Environment*, **64**, 55–65.

Pantelidis, L. 2010. An alternative rock mass classification system for rock slopes. *Bulletin of Engineering Geology and the Environment*, **69**, 29–39.

Paronuzzi, P. 1989. Probabilistic approach for design optimisation of rockfall protective barriers. *Quarterly Journal of Engineering Geology*, **22**, 175–183.

Pedrazzini, A., Jaboyedoff, M., Froese, C. R., Langenberg, C. W. & Moreno, F. 2011. Structural analysis of Turtle Mountain: origin and influence of fractures in the development of rock slope failures. *In*: Jaboyedoff, M. (ed.) *Slope Tectonics*. Geological Society, London, Special Publications, **351**, 163–183.

Ritchie, A. M. 1963. An evaluation of rockfall and its control. *Highway Research Record*. No 17, 13–28.

Robotham, M. E., Wang, H. & Walton, G. 1995. Assessment of risk of rockfall from active and abandoned quarry slopes. *Transactions of the Institution of Mining & Metallurgy*, **A104**, A25–A33.

Salzmann, H., Roth, H. & Haller, B. 2010. Rockfall protection of ore processing facilities in Papua – design and implementation of high energy flexible rockfall barriers. *In*: Williams, A. L., Pinches, G. M., Chin, C. Y., McMorran, T. J. & Massey, C. I. (eds) *Geologically Active. Proceedings of the 11th IAEG Congress, Auckland, New Zealand, 5–10 September 2010*, Taylor and Francis Group, London, 639–644.

Sloan, S. W. 2012. Geotechnical stability analysis. The 51st Rankine Lecture. *Géotechnique*, **62**, in press.

Tsiambaos, G. & Saroglou, H. 2010. Excavatability assessment of rock masses using the Geological Strength Index. *Bulletin of Engineering Geology and the Environment*, **69**, 13–27.

Wyllie, D. C. & Mah, C. W. (eds) 2004. *Rock Slope Engineering*. 4th edn, Taylor and Francis Group, Suffolk.

C5 Retaining structures

T. Hunt, G. J. Hearn* & S. d'Agostino

URS Scott Wilson Ltd, Scott House, Alençon Link, Basingstoke, Hampshire RG21 7PP, UK
*Corresponding author (e-mail: gareth.hearn@scottwilson.com; garethhearn@talktalk.net)

C5.1 Types of retaining structure

Retaining structures are a common feature of road construction in hilly and mountainous areas and can account for up to 20% of the total construction cost. Retaining structures comprise:

- gravity walls, where the weight of the wall and its backfill provide most of the stabilizing force (masonry, gabion and reinforced concrete cantilever walls are typical examples);
- embedded walls, where the soil in front of and behind the structure and anchors (if any) provide the stabilizing force (sheet pile or bored pile walls are typical examples);
- reinforced soil, where the *in situ* soil mass is reinforced with nails or dowels (usually behind a protective face);
- reinforced fill, where steel or geosynthetic geogrids or straps are embedded into the fill during its emplacement.

Due to cost considerations on most low-cost roads, retaining walls are usually designed as gravity structures. Consequently, this chapter focuses on gravity walls constructed from masonry, gabion, mass concrete and reinforced concrete. However, consideration is given to the use of soil nails to strengthen cut slopes (Section C5.2.5) and to the use of reinforced fill structures (Section C5.2.6), as these can provide useful alternatives under certain circumstances.

Walls are constructed in above-road and below-road locations; see Figure C5.1 for illustrations of these terms.

C5.2 Types of earth-retaining structure

Figure C5.2 shows some of the more common retaining structures used in steep terrain for below-road (horizontal backfill) and above-road (sloping backfill) situations. Table C5.1 describes their advantages and limitations in the context of low-cost roads. Dry stone, mortared masonry, composite and gabion walls are labour intensive to construct. This can offer advantages for labour utilization and disadvantages in terms of speed of construction compared, for example, to a mass concrete wall.

The following sections describe the various earth-retaining structures in more detail and further information may be found in, for example, GEO (1993), BSI (1994), Bowles (1996) and Das (2007).

C5.2.1 Masonry retaining walls

Masonry walls are commonly used as slope- and fill-retaining structures along mountain roads, and especially where there is a good local supply of stone for construction. While dry stone walls are sometimes found as both above-road and below-road retaining structures, it is usual to use mortared masonry due to its greater durability. Figure C5.3 shows the use of masonry walls to reinstate a road across a small type 5 landslide (Fig. C3.4) in Nepal. The below-road wall has been founded beneath the failure surface while the above-road wall has been designed to support the remaining failed material in the cut slope. Composite masonry walls (Fig. C5.4) combine the advantages of free drainage and durability provided by dry stone panels within a mortared masonry structure. Despite the illustration in Figure C5.4, they are usually used as slope revetments (Section C7.2.2) rather than as retaining structures. A foundation of suitable strength throughout is paramount for all masonry wall construction.

C5.2.2 Concrete retaining walls

Concrete walls are usually more durable than masonry or gabion, and can be particularly beneficial in below-road locations where the excavation width is constrained. As with masonry walls, concrete walls are unable to tolerate differential foundation settlement without loss of structural integrity. Mass concrete walls are usually only viable at wall heights of up to 4 m, with reinforced concrete cantilever or counterfort walls being preferable for greater heights.

The soil above the base of a reinforced concrete cantilever or counterfort wall is included as part of the weight of the wall in stability calculations. The back of the wall is treated as a vertical face above the wall heel. This is known as the 'virtual back' and is the surface on which the active earth pressures act (Section C5.3.1).

From: HEARN, G. J. (ed.) *Slope Engineering for Mountain Roads*. Geological Society, London, Engineering Geology Special Publications, **24**, 209–229.
DOI: 10.1144/EGSP24.14 0267-9914/11/$15.00

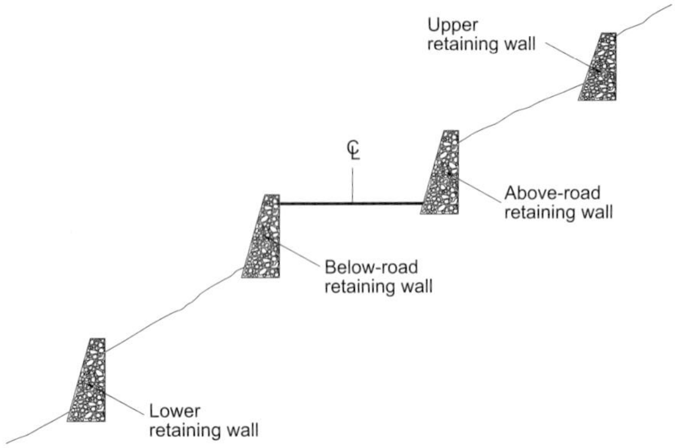

Fig. C5.1. Retaining wall descriptors.

Reinforced concrete grid walls (Fig. C5.5) comprise grids of reinforced concrete infilled with panels of either dry stone or mortared masonry. They have been utilized where the required wall heights are 5 m or greater and where the space for conventional structures is insufficient. Their stability relies on the use of hand-placed rockfill as backfill, and they are effectively part-way between retaining and revetment structures (see Section C7.2.2).

C5.2.3 Gabion walls

Gabion walls are among the most commonly used wall types on low-cost roads. They are often the preferred choice where heavy groundwater seepages and continued earth movements are anticipated. Gabion walls generally have the advantage of requiring a less skilled workforce, without the need for specialist construction equipment. As

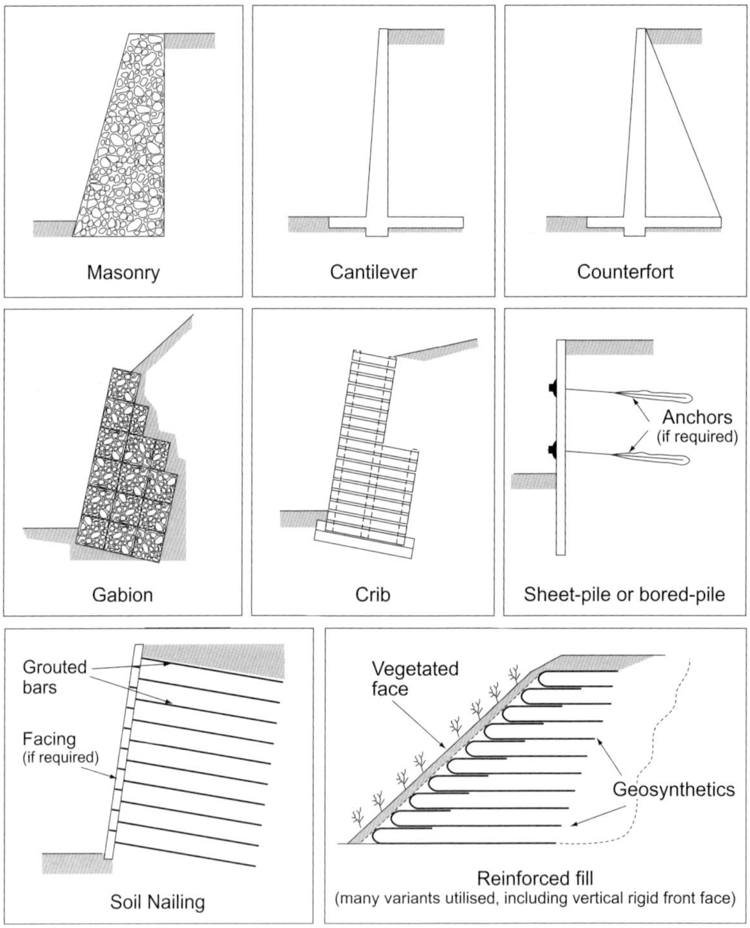

Fig. C5.2. Typical retaining structures.

Table C5.1. *Advantages and disadvantages of various retaining structure types*

Material	*Type*	*Advantages*	*Disadvantages*
Masonry	Dry stone	Stone usually locally sourced; permeable; no specialist equipment needed. Can accommodate varying founding depths and required wall heights	Unable to accommodate very much movement without distress; good quality masonry work necessary; need to protect against washout of fines from backfill
	Mortared	Stone usually locally sourced; higher unit weight and greater durability than dry stone, composite, and gabion walls	Unable to accommodate any movement without distress; requires good foundations due to inflexibility; impermeable therefore drainage measures are required
	Composite	As for mortared masonry but less cement required due to dry stone inserts that are permeable	Unable to accommodate movement without distress; requires good foundations due to inflexibility; need to protect against washout of fines from backfill
Concrete	Mass Concrete	Simple to construct; may be preferred where durable stone is locally unavailable for masonry or gabion walls; durable and 'stiff'	Requires larger cross-section with greater quantities of concrete than reinforced concrete options; requires good foundations due to inflexibility; impermeable therefore drainage measures are required; usually only practicable to heights of 4m
	Reinforced Concrete – Cantilever	Base generally occupies less width than masonry, gabion and mass concrete walls of the same height; durable and 'stiff'	Higher skills required compared to masonry, gabion and mass concrete; requires good foundations due to inflexibility; impermeable therefore drainage measures are required; generally uneconomic above 8m height
	Reinforced Concrete – Counterfort	As RC cantilever, but additional bending restraint allows thinner section	As above, but can be constructed to greater heights
	Reinforced Concrete – Grid	Base occupies less width than other walls for the same retained height	Only provides support against shallow slope failure; hand-placed rock fill required as backfill, contractor experience in the construction of these walls may not exist outside the Philippines
	Reinforced Concrete – Core pile	No excavation required, through-drainage permitted between piles. Can be constructed in situations where conventional gravity walls would not function or would be difficult to construct.	High cost; requires specialist piling rigs; relies on passive earth pressure in front of wall plus bending restraint; significant piling depths may be required
Stone-filled wire boxes	Gabion	Technique well-known; can accommodate limited movement without distress; permeable; stone usually locally sourced; no specialist equipment required	Moderate durability; not recommended for below-road walls due to flexibility and internal settlement; lower unit weight compared to mortared masonry or concrete walls so wider wall necessary for same retained height; difficult to construct on variable founding level unless base is made up in mortared masonry or concrete; protection required against washout of fines from backfill
Timber or reinforced concrete lattice structure filled with gravel	Crib	Rapid to construct; pre-cut or precast sections can be held in stock for emergency works; permeable; easy to adjust height of wall	Concrete base usually required and particularly for varying foundation level; possible problems of durability if timber cribs are used, and cost if reinforced concrete used; some flexibility but less than gabion; protection required against washout of fines from backfill; lower unit weight compared to concrete or masonry walls so larger wall section is required to provide the same retained height
Steel	Sheet pile	Rapid to construct if not anchored; narrow section	Unanchored sheet pile walls rely on passive earth pressure in front of the wall and will fail if slope below fails; anchored sheet pile walls are difficult and costly to construct; impermeable unless through-drainage is provided; may be impracticable in boulder soils and greatly varying rockhead levels
Steel dowels	Soil nails	Often used in road widening to enable steepening of cut slopes	Requires special installation equipment and experienced operators
Steel mesh or geosynthetic and granular fill	Reinforced Fill	Bearing pressures are distributed evenly over width of foundation; some ground movements can be tolerated; free-draining.	Can be expensive if materials have to be imported; selected backfill material often necessary

Fig. C5.3. Use of mortared masonry retaining walls for road reinstatement. Note the illicit tipping of spoil in front of the newly constructed below-road wall.

with masonry walls, the use of locally available stone can have significant cost savings compared to other wall types. Gabion wire must be galvanized to protect against rust. Life expectancy may be significantly reduced in damp conditions, however, especially where the bedrock or soil is acidic and contains iron compounds.

C5.2.4 Crib walls

Crib walls can consist of precast concrete (Fig. C5.6) or timber (Fig. C5.7) members (imported or obtained locally) interlocked and filled with gravel or rockfill to form a gravity structure. A filter fabric is usually provided behind the face of the wall to prevent the fill from migrating through the open structure of the crib. This form of wall is less common as it has only minor advantages over the use of gabion, notably that it is generally heavier and requires a thinner section for the same retained height. Precast concrete crib walls are, however, more expensive than gabion. The use of timber provides a potential low-cost solution, although stringent quality control is required to ensure the long-term durability of these structures. Crib walls are the most free-draining of all wall options and are among the most visually appealing.

C5.2.5 Soil nails

Although they do not constitute a retaining structure as such, soil nails effectively achieve the same purpose and are therefore included here. Soil nails are steel dowels that are installed in soil cut slopes by driving or drilling and grouting and, depending on the nature of the slope materials, may be detailed as discrete units or tied into a protective facing system. They are often used to permit steepening of the design slope angle in order to reduce land take and earthworks volumes in road-widening schemes. They are also used as a retro-fit measure to improve the stability of existing oversteep cut slopes. Upper-bound slope surface angles for soil nails are of the order 60–70°. Pun & Urciulli (2008), for example, describe the load transfer mechanism, nail inclination effects, binding stiffness, design, pull-out resistance, installation and performance of soil nails. Soil nails may also be installed in slopes by hammering the nail (driven nails) into the soil mass or weathered rock. Alternatively, FHWA (1994) provides details of launched nails (ballistic nails) whereby nails, commonly up to 6 m long, are launched into a slope from a nail gun attached to a tracked excavator.

Soil nails are often used in conjunction with a facing system to enhance the local face stability or where slope

Fig. C5.4. Composite masonry retaining wall.

Fig. C5.5. Reinforced concrete grid wall.

Fig. C5.6. Concrete crib wall. Note the mortared masonry protection to the backfill due to anticipated high surface runoff.

materials might otherwise break up and flow around the nail shafts. Common facing systems include reinforced shotcrete (Figs C5.8 & C5.9), reinforced concrete slabs and reinforced concrete grillages. Wire mesh fixed to soil nail heads is also often used when a vegetated slope surface is desired (e.g. Kulkarni 2009).

Instrumented testing of soil nails shows that they have relatively little stabilizing effect in shear across a surface of rupture and instead tend to deform and align themselves parallel to the direction of movement. Soil nails are typically grouted into pre-drilled holes and orientated with a shallow declination (10–15°) into the slope (α on Fig. C5.8). The designer may encounter instances where steeper declinations are desired, for example, to ensure that the nails remain wholly within a site boundary. However, declinations greater than 25–30° are generally avoided due to the larger deformation/strain of the slope face required to mobilize the nail tension, which may lead to serviceability problems.

Soil nails are usually installed in a staggered grid pattern, with spacing of the order 1–3 m and nail lengths 0.7–1.0 times the vertical height of the slope. Nail lengths of 10–20 m are not uncommon and diameters vary, but are typically 25 mm. GEO (2004) suggests that, for slopes steeper than 65°, soil nails should be used in association with

Fig. C5.7. Timber crib wall.

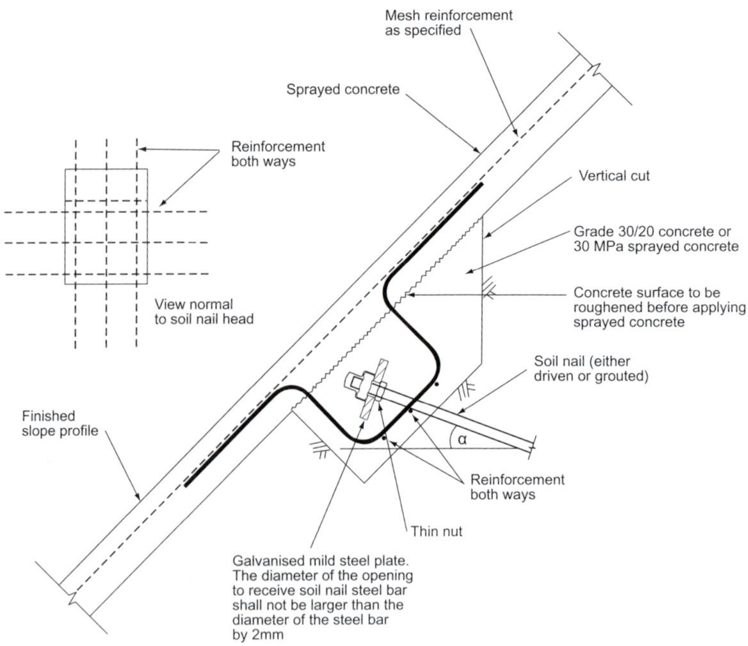

Fig. C5.8. Typical detail of soil nail incorporating shotcrete surface protection (modified from CEDD Standard Drawing C2106/3F, 2008 with permission of Civil Engineering and Development Department of the Government of Hong Kong, Special Administrative Region).

Fig. C5.9. Installation of soil nails in a cut slope.

reinforced concrete grillage beams. The design and specification of soil nails are discussed, for example, in TRRL (1993), Department of Transport (1994), FHWA (2003), GEO (2004, 2008), CIRIA (2005), BSI EN (2006), Pun & Shiu (2007), Pun & Urciulli (2008) and BSI (2009). The numerical analysis of soil nail walls using finite element modelling is discussed, for example, in Javanmard & Ahmadi (2009) and Giri & Sengupta (2009).

Soil nails have been used as reinforcement to fill slopes. However, there remains a concern with this application since the soil nails are largely reliant on dilation of the surrounding soil during shear to ensure the development of adequate pullout capacity; sufficient movement of the fill to cause dilation might be unacceptable as far as pavement rideability is concerned. Where soil nails represent the only feasible solution they should be extended through the fill and into a competent stratum behind, with the use of a rigid slope surfacing to ensure restraint of the fill mass. Soil nails have previously had limited application to low-cost roads due to cost and specialist installation equipment requirements.

C5.2.6 Reinforced fill structures

Reinforced fill structures have also been used relatively infrequently in the low-cost road sector. The most likely reasons for this are the lack of design expertise, the need to import reinforcing strips, geosynthetics or steel mesh and the requirement for good fill control and compaction. However, in many countries, the steel mesh required for gabion walls also has to be imported and so the cost differentials may not be significant. On existing roads, the replacement of failed masonry or gabion walls with reinforced fill structures may prove impracticable given the working area required to construct them. Nevertheless, they have been used in Ethiopia, for example, to reinstate sections of road damaged by landslides (Fig. C5.10). For new roads, the required construction widths are usually available and reinforced fill structures offer the advantage of distributing loads over a wider section, thereby reducing bearing pressures. Ramasamy (2009), for example, describes the use of high-strength polyester geogrid and gabion mesh facing to construct reinforced fill slopes to 25 m in height in mountainous terrain in Malaysia.

C5.3 Design of retaining walls

The design of all retaining structures requires consideration of the interaction between the ground and the structure, and involves two sets of calculations:

- equilibrium calculations to determine the overall proportions and geometry of the structure to achieve equilibrium under the relevant earth pressures and forces;

Fig. C5.10. Reinforced fill wall.

- structural design calculations to determine the size and properties of the structural sections necessary to resist the bending moments and shear forces determined from the equilibrium calculations.

Both sets of calculations are usually carried out in accordance with limit state design. Limit state may be divided into ultimate limit state (e.g. instability or failure by rupture of the structure) and serviceability limit state (including substantial deformation of the structure). This discussion does not consider structural or serviceability limit state failure, but concentrates on equilibrium considerations and the design and construction issues that serve to prevent or reduce the incidence of failure of the ground, not the internal failure of the retaining structure itself.

C5.3.1 Equilibrium calculations for retaining walls

Unless a retaining wall is on the point of failure, the soil against its rear and front faces is not in limiting equilibrium. A relatively small movement or rotation is sufficient to reduce the lateral pressure behind the wall to the active or minimum value, but a larger movement is required to

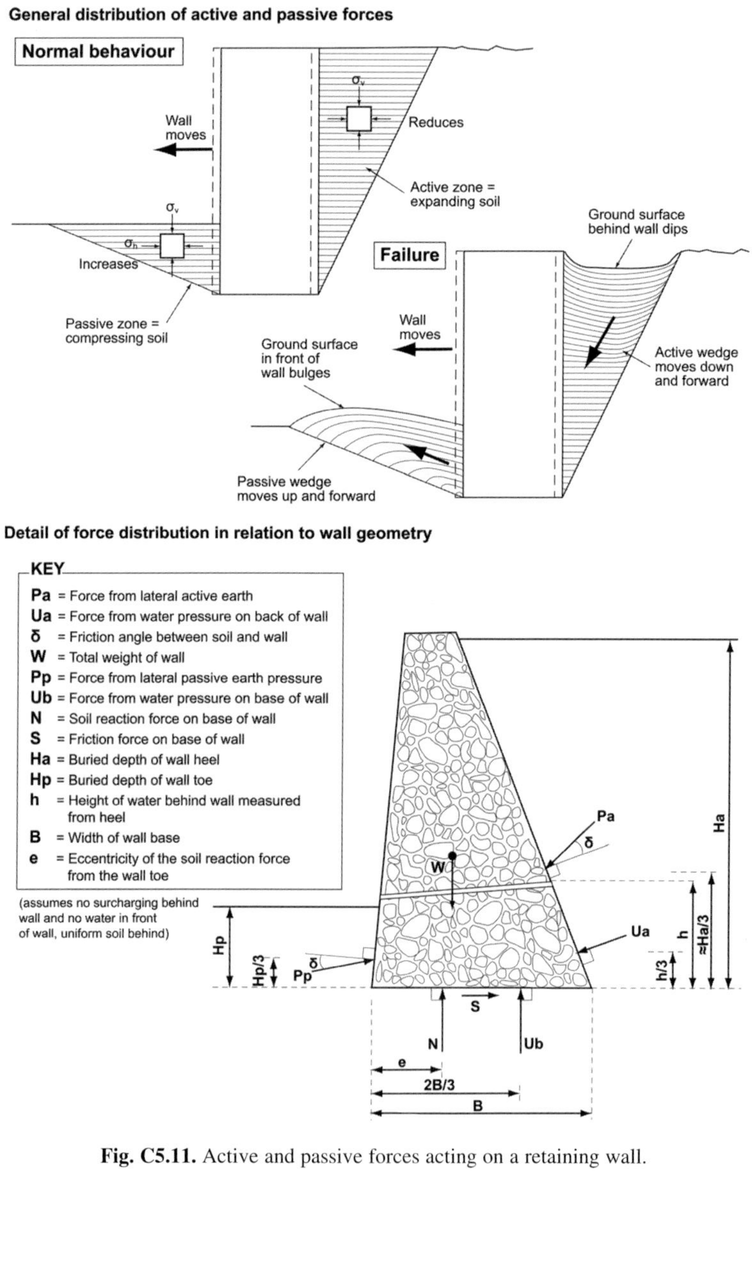

Fig. C5.11. Active and passive forces acting on a retaining wall.

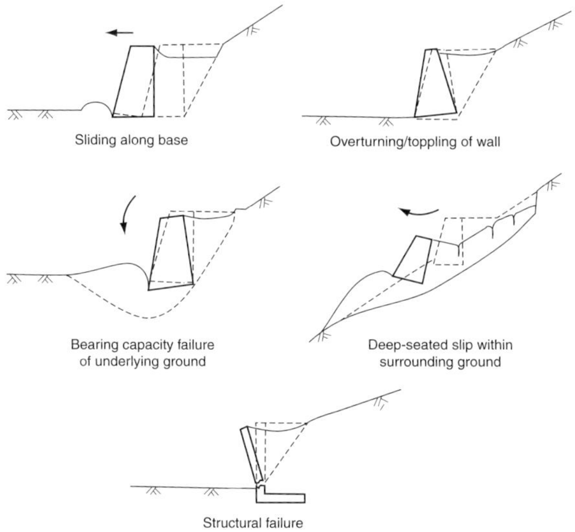

Fig. C5.12. Failure mechanisms for soil retaining walls (modified from GEO 1993 and BSI 1994; BS8002, reproduced with permission from the Head of the Geotechnical Engineering Office and the Director of the Civil Engineering and Development Department, the Government of Hong Kong Special Administrative Region).

increase the resistance in front of the wall to the passive or maximum value. Figure C5.11 shows the distribution of active and passive forces acting on a retaining wall prior to and after failure. In examining the stability of the wall, these two limiting values of minimum active thrust and maximum passive resistance must be analysed.

Retaining walls can fail (Fig. C5.12) by sliding, overturning, bearing capacity failure and due to loss of overall slope stability (i.e. deeper landslide movements of the ground beneath the base of the wall). The calculation of forces acting on gravity retaining walls is not presented here as it is covered in widely available and published standard texts, for example Lamb & Whitman (1979), GEO (1993), Navfac (1986), Bowles (1996) and CIRIA (2000).

Sliding is checked by resolving all of the horizontal forces on the wall and confirming that there is a reasonable margin (factor of safety) between the destabilizing and stabilizing forces. Overturning is checked by taking moments about the front edge of the wall base and ensuring that the restoring moments exceed the overturning moments by the required margin. To check the bearing capacity, the pressure distribution exerted by the wall base on the underlying soil is determined from the overturning calculation and compared to the resistance of the soil to a circular failure beneath the wall. The possibility of deep-seated failure within the underlying ground is checked by undertaking slope stability analyses, as described in Section C3.2. However, for Eurocode 7 compliance (BSI EN 2004), the requirement for a design to have a lumped factor of safety of a certain value greater than 1.0 (such as 1.5 against sliding and 2.0 for wall overturning) is replaced by the use of partial factors. These are applied to external loads, groundwater conditions and soil strength parameters in order to account for variability and uncertainty in the stability calculation (Section C3.2.5).

The main destabilizing force is the active earth pressure acting on the back of the wall. The magnitude and distribution of the active earth pressure depends on many variables including the strength of the soil, the topography of the ground behind the wall and the friction between the wall and the soil. These factors are listed and their effects summarized in Table C5.2.

Methods of calculating active earth pressures are given in standard texts, for example GEO (1993), Navfac (1986) and Bowles (1996).

Equilibrium calculations require the input of the following:

- the dimensions of the wall with respect to final ground levels;

Table C5.2. *Factors affecting active earth pressures*

Factors affecting active earth pressures	*Pa increases if . . .*	*Pa reduces if . . .*
Friction angle of soil behind wall	Soil is weaker	Soil is stronger
Ground topography behind wall	Ground slopes up behind wall	Ground slopes down behind wall
Surface loading behind wall	A large uniform load is applied behind the wall	Not applicable
Inclination of back of wall	Rear of wall leans forward away from the slope	Rear of wall lies back into slope
Friction between back of wall and soil	Wall/soil friction is low	Wall/soil friction is high
Water pressures on back of wall	N/A	Water pressures are higher*
Cohesive soil behind wall	N/A	Soil cohesion is higher†

*Although the active earth pressures reduce as the water pressure increases on the back of the wall, the water pressure must also be separately added to the rear of the wall which causes a significant increase to the overall destabilizing forces and moments.

†When retaining walls support *in situ* clay, lateral earth pressures will be controlled by c_u in the short term (during or immediately after construction). In the longer term the earth pressures are controlled by c' and φ'. Since c_u is usually much higher than c', earth pressures generally increase with time and design should be undertaken using long-term effective stress parameters (c' and φ').

- the shear strength properties of the retained soil (backfill is usually specified to be non-cohesive in which case c' is assumed to be zero);
- the wall friction between the backfill and the material forming the wall;
- the base friction between the base of the wall and the supporting ground beneath;
- whether or not passive pressure acting on the front of the wall is to be taken into account;
- water pressures;
- surcharges; and
- seismic loads.

Output data from the calculations will normally include:

- resisting force, activating force and factor of safety against sliding;
- resisting moment, overturning moment and factor of safety against overturning; and
- maximum eccentricity of normal force, maximum stress at the base of the wall and factor of safety against bearing capacity failure.

It is often appropriate to prepare standard details for retaining walls of up to, say, 6.0 m in height for use in typical locations immediately adjacent to the road without the need to undertake individual stability analyses. If such drawings are prepared, however, the notes on the drawings must make clear the assumptions made and the need to carry out individual calculations if any of the assumptions prove to be invalid during construction. For walls of a greater height than 6.0 m, where the consequences of failure are likely to be more severe and more costly, the designs should always be carried out on an individual basis.

C5.3.2 Passive pressures

It is common practice to ignore passive pressure for retaining wall design for roads in mountainous or hilly terrain. As noted earlier, for full passive pressure to be mobilized a significant movement of the wall has to take place. In addition, allowances may need to be made for the possibility of a trench being excavated in front of the wall (e.g. due to the installation of a roadside drain or other facilities in urban areas). In the case of below-road retaining walls, loss of passive resistance may also occur as a result of failure or erosion of the soil in front of the wall. Furthermore, passive pressure will be significantly reduced where the wall is located at the top of a steep slope (due to the passive wedge potentially daylighting on the slope face).

C5.3.3 Water pressures

Water pressures within the soils behind the wall act to increase the destabilizing force on the back of the wall, increasing its tendency to slide or overturn. Uplift water pressures on the underside of the wall reduce the stabilizing benefit of the self-weight of the wall. This double effect means that wall design is very sensitive to water pressures. Unfortunately, water pressures are often difficult to measure and vary considerably with time and location. It is therefore important that the designer gives careful consideration to water conditions during design, makes a conservative estimate of the likely maximum water pressures and takes measures to ensure that these design pressures are not exceeded. Due to the difficulties in achieving an acceptable factor of safety for an undrained wall, standard details for retaining walls usually specify free-draining backfill material and drainage through the structure and at the base of the backfill (Section C5.5.2); water pressures are therefore often ignored in the analysis.

C5.3.4 Surcharges

These can either be permanent or temporary and uniformly distributed, such as traffic loads, which are usually treated as uniformly distributed loads, or concentrated loads (e.g. building foundations). A surcharge of 10 kN/m^2 is frequently assumed as a suitable traffic load, recommended for example for Approach 1 Combination 2 of Eurocode 7 (Section C3.2.3). The surcharge effects of compaction during wall backfilling might also need to be considered in the design or else minimized in the construction (Section C5.6).

C5.3.5 Seismic loads

These will depend on the magnitude of local seismic events and national codes will usually specify the seismic acceleration force to be adopted in the design of critical structures. These are unlikely to include retaining wall structures on low-cost roads. For example, during the rehabilitation of the Halsema Highway in the Philippines following the 1990 earthquake it was concluded that the addition of seismic loads into the retaining wall designs would substantially increase costs and render the project uneconomic. In that particular case, because the factors of safety against failure of the underlying slopes (geotechnical failure) were judged to be close to unity, there was little justification in building any conservatism into the design of retaining walls as the underlying ground would ultimately fail anyway during seismic loading.

C5.3.6 Factors of safety

There are many different ways of defining factor of safety and these are described in the various codes of practice and standards for retaining wall design, including AASHTO (2007) and BSI EN (2004). Broadly, there are two main types of safety factor: lumped factors and partial factors.

The lumped factor of safety is the ratio of the *stabilizing* forces, moments or pressures to the *destabilizing* forces, moments or pressures. Obtaining a lumped factor of safety of less than 1 indicates that the wall will fail because the destabilizing forces exceed the stabilizing forces. Lumped factors of safety give a measure of the margin by which failure is avoided. Suggested minimum lumped factors of safety are given in Table C5.3.

Table C5.3. *Suggested minimum lumped factors of safety for retaining walls*

Failure mechanism	*Recommended lumped factors of safety (FoS)*	*Stabilizing forces/ moments/pressures*	*Destabilizing forces/moments/ pressures*
Sliding	Resolve forces horizontally to obtain FoS >1.5	Base friction, passive soil pressure	Active earth pressure, water pressure on back of wall and surcharge
Overturning	Take moments about wall toe to obtain FoS >2.0	Wall weight, passive soil pressure	Active earth pressure, water pressure on back of wall and surcharge
Bearing	FoS >2.0–3.0*	Bearing capacity of soil under wall†	Pressure exerted by wall and surcharge on underlying soil‡
Overall slope stability	See Section C3.2		

*The relatively high FoS for bearing capacity is applied to limit wall settlements. Where settlement cannot be tolerated (because it could cause unacceptable damage to the wall, the road or other structures and land uses behind or above the wall), a FoS of 3 is used. Less movement-sensitive situations may justify the use of a lower FoS.
†See Section C5.6.2 for methods of assessing bearing capacity.
‡See Navfac (1986) and GEO (1993) for guidance on assessing bearing pressures.

Instead of applying a single lumped factor of safety, some codes of practice require the application of several partial factors of safety to loads and soil strengths (Section C3.2 and Table C3.4). Although the use of this approach means that it is not so clear what overall safety margin is achieved, larger factors of safety may be applied to those parameters that have the highest uncertainty or greatest influence on stability and wall behaviour (BSI EN 2004).

C5.3.7 Case study

During road reinstatement following landslide damage in Ethiopia, consideration was given to the use of three below-road wall types: masonry, gabion and reinforced fill. In all three cases, the following criteria needed to be satisfied:

- the wall foundation should be beneath the landslide failure surface; and
- bearing pressures should be minimized where foundations were to be formed in the more clayey subgrades (by minimizing wall heights and distributing bearing pressures as evenly as possible across the foundation).

Figure C5.13 illustrates the issues that were taken into account when deciding upon wall type and configuration. Figure C5.14 shows examples of the finally constructed walls; in almost all cases, they have performed well. Some minor movements have been experienced in one of the reinforced fill walls, probably due to localized differential settlement of the foundation.

C5.4 Selection of wall cross-section

Overturning and bearing capacity considerations will often dictate the wall height to base width ratio. However, in the case of walls founded on clayey soils (particularly with steep slopes or large surcharge loads behind them), the base width will be governed by sliding resistance. In respect of masonry, gabion, mass concrete and crib walls there are two other variables to take into consideration:

- the inclination of the front and rear faces; and
- the inclination of the base.

Each of these variables has advantages and disadvantages, which are summarized in Table C5.4.

The resistance to sliding for reinforced concrete walls can sometimes be improved by the introduction of a shear key at the base of the wall as shown in Figure C5.2. This may also permit the width of the wall to be reduced, an important factor when the wall excavation width is critical. In general a shear key is preferably located mid-point in the heel, since a key at the rear of the heel will effectively increase the required excavation depth. The depth of the shear key below the base of the wall should not be less than 0.5 m nor greater than half the width of the base. The shear key may do nothing to improve sliding resistance if, as the shear key is deepened, the active pressures increase more rapidly than the passive pressures and shear resistance of the ground increases.

C5.5 Wall backfill and drainage

C5.5.1 Backfill

The preferred backfill placed and compacted behind a gravity wall is a free-draining, well graded, durable material of high shear strength which is free from any harmful matter. Backfill should not contain:

- peat, vegetation, timber, organic or other degradable material;
- synthetic or combustible material;
- material subject to significant volume change;
- soluble or chemically aggressive material; or
- single-sized material, as compaction will be difficult to achieve.

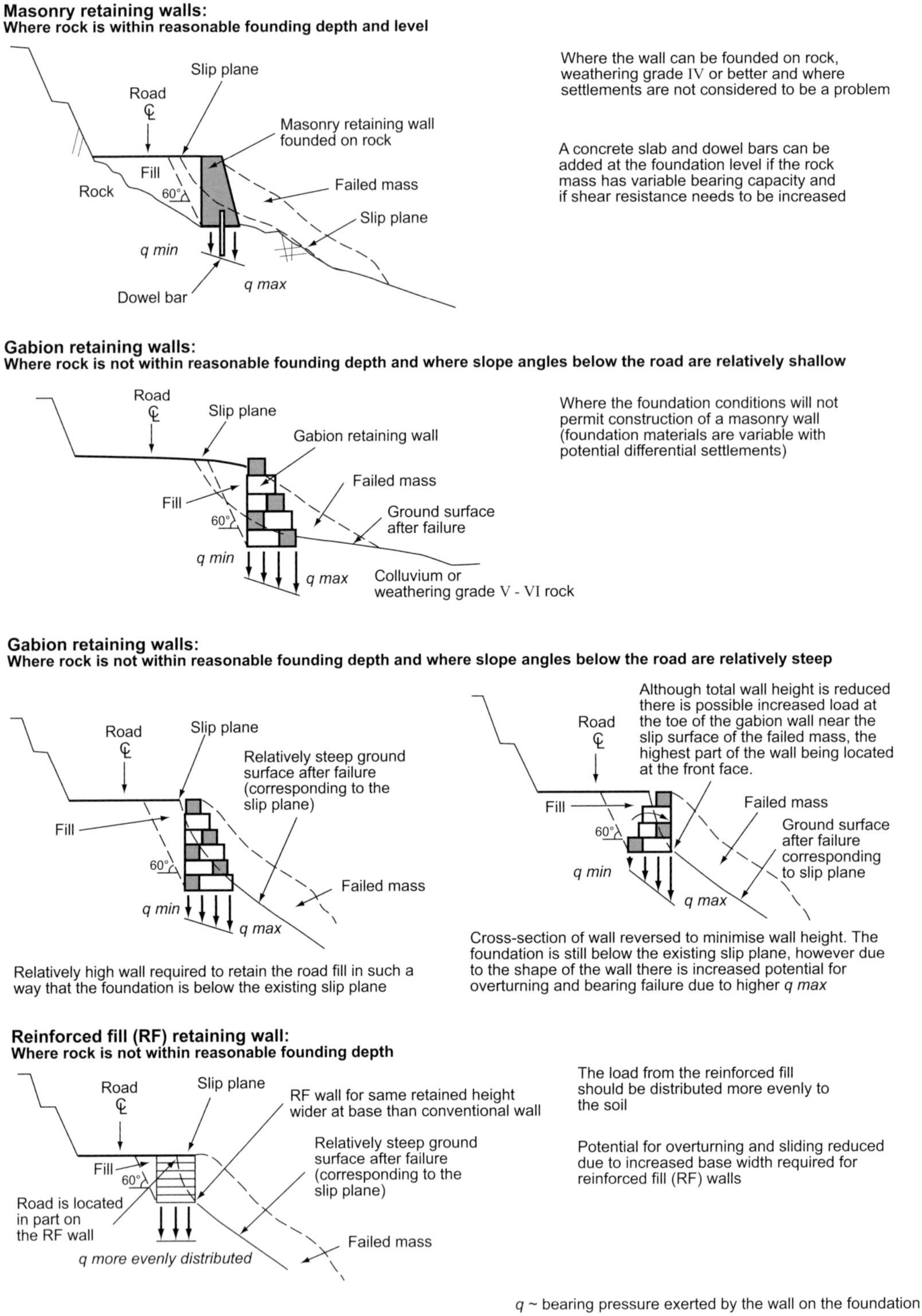

Fig. C5.13. Options for below-road retaining walls and road reinstatement for a road project in Ethiopia.

Fig. C5.14. Illustrations of the three main wall types constructed following selection and design from Figure C5.13.

Clay is not generally recommended as backfill due to its low friction angle and potential problems of swelling, shrinkage and long term consolidation.

Backfill should be compacted in layers not normally exceeding 150 mm. The use of heavy compaction equipment immediately adjacent to a wall is not recommended due to the potential to over-stress it. A hand-guided vibrating plate compactor or pedestrian roller is usually adequate for this purpose.

C5.5.2 Drainage

Weepholes for gabion and crib walls are unnecessary since the materials forming the walls are permeable. Weepholes are usually provided as standard detail, however, for mortared masonry and mass and reinforced concrete walls. Weepholes usually comprise 75 mm diameter uPVC pipes laid across the width of the wall and placed at 2.0 m intervals vertically and horizontally. A filter fabric 'sock'

Table C5.4. *Wall shape advantages and disadvantages*

Shape & Location		*Advantages*	*Disadvantages*
1		Lower toe pressure and greater resistance to overturning compared to 2.	Greater wall height for a fixed retained slope angle compared to 2.
2		Smaller wall height for a fixed retained slope angle compared to 1.	Lower resistance to overturning and higher toe pressure compared to 1.
3	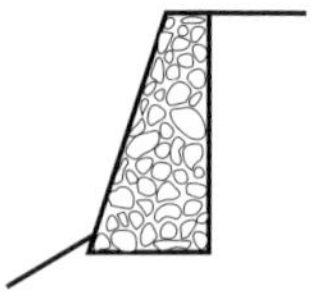	Lower toe pressure and greater resistance to overturning compared to 4.	Greater wall height for fixed height of retained soil compared to 4.
4	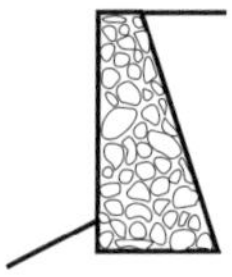	Smaller wall height for fixed height of retained soil compared to 3.	Higher toe pressure and greater extent of excavation into road (more disruption to traffic) compared to 3.
5	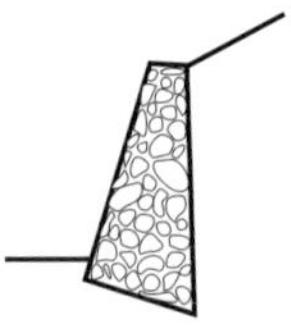	Greater resistance to sliding compared to 1 or 2.	Tilted shape not suitable for gabion construction, increased volume of excavation, positive drainage required at heel to prevent ponding and foundation softening.
6	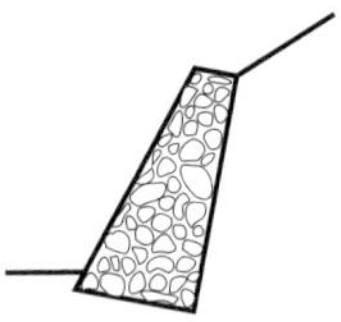	Greater resistance to overturning and ground bearing pressures evened out, compared to 5.	Tilted shape more difficult to construct in gabion, increased volume of excavation, compaction behind wall more difficult, positive drainage required at heel to prevent ponding and foundation softening.

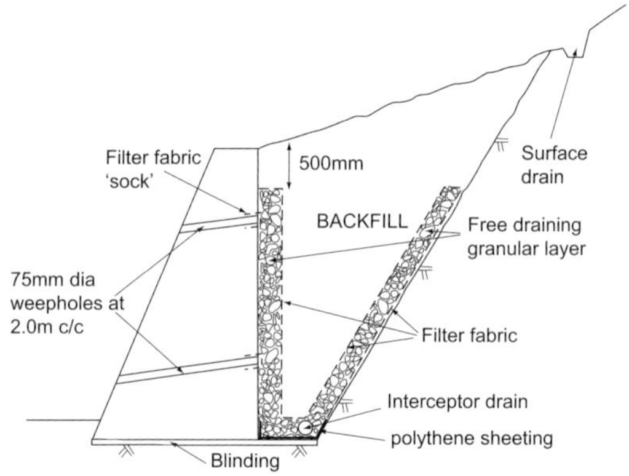

Fig. C5.15. Typical drainage details for retaining walls.

placed over the inlet end will prevent the migration of fines through the weephole. Where uPVC pipe is unavailable, other pipe materials such as bamboo culm are sometimes used.

If the backfill is not free-draining, or if there is a potential for significant build up of water behind the wall due to seepage or a high groundwater table for example, additional positive drainage measures may be necessary. These might comprise:

- a backfill drain (see Fig. C5.15);
- a perforated interceptor drain at the heel of the wall connected to a suitable outlet (see Figs C5.15 & C5.16);
- drainage grips below the base of the wall (see Fig. C5.17); and
- surface drains on the slope above the wall if significant water flow into the backfill is anticipated (see Fig. C5.15).

The use of these positive drainage measures will help prevent the build-up of hydrostatic pressure acting on the

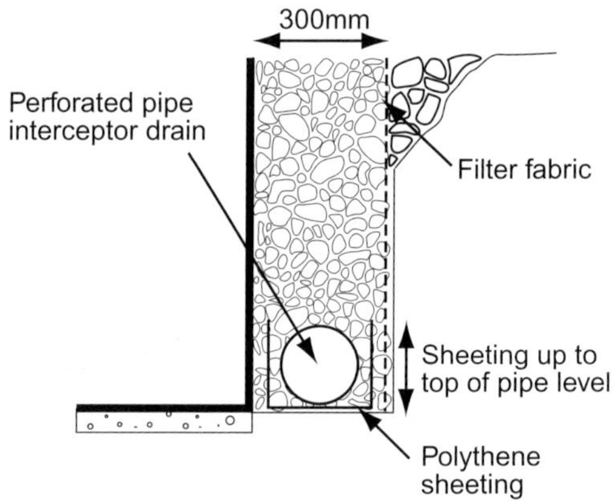

Fig. C5.16. Interceptor drain detail.

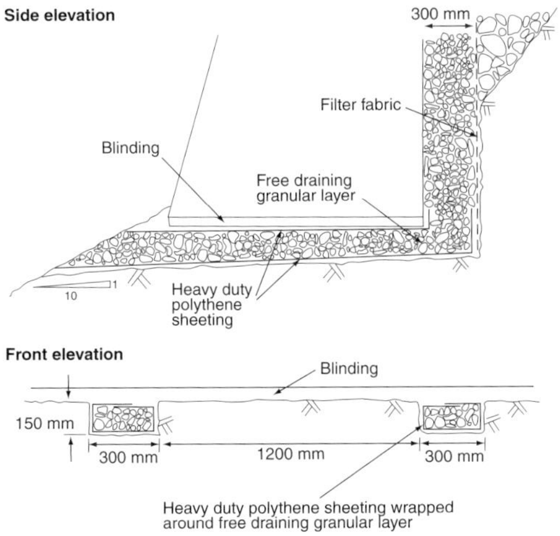

Fig. C5.17. Drainage grip detail.

rear of the wall and upwards on the base, as well as reducing potential softening of the foundation.

C5.5.2.1 Backfill drain

In order to facilitate drainage behind the wall, a free-draining granular layer located immediately adjacent to the rear wall face is strongly recommended. This layer comprises graded unweathered gravel, usually of a minimum width of 300 mm and separated from the backfill by a filter fabric. During backfilling operations the backfill and the graded gravel are placed and compacted in lifts, such that the backfill provides lateral support to the gravel. If seepages are apparent or expected in the base or sides of the retaining wall excavation, the granular layer is often extended along the base of the wall excavation and up the rear face as shown on Figure C5.15. The backfill layer is normally terminated within a minimum of 500 mm from the final road or ground level, and covered with compacted impermeable material to prevent surface water infiltration. For gabion walls, crib walls and composite masonry walls, this drainage layer is usually omitted. However, it is recommended that a filter fabric be placed on the rear face of these walls prior to backfilling in order to minimize the potential for fine material to migrate through the structure.

C5.5.2.2 Interceptor drain

An interceptor drain can be used to help discharge water from the base of the free-draining granular layer. This is especially recommended where the lengthways hydraulic gradient behind the wall is low and where the wall is constructed with an inward-sloping base (Cases 5 and 6 in Table C5.4). The interceptor drain should run the length of the wall and usually comprises a 150 mm diameter uPVC pipe with perforations through the upper half. To ensure that the water enters the pipe, the base of the drainage layer is lined with heavy-duty polythene sheeting that extends up both sides of the excavation (see Fig. C5.16). The pipe should be laid at a minimum gradient of 2% and connected to suitable outlets where the discharge of water will not erode the foundation of the wall.

C5.5.2.3 Drainage grips

Drainage grips can often be used as an alternative to an interceptor drain running the length of the wall, although they are not appropriate for walls with inclined bases or shear keys. Drainage grips are slots, roughly 300 mm wide by 150 mm deep, excavated across the wall foundation with a slight gradient towards the front of the wall. They are then lined with heavy-duty polythene and filled with free-draining granular material (Fig. C5.17). The polythene is then folded on top of the gravel and a concrete blinding layer applied to the top of the folded polythene, thus forming the base of the wall construction. For above-road retaining walls it is usual to discharge these drainage grips via a uPVC pipe into the adjacent side drain. However, drainage grips may be impracticable in many above-road wall locations because the base of the wall may be below the invert level of the side drain. For below-road retaining walls, drainage grips are normally extended via a gravel-filled trench for a sufficient distance downslope in order to facilitate drainage of the ground in front of the wall.

C5.6 Retaining wall construction

C5.6.1 Access for wall construction

The provision of access for wall construction can create problems. Machine access to upper and lower retaining wall positions (Fig. C5.1) at intermediate levels needs to be preplanned with care to minimize ground disturbance that might cause instability.

C5.6.2 Wall heights, founding levels and foundation stability

The designed height of a gravity retaining wall will inevitably have included consideration of a founding level that was either proven from ground investigation (Section B4.3) or assumed from the topography and local exposures of soil and rock. Even when this founding level is based on ground investigation, it still remains indicative until the foundations have been excavated and checked. In addition, particularly in steep terrain, it is unrealistic to assume that the founding conditions will remain the same throughout the width and length of a proposed wall. The excavation should therefore be inspected not only before any wall construction takes place, but also during the course of the excavation itself. A satisfactory founding level may well be reached at a shallower depth than anticipated, reducing both the cost of the wall and the potential for instability to occur in the sides of the excavation. The converse can also apply, whereby the required founding depth proves to be

Table C5.5. *Typical presumed/allowable bearing pressures (modified from BSI 1986)*

Type of material	*Consistency*	*Typical allowable bearing pressure (kN/m²)*
Sandy/silty clay	Soft	50
	Firm	100
	Stiff	150
	Very stiff/hard	300
Silty/clayey sand	Loose	100
	Medium to dense	200
	Very dense	300
Sand/gravelly sand	Loose	200
	Medium to compact	300
	Very compact	400
Sandy gravel	Loose	400
	Medium to dense	600
	Very dense	800
Weak/weathered rock	Weak	1000–4000
Strong/fresh rock	Strong	4000–10 000

greater than that anticipated in the design. Unfortunately, there is often an adherence to what the designer has assumed rather than to the actual soil and rock conditions that become exposed on site during construction. It is strongly recommended, therefore, that:

- construction supervision staff have sufficient engineering geological knowledge to make these observations, and
- there is flexibility in the construction contract to allow design modifications to be made (see Section A2.2 for further discussion).

If a below-road retaining wall is required to cross a low point in the hillside, it will often be appropriate to step-up the foundation towards each end rather than constructing the wall entirely at the maximum height. Gabion and reinforced concrete walls are usually less adaptable to varying depths of foundation than masonry walls, although this can sometimes be remedied by the use of mass concrete stepped foundations upon which the walls are then constructed.

Where adverse bedding, other jointing or sloping rock-head is encountered, it may be appropriate to grout vertical steel dowels (minimum 25 mm diameter) into the foundation in order to secure the wall into the deeper rock mass. The dowels then project into the base of the retaining wall.

The designer of a gravity retaining wall will determine the bearing pressures exerted on the foundation (e.g. Bowles 1996). The confirmation of a satisfactory founding level will need to take into account these design bearing pressures. Allowable bearing pressures for granular soils can be approximated in advance of any field investigations by reference to published values (e.g. Simons & Menzies 1976; BSI 1986). Table C5.5 shows typical allowable bearing pressures for some materials likely to be encountered during foundation excavations for retaining walls. Published or presumed values should not be used for design without site specific verification. Complex soil profiles with weak horizons occurring at depth, or where previous made ground and construction spoil is present, will require careful assessment based on field investigations. For example, during excavation for the replacement of storm-damaged walls in Nepal and the Philippines, it was observed that many had been founded on boulder debris or on uncompacted excavated spoil material placed on sloping ground. In most cases these founding levels were little more than a metre (and frequently less) above *in situ* rock. The original shallow founding levels had been dictated by a reliance on hand excavation and, perhaps, a greater importance placed on achieving adequate bearing capacities than ensuring foundation stability against deeper slope failures.

Fig. C5.18. Dynamic cone penetrometer (DCP).

Table C5.6. *Allowable bearing pressures and the equivalent DCP blow count*

No of blows for 300 mm penetration	*Equivalent mm per blow*	*Allowable bearing pressure (kN/m²)*	
		2 m width foundation	*4 m width foundation*
5	60	90	70
10	30	140	100
20	15	200	160
30	10	270	220
40	7.5	340	290
50	6	400	350

A field assessment of allowable bearing pressures can be made by carrying out probing at the base of the excavation. This method is also useful in identifying soft spots. A typical probe for this purpose is the dynamic cone penetrometer (DCP, Fig. C5.18). Table C5.6 shows a simplified correlation between the blow count and allowable bearing pressures. However, the DCP is of limited use in stony or bouldery soils (e.g. taluvium); results should always be interpreted with caution.

Generally, a nominal 50 mm thick concrete blinding layer is placed on top of the excavated foundation surface before commencing wall construction. This serves three purposes:

- it acts as a levelling course;
- it prevents contamination of the base of the structure during construction; and
- it prevents softening if the foundation would otherwise be left exposed during construction.

Sometimes, particularly for masonry walls, a 200 mm thick concrete footing is also provided. In variable founding conditions this is often reinforced to provide additional load spreading and bridging of soft spots. Shear keys (upstanding blocks) can be constructed in the top of the footing to provide additional resistance to sliding.

For below-road walls, the top of the wall is generally recommended to be at least 50 mm higher than finished road level to prevent road surface runoff from running down the front face of the wall and eroding the soil in front of the foundation. The exception is where the wall forms part of a culvert outlet structure (headwall) and the runoff can be directed down the face of the wall into the outlet.

C5.6.3 Wall length

The required length of wall will often need to be modified from that shown on the design drawings as the excavation proceeds. In Figure C5.19 the wall should have been extended during construction to fit the excavated length. A similar situation occurred during construction of the wall in Figure C5.20 where, to remedy the situation, the structure has been given an angled return into the hillside to retain the fill at the end of the wall.

C5.6.4 Excavation stability

The stability of the sides of an excavation requires careful attention, particularly with respect to workforce safety and for the safety of the public where excavations take place alongside an existing road. There are clear stability benefits to:

- excavating a wall foundation in short lengths (2–10 m depending on circumstances) at a time;
- constructing the wall and backfilling before excavating the next length;

Fig. C5.19. Wall constructed too short.

Fig. C5.20. Angled return into hillside.

- minimizing the time that the excavation remains open; and
- avoiding construction during the wet season.

Alternatively, for long retaining walls, the excavation can be opened up in discrete lengths; intervening ground can initially be left untouched to act as buttresses to any upslope instability until the initial wall sections have been constructed and backfilled.

Fig. C5.21. Unsupported temporary excavation in taluvium for an above-road mid slope retaining wall.

Fig. C5.22. Steep temporary excavation for a below-road retaining wall showing the made ground/original ground interface.

Figures C5.21 and C5.22 illustrate excavations for above-road and below-road retaining walls, respectively, where the precautions outlined above were not taken and a potentially unstable situation was created in each case.

C5.6.5 Construction quality control

Quality control is important to ensure that the completed structure is durable and meets the requirements of the

Fig. C5.23. *In situ* density testing of backfill.

Fig. C5.24. Examples of good (on the left) and poor (on the right) mortared masonry.

specifications. Some of the more important quality aspects are described below.

C5.6.5.1 Backfill

For all wall types, sieve analyses are carried out on both the wall backfill and the backfill drain (free-draining granular layer) to check for compliance with the specification for the grading envelope. Quality control compaction tests are normally carried out for every 1000 m^3 of fill placed (Fig. C5.23). Wesley (2008) discusses compaction methods and quality control for soils of high variability and soils that soften on compaction. The review is especially relevant to residual soils.

C5.6.5.2 Mortared masonry walls

For mortared masonry walls it is important to ensure that all voids are filled with mortar. There is sometimes a tendency for contractors to use the minimum amount of water since this reduces the need for its collection and storage and increases the stiffness of the mortar. However, this makes it more likely that voids will not be completely filled, and may prevent the mortar from setting properly, thus potentially reducing the long-term durability of the wall. The mortar should meet the strength required in the specification (a minimum compressive strength of 17.5 N/mm^2 is often specified). The stones used in the masonry should be of the required size and strength, free from fissures and should be placed with maximum interlock and minimum voids (Fig. C5.24). For dry stone walls it is important to ensure that the quality of the masonry work is such that the volume of voids is minimized. This will usually require that the stones are dressed and laid in brickwork style (Fig. C5.25).

Where castellations (confidence blocks) are required on top of a below-road wall, these should be constructed at the same time and keyed into the wall structure using steel dowels (minimum 25 mm diameter) to provide a greater resistance to vehicle impact.

C5.6.5.3 Gabion walls

For gabion walls, the manufacturer's specifications for mesh size, galvanizing, wire diameter, panel frames, basket connectors and the twisted connections (usually minimum three half turns) need to be adhered to. The wire baskets can either be imported to site as prefabricated units or site-fabricated by hand. Usually, stones should be of an even size that is at least double the mesh size and be of good rock quality. There are large cost savings to be made by contractors in using substandard locally available weathered rock to fill the baskets. This may lead to significant local deformations of the structure under load, putting at risk the performance of the structure itself. Rounded river stone should preferably be excluded but, where necessary, limited to one third of the total stone in any one basket (the remainder being dressed; Fig. C5.26). Long flat stones should be orientated from front to back. Baskets should preferably be filled by hand, arranging stones in dry masonry fashion, with wire cross-ties within each basket.

Fig. C5.25. Dry stone retaining wall in excellent condition after 40 years.

Fig. C5.26. Examples of good (on the left) and poor (on the right) stone filling in gabions.

Fig. C5.27. Masonry buttressing to an above-road gabion wall supporting a failed slope.

Fig. C5.28. Masonry buttressing to a below-road fill retaining wall.

Where gabion baskets are to be used as castellations they need to be well secured with tie-wire to the gabions below.

Successive rows of gabion baskets are preferably staggered, brickwork style, with some baskets placed lengthways from front to back. This will make the wall more integral as a structure and will reduce undue flexing or bulging.

C5.6.5.4 Reinforced concrete walls

For reinforced concrete walls it is important to ensure that:

- the concrete and steel reinforcement meet the specified requirements and adequate cover is maintained;
- the consistency of the concrete is such that it flows freely under the action of a vibrating poker, it meets the minimum specified strength and does not 'bleed' or segregate; and
- the falsework is adequately constrained so that it does not deflect when filled with concrete.

C5.6.6 Improvements to the stability of existing walls

Where existing walls are showing signs of distress and deformation, it may be necessary to completely reconstruct them to a new design. Alternatively, it may be possible to reduce the driving forces behind the wall (e.g. by the provision of additional drainage or by reducing the slope angle) or to increase the resisting forces in front of the wall (such as by the use of buttressing). The practicability of achieving the latter will depend on space and founding conditions. Figures C5.27 and C5.28 illustrate masonry buttresses used to support retaining walls in above-road and below-road situations, respectively. In both cases, and especially the latter, reconstruction of the original wall would have resulted in closure of the road for a considerable period of time with major disruption to traffic.

References

AASHTO 2007. *Load and Resistance Factor Design (LRFD) Bridge Design Specifications*. SI Units 4th edn. American Association of State Highway and Transportation Officials, Washington DC.

BSI 1986. British Standard Code of Practice for Foundations. BSI 8004, (Formerly CP 2004). British Standards Institution.

BSI 1994. Code of Practice for Earth Retaining Walls. BSI 8002:1994, British Standards Institution.

BSI (EN) 2004. Eurocode 7: Geotechnical Design (Part 1 General Rules). British Standards Institution EN ISO BSI 1997-1: 2004.

BSI (EN) 2006. Execution of special geotechnical works. Reinforced fill. BSI EN 14475: 2006. British Standards Institution.

BSI 2009. Code of Practice for Earthworks. 09/30093255 DC. British Standards Institution, United Kingdom.

Bowles, J. E. 1996. *Foundation Analysis and Design*. 5th edn. McGraw Hill, USA.

CIRIA 2000. Modular gravity retaining walls – Design guidance. Construction Industry Research and Information Association Report **C516**, London.

CIRIA 2005. *Soil Nailing – Best Practice Guidance*. Construction Industry Research and Information Association, Report No. **C637**, London.

Das, B. M. 2007. *Principles of Foundation Engineering*. 6th edn. Thomson, Nelson, Canada.

DEPARTMENT OF TRANSPORT 1994. Design Methods for the Reinforcement of Highway Slopes by Reinforced Soil and Soil Nailing Techniques. Design Manual for Roads and Bridges, HA68/94, Part 4.

FHWA. 1994. Applications Guide for Launched Soil Nails. Volumes I and II. FHWA-FPL-93-003.

FHWA. 2003. Soil Nail Walls. Geotechnical Engineering Circular No 7. Report No FHWA0-IF-**03-017**. Prepared by Lazarte, C. A., Elias, V., Espinoza, R. D., & Sabatini, P. J. Washington, DC.

GEO 1993. *Guide to Retaining Wall Design*. Geoguide 1. 2nd edn. Geotechnical Engineering Office, Civil Engineering and Development Department, the Government of the Hong Kong Special Administrative Region.

GEO 2004. *Design of Soil Nail Heads*. Technical Guidance Note No. 21 (TGN 21), Geotechnical Engineering Office, Civil Engineering and Development Department, the Government of the Hong Kong Special Administrative Region.

GEO 2008. *Guide to Soil Nail Design and Construction*. Geoguide 7. Geotechnical Engineering Office, Civil Engineering and Development Department, the Government of the Hong Kong Special Administrative Region.

Giri, D. & Sengupta, A. 2009. Dynamic behaviour of small scale nailed soil slopes. *Geotechnical and Geological Engineering*, **27**, 687–698.

Javanmard, M. & Ahmadi, M. 2009. Numerical analysis of soil nail walls. *International Symposium on Geotechnical Engineering, Ground Improvement and Geosynthetics for Sustainable Mitigation and Adaption to Climate Change, Including Global Warming*, 3–4 December 2009, 361–371.

Kulkarni, T. 2009. Rock fall and debris flow mitigation techniques. *International Symposium on Geotechnical Engineering, Ground Improvement and Geosynthetics for Sustainable Mitigation and Adaption to Climate Change, Including Global Warming*, 3–4 December 2009, Bangkok, Thailand, 129–135.

Lamb, T. W. & Whitman, R. V. 1979. *Soil Mechanics*. John Wiley and Sons, New York.

NAVFAC. 1986. *Foundation and Earth Structures*. Design Manual 7.02. Naval Facilities Engineering Command, Washington DC, 20402.

Pun, W. K. & Shiu, Y. K. 2007. Design practice and technical developments of soil nailing in Hong Kong. *Geotechnical Advancements in Hong Kong since 1970's. Proceedings of the HKIE Geotechnical Division*, 27th Annual Seminar. The Hong Kong Institution of Engineers, Geotechnical Division, Hong Kong, 197–212.

Pun, W. K. & Urciulli, G. 2008. Soil nailing and subsurface drainage for slope stabilisation. *In*: Chen, Z., Zhang, J.-M., Ho, K., Wu, F.-Q. & Li, Z.-K. (eds) *Landslides and Engineered Slopes. From the Past to the Future. Proceedings of the 10th International Symposium on Landslides and Engineered Slopes, 30 June–4 July 2008*, Xi'an, China, Taylor & Francis Group, London, 85–126.

Ramasamy, S. 2009. Construction of 25 m high mechanically stabilized earth structure using Terramesh and Geogrid. *International Symposium on Geotechnical Engineering, Ground Improvement and Geosynthetics for Sustainable Mitigation and Adaption to Climate Change, Including Global Warming*, 3–4 December 2009, Bangkok, Thailand, 253–263.

Simons, N. & Menzies, B. 1976. *A Short Course in Foundation Engineering*. IPC Science and Technology Press, Richmond, Surrey.

TRRL. 1993. *Development of Specifications for Soil Nailing*. Transport and Road Research Laboratory, Research Report TRRL **380**. Crowthorne, Berkshire.

Wesley, L. D. 2008. A simple compaction control method for slope construction. *In*: Chen, Z., Zhang, J.-M., Ho, K., Wu, F.-Q. & Li, Z.-K. (eds) *Landslides and Engineered Slopes. From the Past to the Future. Proceedings of the 10th International Symposium on Landslides and Engineered Slopes, 30 June–4 July 2008*, Xi'an, China, Taylor & Francis Group, London, 1005–1010.

C6 Slope and Road Drainage

G. J. Hearn* & T. Hunt

URS Scott Wilson Ltd, Scott House, Alençon Link, Basingstoke, Hampshire RG21 7PP, UK
*Corresponding author (e-mail: gareth.hearn@scottwilson.com; garethhearn@talktalk.net)

C6.1 The importance of drainage

Good drainage is a key element in the satisfactory functioning of all roads. In hilly and mountainous areas especially, drainage management is critical to the performance and stability of a road and its structures. Within very short periods of time intense rainfall can lead to landslides and extensive erosion on slopes, while sediment-laden storm runoff can cause blockage to drainage structures and scour of foundations and fill slopes. The control of surface water must therefore be uppermost in the minds of those responsible for the design, construction and maintenance of mountain roads.

C6.2 Slope drainage

Slope drainage can be applied both as a routine precautionary measure to improve stability and reduce the incidence of landslides and erosion, and as a reactionary measure to assist in the stabilization of slopes that have already failed. Figures C6.1 and C6.2 and Table C6.1 outline the typical methods of slope drainage found on mountain roads. As a general point, it is recommended that every opportunity is taken to improve the natural drainage of slopes adjacent to roads. This should normally include:

- excavating springs and channelling flows to a safe drainage point;
- ensuring that irrigation systems (e.g. for paddy terraces) and road drainage requirements are as compatible as possible; and
- draining water ponding areas where this has no adverse environmental effects.

In order to maximize the efficiency of surface drainage the following guidelines are recommended:

- design and construct drains so that they have the steepest possible gradient, providing strengthening and scour checks as necessary; and
- use the natural topography (rather than a rigid geometrical design) to maximize the interception of water.

C6.2.1 Routine slope drainage

The main function of drainage in this context is to control surface runoff. A drainage 'cut-off' or catch-water drain is the most common form of routine slope drainage and is often constructed as a Type 1 surface drain (Fig. C6.2) above the crest of cut slopes. However, drain blockage, under-scour from seepage erosion and outflanking by unintercepted surface runoff can render these drains ineffective. This can lead to concentrated runoff and erosion, or slope instability, if not rectified (Text box C6.1). In some instances, such drains may perform the role of a ready-made tension crack allowing the ingress of water into a slope.

The conclusion to be drawn is that the use of cut-off drainage above the crest of a cut slope should only be used where:

- there is a confirmed source of surface water that would otherwise cause the cut slope to become unstable or to erode if it were not safely redirected; and
- there is a long-term commitment to slope drain maintenance by debris clearance and repairs.

On benched cut slopes it is usual practice to provide a surface drainage system (Section C2.3.1); this too can suffer from the problems described in Text box C6.1 if not adequately maintained, as illustrated in Figure C6.3.

Open drains are sometimes used on bermed fill slopes, but these can also be susceptible to erosion and failure. They rely on adequate compaction of the fill slope to minimize settlement and seepage erosion and maximize stability. Regular maintenance is also required once they are operational to ensure they do not become blocked. If their design, construction or maintenance is ineffective, they can quickly become undermined (Fig. C6.4).

C6.2.2 Drainage to stabilize failed slopes

A brief description of each of the slope drainage measures summarized in Figure C6.2 and Table C6.1 is given in the sections that follow in the specific context of slope stabilization. Their main function in this context is to lower groundwater levels within failed material. Additional information

From: Hearn, G. J. (ed.) *Slope Engineering for Mountain Roads*. Geological Society, London, Engineering Geology Special Publications, **24**, 231–242.
DOI: 10.1144/EGSP24.15 0267-9914/11/$15.00

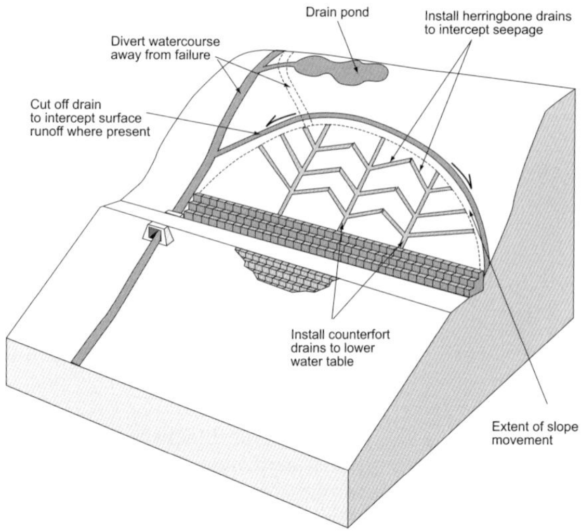

Fig. C6.1. Typical slope drainage measures.

can be found, for example, in Forrester (2001) for subsurface drainage systems and Lau *et al.* (2008) with respect to slope drainage practice in Hong Kong.

C6.2.2.1 Surface drains

Cut-off drains are discussed above with respect to routine drainage measures, and the same issues apply to their use in slope stabilization. Surface drains may prove useful in draining permanent locations of spring seepage on a slope face or in conveying water from a source above an unstable slope to a point at its toe. Surface drains constructed across landslides will however be susceptible to any ongoing ground movements, and where they crack or become dislocated as a result they will almost certainly make matters worse by discharging water into the landslide mass. Articulated drainage structures can provide a solution where the anticipated ground movements are relatively small. The illustration in Figure C6.5 is from Hong Kong where a large articulated drain was designed to convey stream water that drained into the head of an unstable slope to a discharge point at its toe. The drain comprises pre-cast overlapping and interlocking units that are able to tolerate minor ground movements and still function as required.

C6.2.2.2 Herringbone or chevron drains

These drains are constructed herringbone fashion (Figs C6.1, C6.2 & C6.6) on slope faces to collect surface runoff and near-surface groundwater. They are usually constructed to a depth of 0.5–1.5 m (although deeper drains are not uncommon where a greater degree of groundwater lowering is required), and are intended to remain functional even if minor slope movements were to occur. They comprise a central collector trench drain and offset diagonal feeder or branch drains. The upslope face of the diagonal drains should be lined with a filter fabric and the lower face and invert lined with heavy-duty polythene. The drain itself is filled with free-draining granular fill. Care needs to be taken to ensure that the construction of the drain does not lead to further instability. Consequently, diagonal drains are often constructed as surface water interceptor drains. In the event of large anticipated flows, a perforated high-density polyethylene pipe can be installed in the base of the collector drain. Greatest drainage efficiency will usually be achieved by maximizing the number of collector drains, thus reducing the required length of the diagonal drains.

C6.2.2.3 Counterfort drains

These drains are designed to provide deeper drainage and groundwater drawdown and, where constructed to sufficient depth, break up the continuity of the failure surface and provide some frictional buttressing against further movement (Fig. C6.2). These drains are usually constructed along the line of maximum slope and are often dug to a depth of 3 m or more at intervals of 3–10 m depending on the permeability of the subsoil. Ideally, the sides should be lined with a filter fabric and the invert with heavy-duty polythene. A perforated high-density polyethylene pipe is usually installed in the base of the drain if flows are expected to be large.

C6.2.2.4 Sub-horizontal (drilled) drains

Sub-horizontal drains (Figs C6.2 & C6.7) are used to intercept groundwater at depth (e.g. GCO 1985). They require the use of specialist drilling and installation equipment that may not always be available, and they are not easy to install. The drains usually comprise polyethylene pipes, of minimum diameter 40 mm and up to 40 m long, installed in a fan-shaped pattern of pre-drilled holes inclined 5° upwards (maximum 10°). The pipes are perforated and wrapped in a filter fabric to reduce the likelihood of clogging. In theory, if not always in practice, they should be capable of being flushed with water. This type of drain is costly to install and not always successful unless the drain is able to intercept seepage horizons and other flow lines at depth. Additionally, the drains are only able to accommodate very minor continuing slope movements. Although there are a number of sites where such drains have performed very successfully, they are not generally recommended for use on low-cost roads (except in conjunction with other measures at major landslide sites, where they offer the potential for decreasing pore-water pressures at depth).

C6.2.2.5 Toe drains

In many situations the drainage of a slope can be improved by increasing the depth of drainage in the roadside drain at the toe. A trench drain containing a perforated pipe and a gravel backfill is constructed beneath the lined roadside drain (Fig. C6.8) and the perforated pipe discharged into the inlet of the next culvert down-gradient. Where the topography does not allow this, it may be necessary to discharge the water by means of a relief drain across the road. The cost and disruption associated with the construction of these drains can be significant if they are installed as an afterthought, or in response to ground movements and groundwater levels that do not become apparent until after

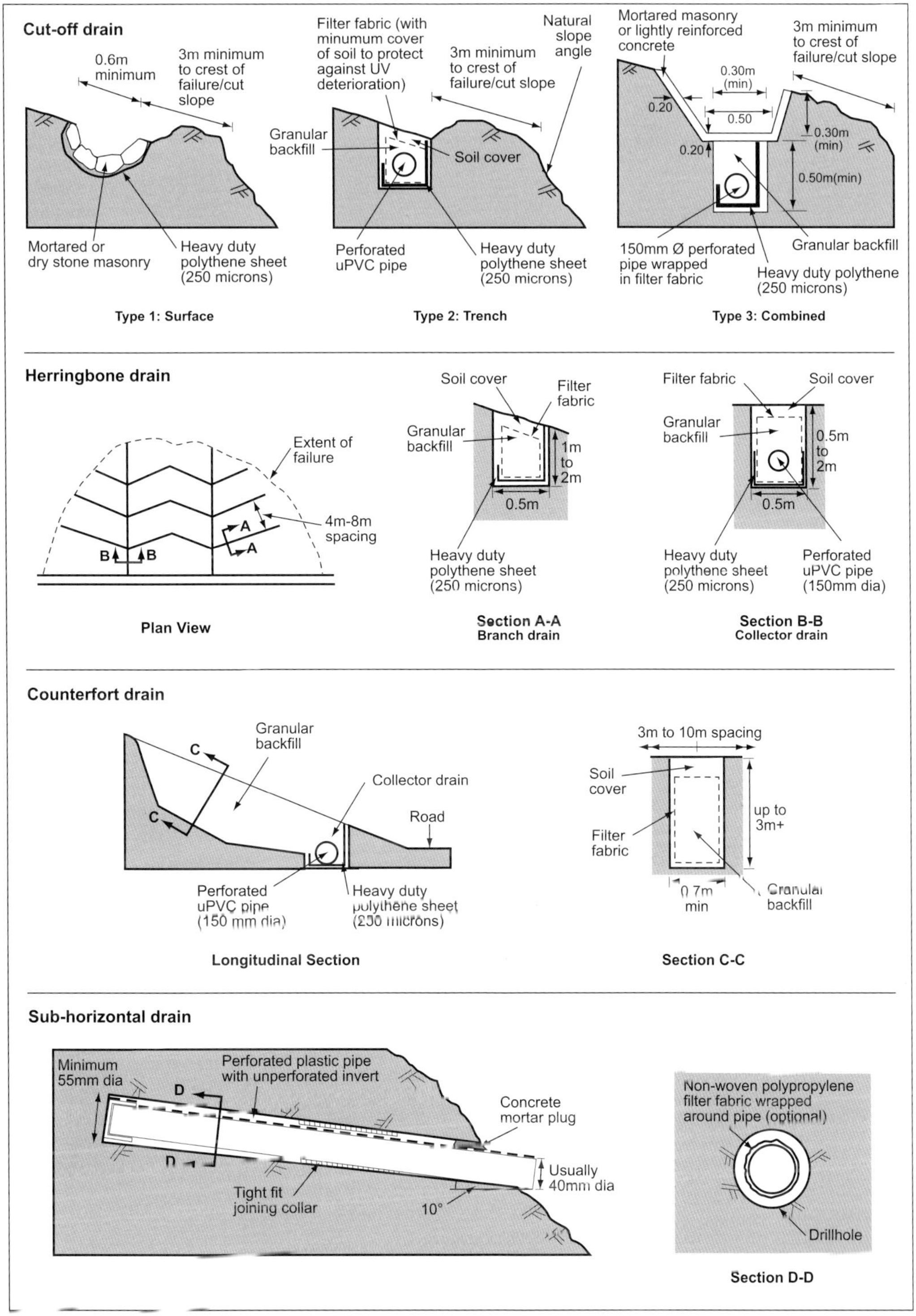

Fig. C6.2. Typical slope drainage details.

Table C6.1. *Types of slope drainage, their function and limitations*

Function	*Type*	*Advantage*	*Limitation*
Interception of surface runoff	Unlined cut-off drain (open ditch)	Cheap, can in some instances prevent ingress of surface runoff into landslide masses	Prone to leakage and erosion; may act as incipient tension crack beyond slope crest; requires frequent inspection for damage/blockage; access may be difficult for maintenance
	Lined cut-off drain (type 1 on Fig C6.2)	As above, though less prone to erosion and leakage	Requires frequent inspection for damage/blockage; access may be difficult for maintenance; concentrates flow and erosion if ruptured by ground movement
	Sub-soil drain (type 2 on Fig C6.2)	Usually not prone to erosion and leakage and can tolerate some ground movements while continuing to function	May become clogged with silt; can be surcharged during large surface flows; may encourage water to enter the slope where excessive ground movements create 'sags' in vertical alignment or tears in the polythene; access may be difficult for maintenance
	Lined cut-off drain with subsoil drain (type 3 on Fig C6.2)	Combines surface and subsurface drainage. Can accommodate large surface flows	Requires frequent inspection for damage/blockage; may encourage water to enter the slope where excessive ground movements create 'sags' in vertical alignment or tears in the polythene; access may be difficult for maintenance
	Bench drain on cut slopes Berm drain on fill slopes	Collects and discharges surface runoff from one bench or berm to the next. Reduces the tendency for large quantities of water to pond and seep into the slope material	Will crack and dislocate following any ground movements, may become blocked by falling debris or silt if not properly maintained
Reduction of shallow sub-surface water and drainage of seepages	Herringbone drain	Depending on depth, usually able to intercept water up to 1.5m below slope face; can be used to drain seepage areas; can accommodate some slope movement; can be used to help stabilise shallow slope failures up to 2m deep	May have very limited effect on overall stability of deep-seated failures; may create shallow instability during construction, hence preference to minimise branch length
	Counterfort or trench drain	Generally, able to intercept water up to 3m depth below slope face; can act as a 'buttress' if base is below slip surface	Usually needs to be machine dug; difficult to construct in boulder material
Interception of deep water table	Drilled sub-horizontal drain	Only feasible method of intercepting groundwater at depth	Relatively high cost; drilling equipment required; may not always be successful

Text box C6.1. Experience with cut-off drains in Nepal

Cut slopes formed along mountain roads constructed in the 1970s and 1980s in Nepal were often provided with cut-off drains located above the slope crest as standard practice. These were predominantly lined with mortared masonry (Type 1 on Fig. C6.2) and were intended to intercept surface runoff before it was able to flow over the cut slope and cause erosion or shallow slope failure. Where some of these drains had been excavated into residual soils, they were quickly undermined by seepage erosion caused by shallow subsurface flow in the saturated soils above rock head level. This was most common in the silty, sandy soils derived from *in situ* weathered high-grade metamorphic rocks. In the more clayey residual soils, the drains tended to be less susceptible to this mode of failure. Where cut-off drains intercepted surface runoff in the manner intended, they often became blocked by sediment and vegetation debris and thus ceased to function due to irregular maintenance. Unlined cut-off drains were sometimes used instead. However, these often became subject to invert scour and ultimately gullying on the steeper gradients. They also led to leakage into the surrounding soil as a result of seepage on shallower gradients and at low points in the topography.

road construction. Care is also needed during their installation to prevent or minimize further instability. Once a stability assessment has been carried out, these deeper drains should be constructed over short lengths at a time to minimize the length of slope temporarily unsupported. Figure C6.9 illustrates how excavation for a 1-m deep subsoil drain triggered ground movements in the adjacent slope over relatively short excavation lengths.

C6.3 Road drainage

C6.3.1 Design criteria

Inadequate control of road side drainage and cross drainage can lead to significant erosion in streams and gullies, creating instability on adjacent slopes. The underlying principle should be to provide sufficient drainage infrastructure to

Fig. C6.3. Typical benched cut slope erosion and drain blockage problems.

convey the design flows without surcharge or erosion. In the development of this design, the designer is often required to make engineering decisions based upon very limited data. It is likely that there will be few, if any, records of flow or flood levels in the streams and rivers that a mountain road is to cross. There should therefore be a strong reliance on visual observation of conditions on the ground during design, and the designer may need to rely on geomorphological evidence

Fig. C6.4. Slope failure on fill embankment with berm drain at immediate risk. Note that when the drain fails water will become concentrated into the failed mass below.

Fig. C6.5. Construction of an articulated drain across an unstable slope.

Fig. C6.6. Herringbone subsoil and surface water drainage of a landslide.

Fig. C6.7. Installation of sub-horizontal drains.

Fig. C6.8. Installation of a subsoil drain beneath a mass concrete U drain.

The number of bridges and culverts required along a road will vary according to the frequency of watercourses and where the alignment is located in the landscape. Along all roads, but especially those located in hilly and mountainous terrain, it is important to maintain the natural drainage pattern as much as possible and not to truncate any significant drainage paths. In areas of high rainfall and surface water runoff, frequent culverting is therefore important and it is not uncommon to require more than ten culverts per kilometre of alignment. Generally, the frequency of drainage crossing structures will be higher in the lower parts of the landscape, such as along lower valley sides, and lower in the more elevated parts of the landscape such as along or close to ridge lines.

C6.3.2 Field inspection

Field inspection is an essential part of the drainage design for a new road and the assessment of drainage structures on existing roads. The inspection is usually carried out in for much of the interpretation (TRL 1997). The designer should also work closely with engineering geological personnel during both design and construction supervision.

The drainage design adopted for a low-cost road should be consistent with the need to avoid unwarranted expenditure in either construction or maintenance. The design should take into account the following:

- climate and, in particular, rainfall regime;
- geological and geomorphological considerations, including rock and soil types and their permeability, catchment shape and topography;
- land use type and land use practices, such as irrigation systems;
- vegetation cover;
- social considerations, including the water needs of the public and adjacent landowners and farmers; and
- road safety, in terms of drainage surcharge effects.

Some general guidance on the drainage design standards appropriate to low-cost roads is given in Table C6.2. In situations where there is uncertainty over design discharge calculations and/or where surcharge could result in significant damage to slopes and structures, increased safety margins should be built into the figures given in the table.

Fig. C6.9. Excavation for a subsoil drain triggers movement in the adjacent slope.

Table C6.2. *General guidance on the design of drainage structures for low-cost roads*

Culverts and bridges	
Design flow	1 in 5 years (side drain relief culverts*) 1 in 10 years (pipe culverts and drifts) 1 in 25 years (box and slab culverts and short-span bridges)
Freeboard	Culverts are usually designed to flow full during a 1 in 10 year flood, and no freeboard above this level is usually provided Short-span bridges are usually designed with a freeboard 600 mm above the 1 in 25 year flood level
Minimum gradients	1.0% (culverts)
Maximum velocity	1.2 m/s (discharge to existing unlined (non-bedrock) watercourse). For lined watercourses† the lining is generally designed for the predicted exit velocity so the maximum velocity is not relevant
Minimum sizes	For stream crossings, 1000 mm diameter pipes; for other shapes an equivalent area should apply subject to a minimum dimension (height or width) of 600 mm; for access crossings and side drain relief culverts 900 mm diameter pipes are appropriate
Drifts	
Maximum depth of water	0.6 m
Maximum velocity	1.5 m/s; criterion is based on vehicles crossing the drift; drift design should include appropriate scour protection† on the downstream side so the maximum velocity is not relevant
Side drains	
Design flow	1 in 2 years
Minimum freeboard	100 mm above the 2 year flood level
Minimum sizes	0.3 m^2 (unlined side drain cross-section); for lined drains there is no minimum specified
Minimum velocity	0.75 m/s
Maximum velocity:	1.2 m/s (unlined side drains)† 3.0 m/s (lined side drains)†

*Side drain relief culverts should be scheduled in locations where natural water courses do not exist but where it is necessary to convey side drain runoff from one side of the road to the other to avoid surcharge.
†The erodibility of the subgrade and channel material will dictate lining and scour protection requirements.

association with drainage network and catchment delineation from topographical maps and aerial photographs or LiDAR. The principal aim of this work is to establish the pattern of drainage and to determine peak flood levels, flow velocities and sediment transport regimes in the main streams and river crossings. For example, TRL (1997) describes the hydrological investigations and analyses usually carried out.

C6.3.3 Roadside drains and turnouts

Roadside drains channelize water and concentrate flows (Fig. C6.10). Scour of unlined side drains, if not controlled, can lead to the formation of gullies that eventually become sufficiently deep that they undermine the adjacent road shoulder and present a danger to road users. Erosion in side drains can also cause failure of the adjacent cut slopes. Figure C6.11 illustrates erosion and incipient slope failure adjacent to an unprotected section of side drain. The situation is remedied by the use of a lined channel, either mortared masonry or concrete. In some cases dry stone pitching can be used as an inexpensive means of protecting the side drain, especially where low-grade metamorphic rock (phyllite and slate) is available locally that can be cut readily into tabular pieces. Checkdams are frequently used to slow down the velocity of water and hence reduce incision in the drain invert (Fig. C6.12).

In most cases, roadside drains are designed to discharge directly into stream channels or culvert inlets. However, turnouts are sometimes used as intermediate side drain discharge points. Turnouts should be sited where they minimize the potential for slope failure and erosion (Figs C6.13 &

Fig. C6.10. Concentrated flow in side drain.

Fig. C6.11. Side drain erosion leads to failure of adjacent cut slope.

C6.14). If it is necessary to locate turnouts on steeper slopes, consideration should be given to constructing a contour drain that conveys runoff to a point where it can be safely discharged (e.g. either into an adjacent stream or onto a stable slope with underlying materials capable of resisting erosion). On embankments, turnouts should be discharged via cascades or chutes to the toe of the embankment slope.

Generally, wherever there is a potential for significant erosion, turnouts should not be used and all roadside drainage should be discharged into culverts and other cross-drainage structures. On hairpin bends, turnouts can be avoided by conveying water via a culvert from the uphill to the downhill side of the bend. However, road side drain capacities and scour protection at the ultimate discharge

Fig. C6.12. Use of checkdams in side drains vulnerable to erosion.

Fig. C6.13. Slope failure exacerbated by water discharge from a roadside drain turnout.

Fig. C6.14. Erosion beneath side drain turnout. Note that this location is approximately 30 m from the road bend on side-long ground formed in closely jointed phyllite.

point should be increased to accommodate the additional flow. Where there is no choice but to use a turnout, it should be flared to spread the flow and a flexible apron (e.g. a gabion mattress) provided. In some cases, erosion protection measures may be required to extend for several tens or (in extreme cases) even hundreds of metres downslope from turnout locations (Fig. C6.15).

Where a road is on a gradient and surface runoff is directed towards its outer edge, water can become concentrated onto the slope below causing significant erosion. If the road is supported by a below-road retaining wall, road runoff will discharge over the slope at the down-gradient end of the wall. This concentration of runoff may cause erosion, leading to the exposure of the adjacent wall foundation and the fill behind (Fig. C6.16). Where it is apparent that concentrated road runoff is likely to result in scour, either the road edge should be kerbed until the water can be safely discharged elsewhere or provision made for the safe discharge of water at the end of the wall (e.g. via a lined channel sufficiently founded into the underlying slope material).

C6.3.4 Culverts

Three forms of construction are commonly used for cross-road drainage: pipes, boxes and slabs. Pipes are usually used for smaller crossings and boxes for larger watercourses. Double and triple box culverts in larger sizes are often used instead of short-span bridges on low-cost roads. The velocity of stream flow through pipe and box culverts is usually increased by the constricted flow width and because the

Fig. C6.15. Continuous outlet protection below side drain turnout from a hairpin stack.

Fig. C6.16. Severe scour at lower end of retaining wall.

invert of these structures is formed in concrete or masonry, both of which have a lower hydraulic roughness than the natural stream bed. This can increase the tendency for stream flows to erode the natural stream bed at the outlet. Slab culverts comprise reinforced concrete slabs supported at each end by concrete or masonry abutments. Slab culverts are easier to construct to the width of the natural channel than pre-cast pipes or boxes, and their impact on stream flow is usually correspondingly less.

If possible, cross-road drainage structures should be designed to match the dimensions of the natural channel at the design discharge. A crossing should be designed so that it spans the natural course of the stream with the abutment walls aligned parallel to the direction of flow. Guidance on the operating capacity of culverts with respect to the design level flood is given in Table C6.2. The hydraulic design of culverts in relation to the design discharge is described in Chow (1964), for example. However, in many mountain areas, the ability of culverts to convey the size and volume of transported debris is often equally if not more important (Cross 1982). Single-span slab or box culverts and drifts are much better at allowing debris to pass unhindered than pipe culverts. Wherever practicable, single openings are preferred to multiple culverts as these are more capable of transmitting high sediment loads, and particularly large-sized boulders during floods.

A chute inlet, such as that illustrated in Figure C6.17, can also allow efficient transport of debris through a culvert by channelizing and increasing the velocity of flow. This will lead, however, to increased scour potential; the lining of the culvert invert and the protection provided to its outlet (Section C7.4) will need to be designed accordingly.

Figure C6.18 illustrates the manner in which different culvert configurations influence the potential for inlet

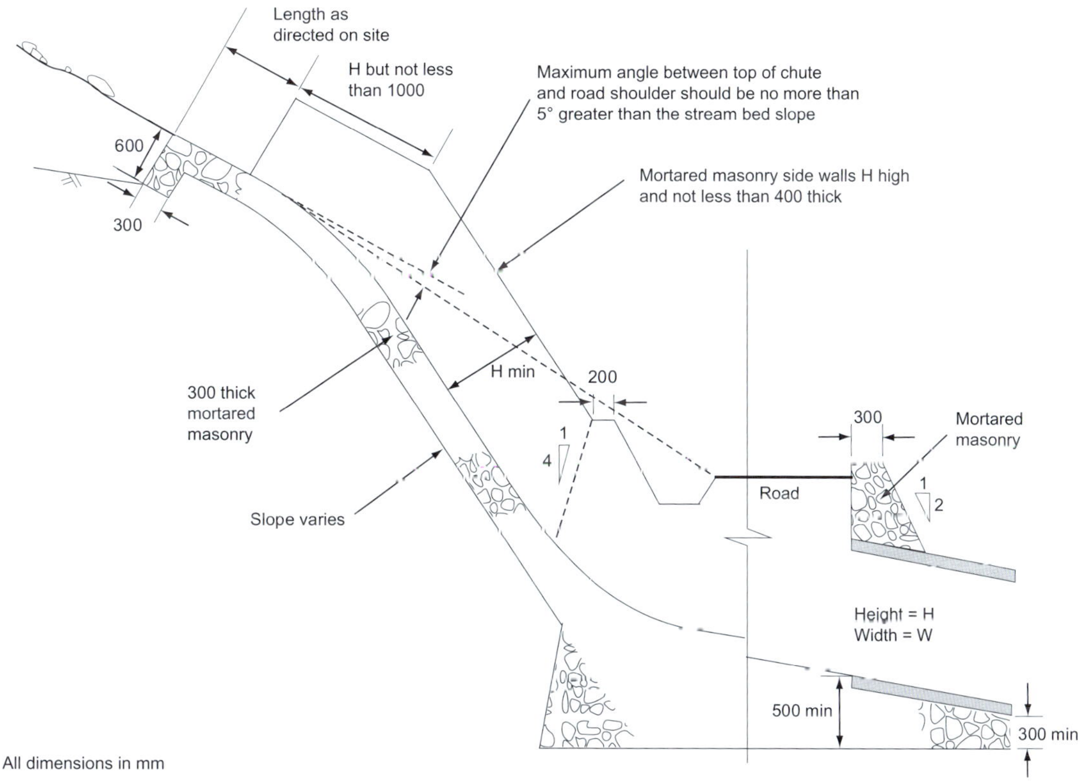

Fig. C6.17. Chute inlet.

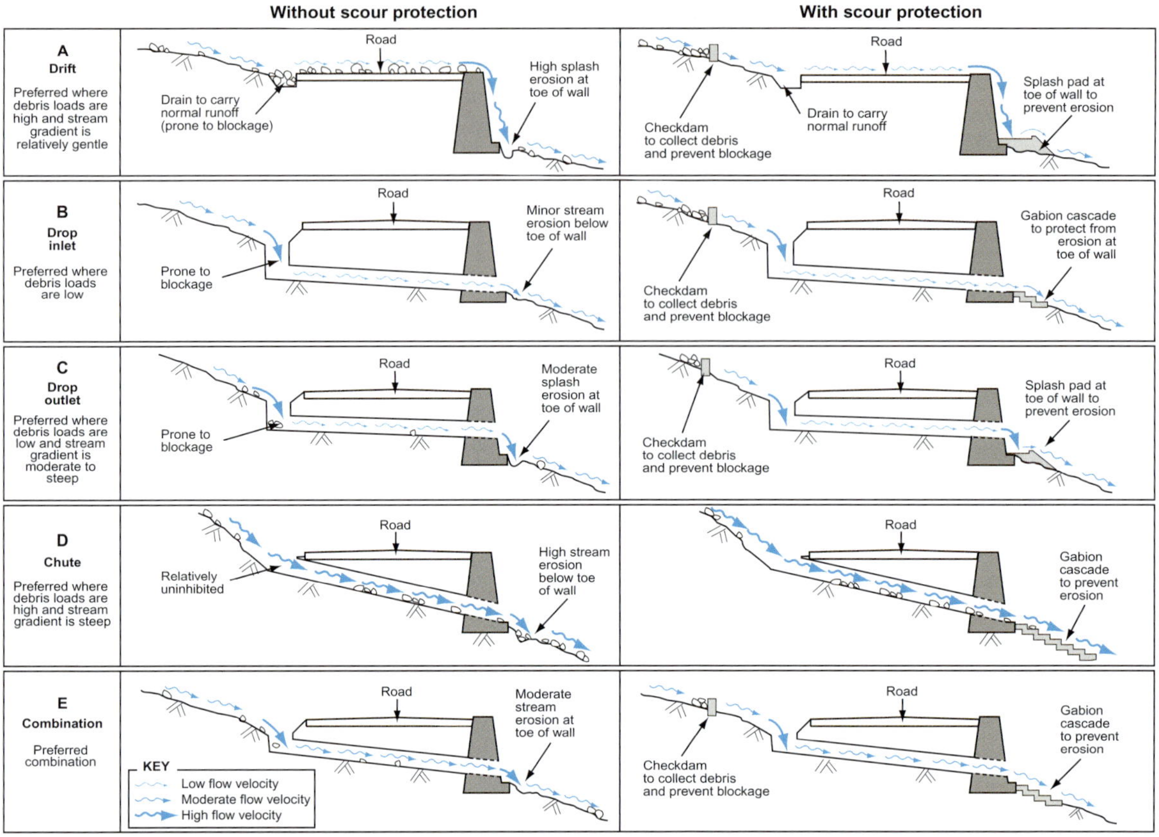

Fig. C6.18. Cross drainage configuration and its influence on inlet blockage and outlet scour potential.

blockage and outlet scour. The preferred culvert configuration is one that minimizes both the potential for blockage at the inlet and erosion at the outlet. The design requires an assessment of the size and likely volumes of sediment load and the scour potential of the soil or rock that forms the channel at and downstream of the outlet. It also requires an assessment of the topography of the drainage crossing to determine the most practical form of construction. For example, culverts constructed to a shallow gradient compared to the natural stream bed will either require a drop-inlet, a drop-outlet or a combination of the two. A drop inlet will encourage blockage while a drop outlet may cause splash erosion of the stream bed and the headwall foundation immediately below the culvert. By contrast, culverts constructed to a gradient parallel to the stream bed will facilitate debris transport through the structure but will lead to increased outflow velocities that may cause abrasion of the channel bed. The right-hand column in Figure C6.18 indicates the form of mitigation that might be used to combat blockage and scour. Case 'E' depicts the optimum combination of inlet, outlet and culvert gradient to minimize these effects.

References

Chow, V. T. 1964. *Handbook of Applied Hydrology*. McGraw-Hill, New York.

Cross, W. K. 1982. Location and design of the Dharan-Dhankuta low-cost road in eastern Nepal. *Proceedings of the Institution of Civil Engineers*, Part 1, **72**, 27–46.

Forrester, K. 2001. *Subsurface Drainage for Slope Stabilization*. American Society of Civil Engineers Press, Reston, Virginia.

GCO 1985. Groundwater Lowering by Horizontal Drains. Geotechnical Contol Office Publication No 2/85. Craig, D. J. & Gray, I.

Lau, T. M. F., Sun, H. W., Hui, T. H. H. & Ho, K. K. S. 2008. Review of slope surface drainage with reference to landslide studies and current practice in Hong Kong. *In*: Chen, Z., Zhang, J.-M., Ho, K., Wu, F.-Q. & Li, Z.-K. (eds) *Landslides and Engineered Slopes. From the Past to the Future. Proceedings of the 10th International Symposium on Landslides and Engineered Slopes, 30 June–4 July, 2008*, Xi'an, China, Taylor & Francis Group, London, 1769–1774.

TRL 1997. *Principles of Low Cost Road Engineering in Mountainous Regions*. Overseas Road Note 16, Transport Research Laboratory, Crowthorne.

C7 Erosion control

G. J. Hearn[1]*, J. H. Howell[2] & T. Hunt[1]

[1] URS Scott Wilson Ltd, Scott House, Alençon Link, Basingstoke, Hampshire RG21 7PP, UK
[2] Living Resources Ltd, Lukesland House, Ivybridge, Devon, PL21 0JF, UK
*Corresponding author (e-mail: gareth.hearn@scottwilson.com; garethhearn@talktalk.net)

C7.1 Sources of erosion

An important element in roadside slope management is the preservation and encouragement of the natural vegetation. However, the effects of road construction on erosion can often extend well beyond the right of way. In mountainous terrain, road construction itself often facilitates access to areas of forest that then become logged for timber as development takes place within the wider road corridor. The gradual depletion of forest cover can lead to increased runoff, causing slope erosion, gullying and downcutting in stream channels, which eventually increases the incidence of landslides. This process may be exacerbated by the indiscriminate dumping of excavation spoil during construction, creating debris hazards downstream (e.g. Kojan 1978; Hearn 1987; Zurick & Karan 1999; Hart *et al.* 2002; Campos *et al.* 2010 with respect to road corridors in Nepal, the Philippines and Brazil).

However, even where these effects are minimized through good land use planning and engineering management practices, there is usually a significant potential for erosion to occur on unprotected cut and fill slopes, in stream channels and at other drainage discharge points (Section C6). Erosion on roadside slopes is sporadic and difficult to predict since it is related to rainfall events, slope drainage patterns and the characteristics of the ground. The largest volumes of erosion are usually generated from short-duration, intense rainfall events. Methods used to combat this erosion are described and discussed below in the context of slopes and streams.

C7.2 Slope erosion control

C7.2.1 Options

Broadly, there are two main options for erosion control:

- physical measures such as revetments and slope coverings; or
- bio-engineering (i.e. vegetation-based) measures.

A combination of the two is also often used. On low-cost roads, the desire to minimize costs may lead to an emphasis on the use of bio-engineering (Section C7.2.4). However, planting schemes can take some time to establish and should only be used on their own when dealing with localized and shallow erosion that does not involve any significant depth or slope instability. In many circumstances a more substantial and immediate means of slope protection is required through the use of revetments.

C7.2.2 Revetments

Some walls adjacent to the road, usually located in cut slopes, are designed to lean against the slope as revetments (Fig. C7.1). Revetments are not intended to offer slope support; they act merely as slope face protection and are usually only appropriate to protect soils, or weak or weathered rock. They are particularly beneficial in reducing seepage erosion at the toe of cut slopes and preventing the subsequent softening and leaching of materials that might otherwise lead to progressive erosion and failure of the entire slope. Along mountain roads, revetments are usually constructed from masonry (although gabions can be used) and are inclined at the same angle as the cut slope; they can be as little as 1 m high although they are commonly constructed 2–4 m in height.

Revetments can either be used on their own (Figs C7.2 & C7.3) or in combination with bio-engineering works (Fig. C7.4) where planting and surface drainage schemes are carried out on the slope above. This combination can be cost-effective in tackling the majority of shallow slope instability and erosion problems along mountain roads. For example, the cost of the combined revetment and planting scheme in Laos shown in Figure C7.4 amounted to slightly less than US$ 5/m^2 in 2008 using local labour and materials.

Other forms of revetment include reinforced concrete slabs cast onto a slope and supported either through the use of dowels into rock or a basal support beam bolted or anchored in place. An example from Laos (Fig. C7.5) shows the use of this approach in a gully head immediately

From: Hearn, G. J. (ed.) *Slope Engineering for Mountain Roads*. Geological Society, London, Engineering Geology Special Publications, **24**, 243–268.
DOI: 10.1144/EGSP24.16 0267-9914/11/$15.00

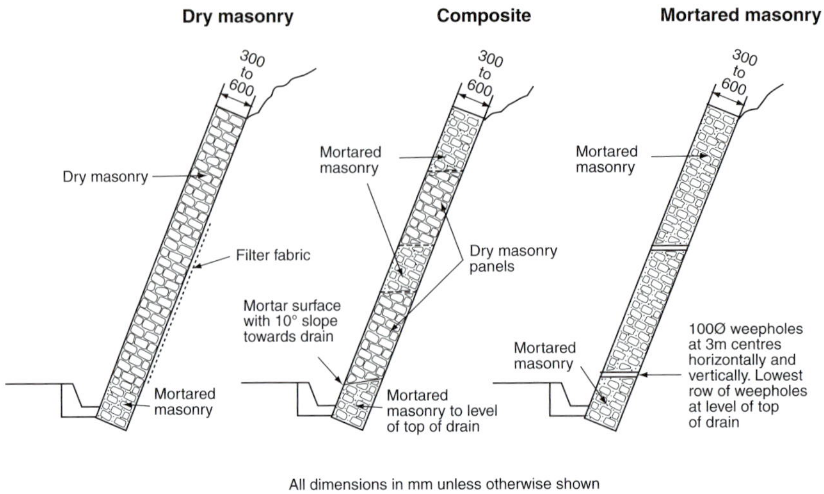

Fig. C7.1. Masonry revetments.

below a road. In this example, the structure is concave in plan so that it is keyed into the sides of the gully head.

C7.2.3 Surface coverings

Surface coverings can include:

- shotcrete;
- chunam; or
- mattresses and matting made out of geosynthetic or natural materials.

The following points should be considered before a decision is made to use shotcrete and other impermeable coverings, such as chunam (see below).

- Water pressures may build up behind the covering unless adequate drainage, usually in the form of weepholes, is provided.
- The use of dowels to support a shotcrete mesh will require embedment into a rock mass of sufficient strength. Shotcrete should not normally be applied to completely weathered rock or soil slopes.

Fig. C7.2. Mortared masonry revetment used to protect the lower portion of a cut slope against erosion and shallow failure.

Fig. C7.3. Combined concrete and mortared masonry revetment.

- Direct rainfall or slope runoff that runs down an impermeable surface will gather speed and will pose a potential erosion hazard at the base of the covering, possibly leading to loss of support and failure. It is usual, therefore, to provide additional slope or stream protection below the covering, including a downturn to the covering into the slope at its lowest point.
- The presence of concrete coverings on cut slopes or hill-sides can be unsightly.

Fig. C7.4. Composite masonry revetment used in combination with planting.

Fig. C7.5. Reinforced concrete slope revetment.

C7.2.3.1 Shotcrete

Shotcrete is a term used to describe a sprayed concrete that is usually applied to a steel mesh secured to a slope by dowels. Shotcrete is normally used on weathered and closely jointed rock slopes that are being eroded through progressive ravelling, erosion and dislodgement from the rock face. It provides more or less instant protection to a newly excavated rock face or to an area of slope that is experiencing shallow failures due to rock ravelling and rockfall, either below or above a road. Shotcreting is used most commonly on weathering grade III and IV slope materials (Section A3.1); lower weathering grades (fresh to slightly weathered rock) usually do not require slope protection while higher weathering grades (completely weathered rock and residual soil) should be treated as a soil and protected by other methods such as geosynthetic coverings (mattresses and matting) and/or vegetation.

Shotcrete is often used to improve the face security of blocky rock masses in conjunction with dowels, bolts and anchors (Section C4.4). Shotcrete can be unreinforced or reinforced, although it is considered beneficial to provide at least nominal reinforcement in the form of a light gauge mesh located centrally in the layer that can be moulded to the slope face prior to shotcrete application. Although early practice used chain link mesh as the reinforcing material, 6 mm diameter welded mesh has become more common. The latter is more rigid, while the former tended to rebound while being sprayed. The use of 'tell-tales', or wire gauges attached to the mesh, provide a convenient means of indicating when the required spray thickness has been achieved. This is usually between 50 and 75 mm, but can be up to 100 mm where required structurally. Steel-fibre-reinforced shotcrete is also becoming increasingly used. In this case, although the rock face has to be dowelled, the mesh is omitted (Fig. C7.6).

Design of the shotcrete face is based on an assessment of the jointing pattern and stability of the rock mass. At low-risk sites it is usual to apply the shotcrete to a steel mesh secured to the slope by close-centred dowels that can usually be hammered into place rather than drilled and grouted. At higher risk locations, surface protection may be required to be integral with the reinforcement of a deeper mass of rock behind the rock face. In such situations it is more common to grout steel dowels of up to 4 m in length and 25–32 mm in diameter into drilled holes at 1.5–2 m centres, depending on site conditions. The uneven surface of the slope face adds significant rigidity to the shotcrete skin which can contain smaller rock blocks without the need for additional anchorage.

Fig. C7.6. Use of fibre-reinforced shotcrete for rock slope protection.

Fig. C7.7. Hand-applied mesh-reinforced concrete.

Figure C7.7 shows a slope below a mountain road in Laos. The slope was formed in closely jointed phyllite, and was located at the head of a large and deep erosion gully. The slope had progressively ravelled back, causing loss of the outside edge of the road. Although the original contract documents specified mesh-reinforced shotcrete, specialist equipment was unavailable in the country at the time. Consequently, the specification was relaxed to permit a concrete surface protection to be applied by hand in panels onto the dowelled mesh. The cost of this application was a little less than US\$ $10/m^2$ in 2008 using a local contractor.

C7.2.3.2 Chunam

Chunam is a lime–cement–soil mortar which can be applied to a slope by trowel, and has been used widely in Hong Kong to protect against erosion on residual soil slopes. Chunam surfacing can crack easily as it contains no reinforcement, thus concentrating runoff and erosion potential. For this reason, chunam is now seldom used.

C7.2.3.3 Mattresses and matting

Geosynthetic materials are becoming increasingly used for erosion control as well as for fill reinforcement (Section C5.2.6) and as filter fabric in retaining wall (Section C5.5.2) and drainage (Section C6.2.2) applications. For erosion control on slopes these materials usually comprise thin mats or thicker mattresses that are polymer-based and have a void space of up to 90% that allow them to be filled with gravel or soil. Their important attributes as far as slope erosion is concerned include their flexibility, durability, permeability and ability to allow vegetation to grow through their structure. For example, Kulkarni (2009) describes the use of turf reinforcement matting combined with a steel confinement mesh bolted in place with grouted soil nails to stabilize and protect a residual soil cut slope in Malaysia. The slope was then hydroseeded and the cost of the combined slope stabilization and erosion protection was of the order of US\$ $150/m^2$ (Kulkarni pers. comm. 2009), i.e. significantly higher than surface protection measures alone.

Where rapid plant growth is anticipated, degradable geotextiles can be used and the covering regarded as a temporary measure. Most geotextile applications for erosion control on low-cost roads are of the degradable type, using locally or regionally derived materials (e.g. jute netting). Further information on the use of geosynthetics (including geotextiles) in erosion control can be found in Morgan & Rickson (1995), Gray & Sotir (1996), Saathoff (2003) and Barker *et al.* (2004).

C7.2.4 Bio-engineering measures

Bio-engineering measures can be applied to prevent or reduce erosion on newly constructed cut and fill slopes, in landslide back scarps and on failed debris. There is a range of standard techniques that can be applied but, as a general rule, the physical rooting conditions for plants mean that landslide debris can be revegetated much more quickly than landslide scarps. The speed of plant growth depends on many factors but particularly on the density and permeability of the soil and the micro-climate of the locality.

Through long-term and widespread practice, vegetation has been found to be effective in slope protection along low-cost roads. However, its use in engineering lacks the numerically based predictability of calculations that are possible with civil engineering structures using materials with defined properties. Although the role of vegetation in increasing soil shear strength, soil suction and other factors relating to slope stability has been numerically assessed in detail (e.g. Wu 1995; Norris & Greenwood 2003; Greenwood 2006), these assessments are heavily dependent on both species type and site conditions. Norris & Greenwood (2003), for example, quote increases in factors of safety of 10% due to the presence of plant roots for potential slope failures of 0.5 and 1 m depth in Greece and Italy. However, actual root development is frequently very different from that hypothesized in numerical assessments. Little is still known of the behaviour of different plants under a diverse micro-climatic environment and in the wide range of soil conditions that are often encountered. The plant root geometry required to fulfil specific engineering functions therefore cannot yet be relied upon.

In the context of mountain slopes, many of the claims made for deep rooting and wide geographical site versatility for certain species have to be viewed with much caution. A particular example is provided by vetiver grass, where studies have shown root penetration to more than 5 m (National Research Council 1993). While these are undoubtedly correct, they were observed only in lowland or valley situations, and not on poorly developed hillside soil profiles.

Studies of plant roots are relatively few and tend to be biased towards:

- commercial tree species growing on wet soils (e.g. Ray & Nicoll 1998);
- examination of trees which have blown down (e.g. Cutler *et al.* 1990; Sthapit 1996); and
- investigations into the liability of windthrow (e.g. Fraser & Gardiner 1967).

Among those studies which have been conducted specifically for engineering purposes, the research by Phillips & Watson (1994) is among the most comprehensive. In addition, Norris & Greenwood (2010) describe methods for investigating the properties of vegetation as a contributor to slope stabilization.

Fig. C7.8. Natural root development of a tree in shallow soil overlying laterite.

Table C7.1. *The main plant classes and their anticipated effective depths of rooting (adapted from Howell 1999)*

Plant type	*Anticipated maximum effective rooting depth in good conditions (m)*	
	Cut slope in original ground	*Unconsolidated debris*
Small grass	0.1	0.1
Large grass	0.5	0.5–1.0
Large bamboo	0.5	1.0
Shrubs	0.5	1.5
Trees	0.5–1.0*	1.5–2.0*

*This is a simplification but can be safely assumed to be nearly always the case in designing engineering applications.

Observations from long-term research in Nepal undertaken by Howell (1999) and others led to the conclusion that the actual or effective depth of rooting is in any case less than the maximum penetration depth of plant roots under ideal growing conditions. Furthermore, root development is closely linked to the presence of a penetrable stratum; on steep slopes, shallow soils and weathered rock usually pose significant limits to root penetration. This is illustrated by Figure C7.8 which shows the roots of a tree that grew in shallow soil over laterite material to a depth of 150 mm. Penetration up to approximately 500 mm was found to occur only along discontinuities in the laterite. In practice, therefore, the rooting depths given in Table C7.1 should be taken as the maximum that are likely to be achieved on most mountain slopes in the humid tropics and subtropics for different categories of plants.

The role played by bio-engineering plants is mainly in the protection of the surface against erosion (i.e. the detachment of soil particles), and this is controlled by the microscale architecture of plant stems and roots. Grasses, with their multiple stems and dense networks of shallow roots, perform this function most readily. Soil reinforcement can also be achieved through dense grass root systems, as well as by other plants. In particular, it appears that woody plants propagated from cuttings may well produce dense but shallow rooting systems for reinforcement, whereas those grown from seed may produce roots with a more vertical orientation, that is, a deeper penetration. The latter, in the right conditions, would provide something of an anchoring function (Fig. C7.9).

Many bio-engineering techniques depend on a certain method of plant propagation; for example, brush layering, palisades and fascines are all constructed using hardwood cuttings. Grasses, however, appear to be the most straightforward in the development of their rooting patterns. Whether they are grown from seed or from cuttings, they seem to develop a similar full root pattern which is determined by the species rather than by the method of propagation (Fig. C7.10).

The capacity of well-established and robust ground cover plants to protect soil surfaces against erosion is not in

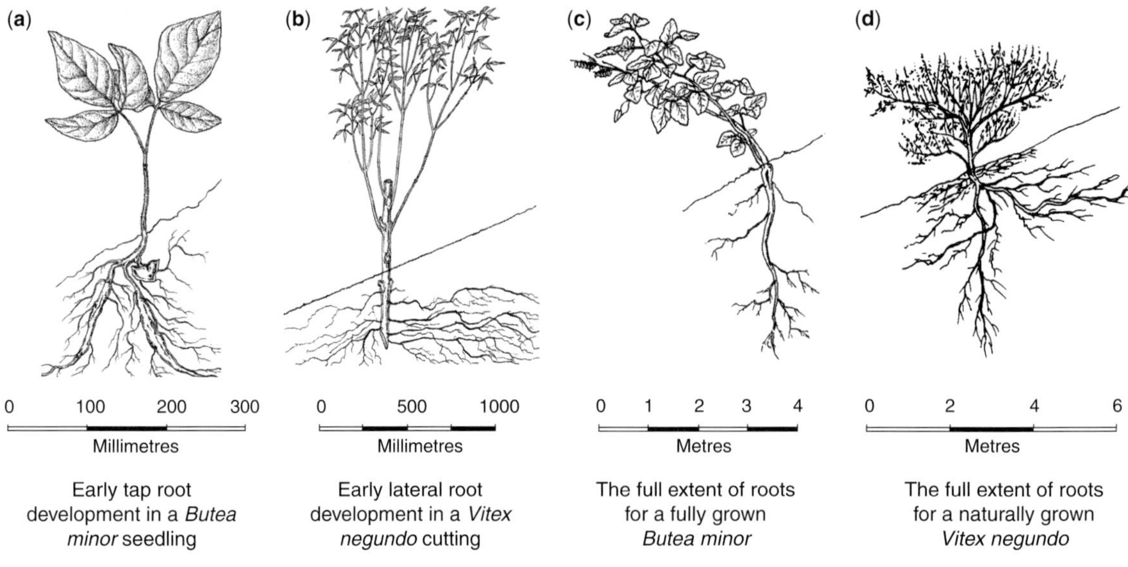

Fig. C7.9. Differences of root development in two shrubs used for bio-engineering in Nepal (modified from Howell 1999).

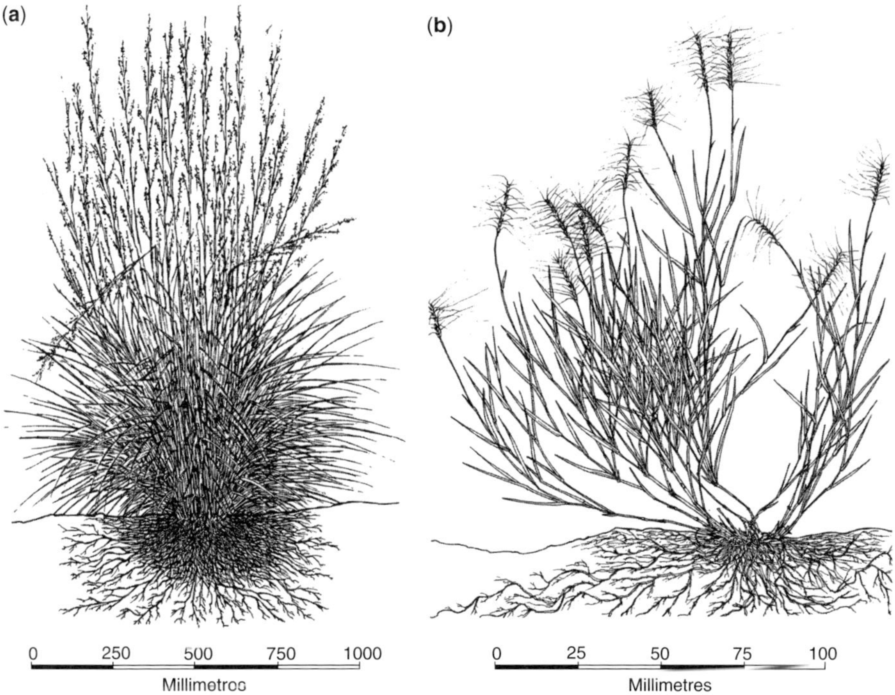

A mature clump of the large grass *Cymbopogon microtheca*

A plant of the small grass *Pogonatherum paniceum*

Fig. C7.10. Differences of rooting in two grass species (modified from Howell 1999).

question (for example Fig. C7.11). Other engineering functions are certainly served by vegetation, but to lesser degrees. In particular, plant roots contribute to soil strength and resistance to deformation, but the extent to which this is achieved can be considered to range between two essentially unmeasurable extremes:

- very high in a well-drained, porous, stony soil penetrated by plant roots with multiple bifurcations; and
- very low in a poorly drained, silty clay soil penetrated by plant roots with a tendency towards straight, unbranched roots.

For these reasons it is safe to assume that plants can be relied upon to contribute to erosion protection but cannot yet be relied upon for even shallow landslide stabilization.

C7.2.4.1 Using vegetation in engineering

Different types of plants and planting materials (seeds and cuttings) give rise to a variety of rooting patterns, with the result that the surface layer of soil will be bound together and its resistance to deformation increased. It is often important that vegetation is 'engineered' to achieve a desired cover, since many of the plants that naturally colonize bare ground are not the best for controlling erosion.

The poor predictability of vegetation growth means that it cannot be guaranteed to provide an immediate solution. Furthermore, there are situations in which vegetation can actually reduce slope stability if used wrongly. As a result of this, the following general rules should be adopted.

- Grasses that form large, dense clumps generally provide the most robust slope protection in humid tropical and subtropical areas, where rainfall can be particularly intense. This type of plant is usually best for erosion control. However, most grasses will not grow under the shade of a tree canopy.
- Shrubs (i.e. woody plants with multiple stems) and small trees (i.e. woody plants with single stems) can often be grown from cuttings taken from their branches (e.g. Fig. C7.9b). Plants propagated by this method tend to produce a mass of fine, strong roots. These are often better for shallow soil reinforcement than the natural rooting systems developed from a seedling of the same plant.
- On steep or fragile slopes trees should not be allowed to grow to more than 10 m in height, nor large bamboo clumps permitted to grow at all. These big plants should also be restricted to slopes that are less than 1V:1.5H

Fig. C7.11. Dense surface protection provided by the large grass *Imperata cylindrica*, 16 months (including two wet seasons) after planting on landslide debris composed of WG V and VI soil.

and should not be planted or allowed to grow in the upper parts of cut slopes that are any steeper than this.

- Maintaining a line of large trees or bamboo clumps at the base of a slope can help to buttress the slope and reduce undercutting by seepage or drainage lines.
- There is no single plant species or technique that can resolve all slope protection requirements.
- Plant roots cannot be expected to contribute to soil reinforcement below a depth of 500 mm.
- Plants cannot be expected to reduce soil moisture significantly at the critical periods of intense and prolonged rainfall when slopes are most likely to fail.
- Grazing by large numbers of domestic animals can devastate a planted site if it occurs before the plants are properly established. Some form of protection may be required.

In most cases the establishment of a full vegetation cover on fill slopes and colluvium slopes can usually be achieved in one or two wet seasons. However, plants may take three to five wet seasons to develop fully in cut slopes formed in clayey residual soil. More permeable soils will have faster plant growth rates, but drier locations will lead to slower rates. Once established, plants are flexible and robust. They can recover from significant levels of damage (e.g. flooding and debris deposition), although recovery may not occur fully until the following growing season. Many trees and shrubs are killed by fire, although in certain habitats fire is necessary to trigger the germination of some seeds. Also, some shrubby species regrow after fire. Once grass is established, it will normally survive a fire and put out new shoots. For this reason fire is frequently used by graziers to improve pasture. However, if a recently planted site is burnt, the grass plants may be killed.

Since plant growth remains dependent on indeterminable factors (as discussed earlier), care must be taken in selecting both the planting technique and the actual species used. The most reliable approach is to use the most robust species available in a locality. With experience, the practitioner will be able to predict likely growth characteristics on different sites.

The techniques recommended in Table C7.2 will always improve slope stability and, in certain situations, enhance drainage as well. For this reason, it is 'safe' for them to be added to any other design of slope treatment.

C7.2.4.2 Bio-engineering technique selection

Bio-engineering measures have been used in many parts of the world. As a result, there are numerous methods adapted to different environments and specific site requirements. Only some of these are appropriate to roadside slopes within the humid tropics and subtropics; others have been developed for wholly different environments, such as the temperate and alpine climates of Europe and North America. Greater care must be taken in adapting methods from one place to another than with purely physical engineering measures.

Table C7.2 summarizes the available techniques that have already been proven through field trials and widespread implementation in a number of countries. Between them, they provide a range of options that allows slope protection works to be undertaken in most situations found along a typical mountain road within the humid tropics and subtropics. Some of these techniques are illustrated in Figures C7.12 and C7.13.

Table C7.3 shows how an understanding of the characteristics of the site allows the most relevant technique to be identified. An important feature of bio-engineering works is that the design should be at the subsite level, that is, individual slopes and sections of slope within a wider application site. The engineer should therefore expect to use a range of different techniques on the same site.

C7.2.4.3 Slope preparation

Before bio-engineering slope treatments can be applied, the site must be properly prepared. The slope surface should be clean and firm, with as much loose debris removed as possible.

Slopes must be trimmed to the final desired profile, with a slope angle of between 1V:1.75H and 1.75V:1H (30° and 60°) depending on slope material. Excessively steep sections of slope must be removed by trimming and regrading. In particular, a small failure in an over-steep lower part of a slope can regress and destabilize the entire slope. These steeper sections should therefore be removed. However, if this were to require a large excavation volume, buttressing or a revetment wall may offer suitable alternatives.

C7.2.4.4 Selection of the appropriate plant species

Plants must be of the right type to perform the bio-engineering role that is required. The possible categories include:

- grasses that forms large clumps;

Table C7.2. *Bio-engineering techniques*

Technique	*Advantages*	*Limitations*
Linear grass planting (Fig. C7.13): rooted slips of large grasses are planted in lines across a bare or eroded soil slope. Slips are made by splitting out the clumps to give a small section of both root and shoot. Lines are usually horizontal (dry, well-drained soils), vertical (moist, poorly-drained soils) or diagonal (variable moisture/soil properties)	• The best and quickest way to create a surface vegetation cover on a steep, bare slope. • Effective on almost all soil slopes up to 2V:1H. • Robust protection and shallow reinforcement of the surface soil.	• Requires a slope with at least 30% soil cover. • Slow to establish on rocky cut slopes. • Where horizontal lines are used, slowed runoff can increase infiltration to cause shallow slumping. • Where diagonal or downslope lines are used on non-cohesive soils, erosion may take place between planting lines.
Direct seeding: the seeds of shrubs and small trees are inserted into crevices in slopes composed of weathered rock.	• The best way to establish vegetation on rocky slopes.	• Slow to provide a coverage good enough to resist erosion.
Brush layers (Figs C7.12 & C7.13): woody cuttings from shrubs or small trees are arranged in lines in shallow trenches running horizontally across slopes formed in unconsolidated debris. These can be installed on slopes up to *c.* 1V:1.25H.	• Instant physical barrier that interrupts runoff. As the plants root and grow, they protect and reinforce the soil. • Stronger than grass. • Often successful on loose stony debris. • Most shrubs tolerate shade, so this method can often be used under tree canopies where grasses will not grow.	• Can only be installed on slopes of up to 1V:1.25H on unconsolidated materials. • Construction causes considerable disturbance to the slope.
Fascines (Fig. C7.12): bundles of long woody cuttings are arranged horizontally in shallow trenches across slopes formed in unconsolidated debris. These can be installed on slopes up to *c.* 1V:1.25H. After burial in the trenches, they put out roots and shoots, forming a strong line of vegetation. The technique is sometimes called live contour wattling.	• Provide surface protection and shallow root reinforcement. Once established, they can also catch debris. • In certain locations, fascines can be angled to provide drainage.	• Brush layers are quicker and easier to construct than fascines. • Can only be installed on slopes of up to 1V:1.25H on unconsolidated materials. • Construction causes considerable disturbance to the slope.
Palisades (Figs C7.12 & C7.13): woody cuttings are planted in lines across the slope, usually following the contour. This can be done on a wide range of sites up to *c.* 1.75V:1H.	• Form an immediate barrier that traps small debris moving down the slope; after some time, a small terrace will develop. • Less disturbance to the slope than brush layers, so they can be installed on steeper slopes. • In certain locations, palisades can be angled to provide drainage.	• Not suitable on soils that are poorly drained and are subject to high rates of small-scale slumping. • Not as strong as brush layers.
Truncheon cuttings (Fig. C7.13): big woody cuttings from trees are inserted upright at intervals in slopes formed in deep or poorly stabilized and unconsolidated debris.	• Relatively strong plant material on slopes that are still unstable. • Withstand damage from moving debris.	• Take a long time to establish a complete cover. • Need a lot of planting material.

(Continued)

Table C7.2. *Continued*

Technique	*Advantages*	*Limitations*
Live checkdams (Fig. C7.13): small checkdams with structural elements made from the woody cuttings of trees are placed at intervals in erosion gullies.	• Low-cost flexible structures to reduce erosion where water flow is concentrated. • Relatively limited disturbance to the slope. • Can be used on weak, unconsolidated materials.	• Not as strong as checkdams of gabion or masonry. • Require careful supervision and maintenance.
Tree planting: potted seedlings from a forest nursery are planted at intervals across a soil slope.	• Restoration of a forest mix of trees in the long term.	• Takes a relatively long time (5 years or more) to contribute significantly to slope strengthening or to establish a complete cover. • Seedlings are vulnerable to grazing; protection is required during the first 3 years or longer.
Large bamboo planting: a section of the stem and root of a large bamboo is planted, usually at the base of a slope or along a stream bank. The cutting is about 2 m in length and has to be excavated from the mother clump carefully.	• Large bamboos support the base of a slope or strengthen river banks by creating a very strong line of plants. Once they are well established, they are highly flexible, immovable barriers. • With their multiple stems, they catch debris moving down the slope.	• Bamboos take about 5 years to contribute significantly to slope strengthening. • Cannot be used in hot, dry sites, since most bamboos require cool, moist conditions. • Bamboos planted in steep upper slope situations develop deep roots slowly and so are prone to slumping for some years (up to 7 or more) after planting.
Random grass planting: grass slips are planted at random, usually with a specified number of plants per square metre. This is usually in association with a temporary geogrid surface covering, such as woven jute, coir netting or a synthetic netting.	• Rapidly forms a strong and complete surface covering. As with the other grass systems, the roots provide strengthening to the surface soil layers.	• This should not be used where the specific erosion control or structural advantages of line planting patterns are important.
Downslope grass lines of planted slips: grass slips are planted in lines running straight down the line of maximum slope.	• This arrangement provides the maximum amount of surface drainage by channelling runoff and minimizing infiltration. • It protects against erosion and reinforces the surface. • It can be important on very steep and poorly drained soil slopes cut in silty clays, where the weaker surface layer can become saturated to the point of slumping if drainage is impeded using horizontally configured systems.	• On drier sites, grass plants can suffer from drought due to the increased drainage. • On some weak materials rills can develop down the side of the plant line, damaging the grass slips and reducing their growth. These rills could develop into gullies on deep, erodible soils.
Grass seeding: seeds of grass plants are spread across a soil slope. On slopes steeper than *c.* 1V:1.5H, a mulch or other temporary surface covering is usually required.	• Creates an even cover over bare slope surfaces. • Fully protects and reinforces slopes after a few years of growth. Reinforcement depends on the character of the plant used. • Good where soil cover is thin or discontinuous.	• Plants take longer to develop from seeds than from slips. • Very heavy rain in the days immediately after sowing can lead to seeds being washed off the slope or damage to the very small seedlings.

Technique	*Advantages*	*Limitations*
Vegetated stone pitching: stone pitching is where small cobble-sized pieces of rock are hand-laid to protect a soil slope surface; grass slips or woody cuttings can be inserted between the cobbles to establish reinforcement.	• A very strong surface protection in gully beds or other locations where periodic water flows may occur. • Stronger than dry stone pitching but more flexible than mortared pitching.	• More costly than other forms of bio-engineering surface protection.
Wire bolsters (Fig. C7.12): usually 0.3 m diameter and in 5 m long sections joined end-to-end, a stone-filled roll of gabion arranged in a shallow cross-slope trench and secured with steel pegs driven into the ground.	• Provide an immediate check on scour potential. • A very strong surface protection in weak and highly erodible soil	• More costly than planting schemes. • Construction can cause considerable damage to the slope.
Geotextile coverings: permeable coverings are spread on the surface and used as an aid to starting vegetation growth. These may be of natural fibres (e.g. jute or coir) or synthetic (e.g. proprietary products).	• Immediate surface covering to reduce erosion while vegetation gets established. • Some types help vegetation to establish, allowing bio-engineering to be undertaken on very harsh or very steep slopes with little or no soil cover.	• A significantly greater cost than simply using vegetation. • It can be difficult to place geotextiles on uneven ground, and specialist supervision is needed. • Some coverings can create added problems, usually by increasing soil moisture under the covering, to the point where shallow movement occurs.
Wattle fences (Fig. C7.13): fences made of woven branches are used to retain small volumes of debris, forming mini terraces.	• A rapid temporary measure on slopes with loose surface debris.	• Only a temporary measure, as the branches do not grow but gradually degrade. • Wattle fences built in the dry season are liable to collapse in the wet season, when the pressure from retained debris increases.
Hydro-seeding: a mixture of seeds, binder, mulch and fertilizer is sprayed onto the slope surface. Various proprietary systems are available.	• Rapid surface covering of vegetation over large areas. • Good where there is little or no topsoil or where topsoil is poor quality.	• Capital-intensive, relying on specialist machinery, supplies and skills. Much more expensive than hand-planted methods and not normally viable on low-cost roads. • Seed mixes of locally appropriate species can be difficult to obtain, and introduced seed mixes either do not perform as intended or create weed problems. It is also not good environmental practice to import seeds from elsewhere. • Adhesion of spray material can be poor on steep roadside cuts under intense tropical rainfall, leading to incomplete coverage. • Some systems using thick mulch applications do not encourage rooting beneath the mulch layer into the underlying soil, so that rapid early plant growth is followed by a sudden decline or failure of the mulch layer over the original slope surface.

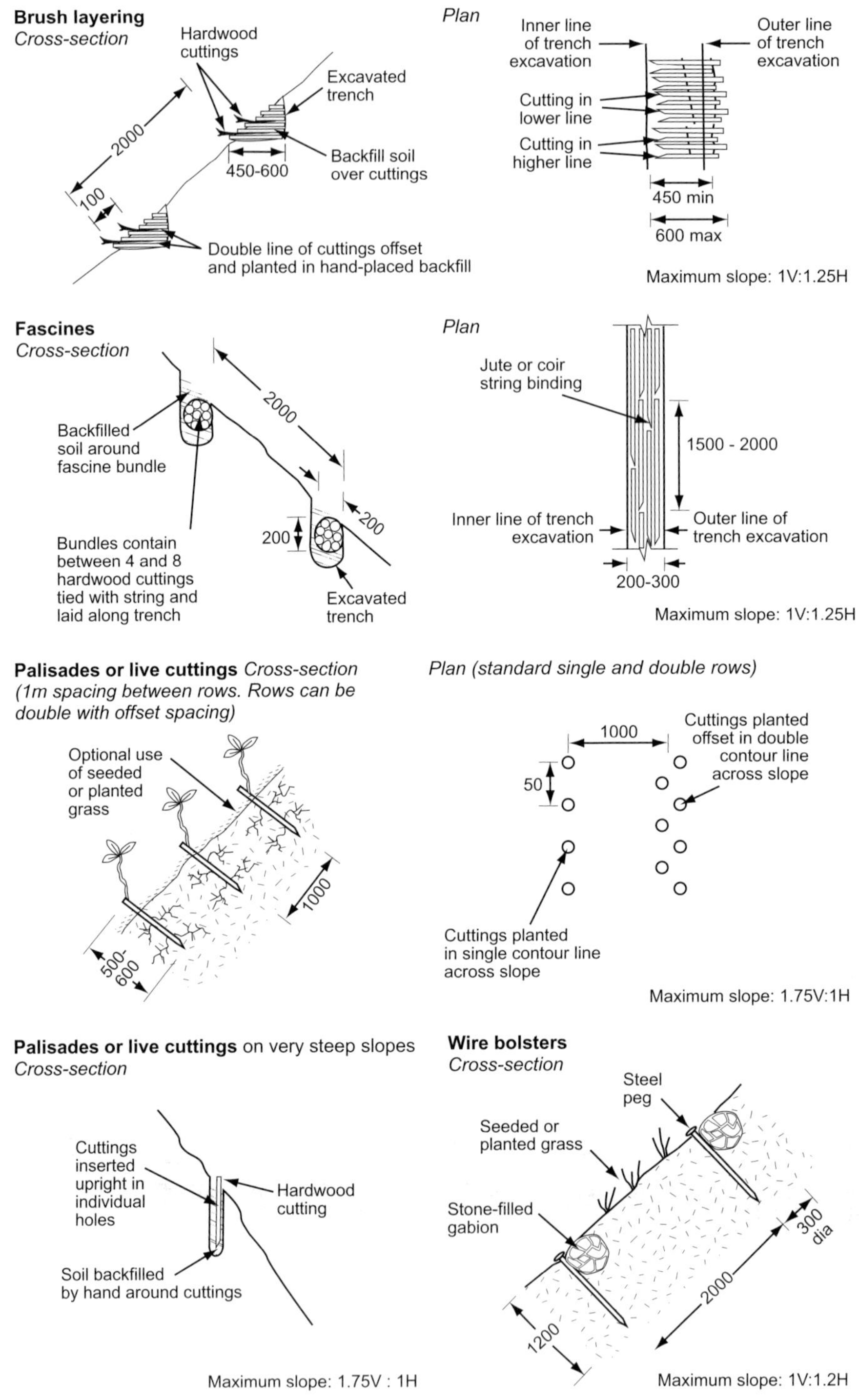

All dimensions in mm

Fig. C7.12. Typical bio-engineering details.

Fig. C7.13. Some planting applications.

Table C7.3. *Selection of bio-engineering technique according to site characteristics*

Site characteristics	*Recommended techniques*
Sites mainly above roads	
Cut slope in highly to completely weathered rock or *in situ* weathered soil, at up to 2V:1H	Grass planting in lines, using slip cuttings
Cut slope in taluvium/colluvium debris, at any slope up to 1V:1H	
Trimmed landslide back scarps in suction-controlled residual soil, at up to 2V:1H	
Roadside shoulder in soil or mixed debris	
Slopes in highly weathered rock at any slope up to about 4V:1H	Direct seeding of grasses and shrubs
Trimmed landslide back scarps in highly weathered rock at any slope up to about 4V:1H	
Backfill behind slope-retaining walls	Brush layers using woody cuttings from shrubs or trees
Sites mainly below roads	
Dry fill slopes and backfill behind walls: these should first be regraded to be no steeper than about 1V:1.5H	Brush layers using woody cuttings from shrubs or trees
Slopes underlain by rock structure and *in situ* weathered soil, at angles between 1V:1H and 1.75V:1H	Palisades using woody cuttings from shrubs or trees
Other debris-covered slopes at angles between 1V:1.5H and 1V:1.25H	Brush layers using woody cuttings from shrubs or trees
Fill slopes and backfill behind walls showing evidence of regular water seepage or poor drainage: these should first be regraded to be no steeper than about 1V:1.5H	Fascines using woody cuttings from shrubs or trees, configured to contribute to slope drainage (e.g. in a herringbone pattern - more permeable backfilled soil within the fascine trenches)
Large and less stable fill slopes more than 10 m from the road edge (slope angle not necessarily important, but likely eventually to settle naturally at about 1V:1.5H or less)	Truncheon cuttings (big woody cuttings from trees)
The base of fill and debris slopes	Large bamboo planting or tree planting using potted seedlings from a nursery
Other sites	
Stream banks where minor erosion is possible	Large bamboo planting
Gullies or seasonal stream channels with occasional minor flow	Live checkdams using woody cuttings of shrubs and trees
Gullies or seasonal stream channels with regular and large flows	Stone pitching, probably vegetated; gabion checkdams may also be required (Section C7.3.1)
Other bare areas such as on the slopes above landslide back scarps or on large debris heaps and stable fill slopes	Tree planting using potted seedlings from a nursery

- shrubs or small trees that can be grown from woody cuttings;
- shrubs or small trees that can grow from seed on rocky sites (note however that tree root penetration can sometimes open up rock joints, reducing stability);
- trees that can be grown from a potted seedling; or
- large bamboos that form clumps.

Plants must be capable of growing in the location of the site. This means that they must be suitable for the:

- site temperature, which in mountain areas is often linked to altitude;
- site moisture, which is a function of rainfall, soil characteristics, topography and climatic factors (rainfall, slope aspect and temperature); and
- soil conditions, both physical and chemical.

Although it is often possible to derive information on the site preference of different plants from forestry and agricultural sources, the complex and often difficult growing conditions on many roadside slopes means that this can be one of the most critical factors in implementing a bio-engineering programme. Observations of existing growth patterns and the use of trials are therefore recommended (e.g. Hunt *et al.* 2010).

C7.2.4.5 Timing of bio-engineering works

In most countries within the humid tropics and subtropics, bio-engineering should be undertaken in the first half of the wet season. In practice this might vary regionally or even locally within a country, and so there are rarely firm rules. There will be strong similarities between the timing of bio-engineering works and those of forestry and agriculture in the vicinity of the sites. In some instances a good indicator is to implement bio-engineering planting at the same time as local farmers plant out rain-fed rice. As a rule, planting work should not be done too late in a wet season since there may not be sufficient time for plant roots to develop adequately to survive the following dry weather. Further details are given in Part D.

Whatever the precise timing, the soil must be moist when the planting is done. If the wet season is also the hottest period, then an absence of rain for a few days after planting can be very damaging. In some cases, plants should be watered by hand every day until it does rain.

C7.3 Stream erosion control

On mountain roads, stream erosion can cause slopes to fail as a result of incision (bed scour) and channel widening and migration (lateral scour). Furthermore, the deposition of sediment load in streams can also lead to slope instability by blocking culverts and river channels, thereby deflecting flood flows against adjacent slopes causing toe erosion and eventual failure.

Erosion protection is usually focused on the following elements of the drainage system:

- in gullies formed on slopes above or below the road;

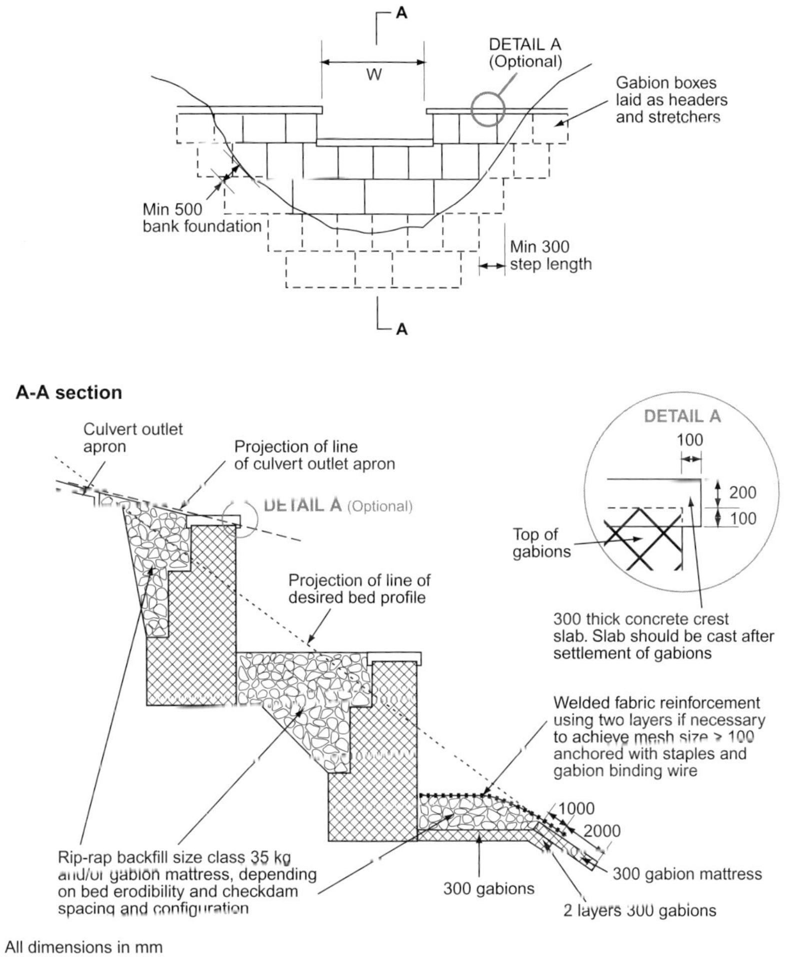

Fig. C7.14. Typical gabion checkdam details.

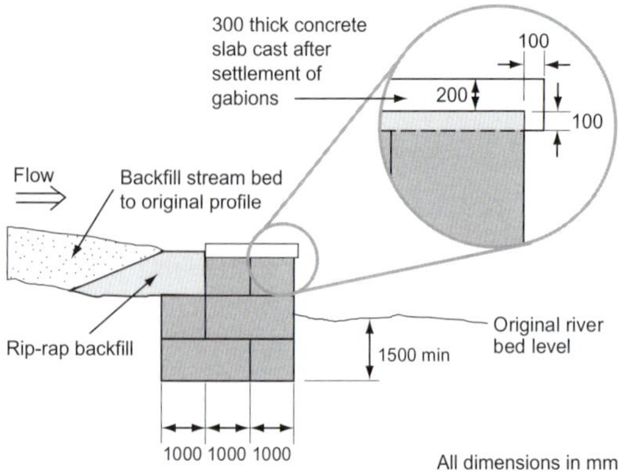

Fig. C7.15. Typical gabion scour check detail.

- at stream crossings, that is, upstream and (mostly) downstream of culverts and bridges; and
- alongside actively eroding river banks or lower valley sides subject to potential scour effects from flooding (covered more fully in Section C7.5).

Fig. C7.16. Dry stone cascade where flow rates are low.

Stream erosion protection measures most commonly comprise checkdams, cascades and channel linings.

C7.3.1 Checkdams

Checkdams can be constructed from live plant cuttings (Section C7.2.4), mortared masonry or concrete where the gully bed is formed in rock or, more commonly, using gabions where it is absent. In all cases it is important to key the structure into the adjacent slopes or gully walls in order to prevent outflanking and side scour (Fig. C7.14). For gabion structures, a gabion mattress is usually constructed at the toe of each checkdam on the downstream side to protect its foundation from scour. The mattress (nominally 0.25–0.3 m thick) is able to fold into scour holes and other irregularities on the stream bed.

Scour checks offer an alternative to checkdams where their function is to introduce hard points in the stream channel long-section. Figure C7.15 shows a typical detail.

C7.3.2 Cascades

In situations where more significant flows are anticipated or observed, cascades may be preferable to checkdams. Cascades are also commonly constructed to convey water down cut slopes and other steep sections of slope where

Fig. C7.17. Gabion cascade where flow rates are high.

Gabion cascade with concrete lining

Masonry cascade

Fig. C7.18. Gabion and masonry cascades.

runoff is either introduced or significantly increased as a result of the design. Typical applications include the conveyance of:

- cut-off drain water down the face of cut slopes to the road side drain;
- stream flow from a truncated watercourse above a cut slope to a culvert inlet at road level;
- flow in stream channels between culverts on hairpin loops and stacks; and
- diverted drainage, either temporary during construction and emergency maintenance or permanently as part of long-term drainage improvements.

Cascades can be constructed in dry stone (Fig. C7.16), gabion (Figs C7.17 & C7.18a), mortared masonry (Fig. C7.18b) or concrete.

Dry stone and gabion cascades can be problematic due to their inherent permeability. Percolating water may cause seepage erosion beneath the structure or softening of the foundation. This can lead to deformation and eventual loss of function, particularly where the cascade is not constructed in a natural watercourse. Heavy-duty polythene sheeting can be laid on the foundation of the structure to contain any seepage water, or an impermeable surface applied to the gabion cascade itself using concrete for example (Fig. C7.18a). However, any significant ground movements will cause tearing to the sheeting and cracking of the concrete, rendering these measures ineffective (Fig. C7.19). Mortared masonry cascades are essentially impermeable but will crack and allow water ingress following only minor settlement or ground movement.

Both gabion and masonry cascades are also prone to abrasion when stream flows contain large-sized debris. A concrete screed can be applied to the steps of the cascade in order to provide some protection. Typical details for gabion cascades with this protection are given in Figure C7.20. Where abrasion is a particular concern, and where founding conditions permit, consideration should be given to constructing the entire structure in concrete.

C7.3.3 Channel linings

These are usually constructed in gabion, masonry or concrete and are designed to convey water from a source to a discharge point either where there is no existing stream channel or where an existing channel has a gradient that is too shallow for a cascade. Gabion-lined channels suffer from the same problem of seepage described above for gabion cascades, especially where anticipated low-flow velocities on shallow gradients will result in significant infiltration,

Fig. C7.19. Gabion cascade constructed on an unstable slope.

thereby leading to softening or erosion of the underlying material. Masonry structures will suffice wherever ground movements are likely to be negligible, and it is usual to apply a concrete screed to the channel invert to protect against scour. Concrete structures are more robust in terms of scour and may be able to accommodate minor ground movements but, as with all drainage structures, they will require regular maintenance and repair if larger displacements occur. Significant ground movements cannot be tolerated by any rigid structure, even with the benefit of regular maintenance (Fig. C7.21). In this case a flexible, though impermeable, articulated structure should be sought, such as that illustrated in Figure C6.5. Bolted together corrugated steel pipes or half-drums may suffice for channels with low flows. If none of these options are likely to work, then consideration should be given to redirecting the drainage elsewhere.

Channel linings (especially those constructed in concrete) will tend to increase flow velocities due to reduced flow resistance, thus increasing scour potential on the downstream side. Provision will need to be made for this by keying the end point into scour-resistant rock or, where this is not possible, constructing a downturn into the stream bed or a scour check across the channel. The use of stilling basins and checkdams should also be considered as a means of slowing down the flow of water.

C7.4 Culvert outlets

Figure C6.18 illustrates how different culvert configurations influence outlet scour potential. Unless they are constructed on erosion-resistant bedrock, some form of protection will usually be necessary at all culvert outlets in mountainous terrain whatever their configuration. As a minimum, a flexible apron should be provided downstream of the outlet using gabion or riprap. Mortared masonry should only be used where it can be constructed directly onto rock or weathered rock, and even then concrete would be more durable in this situation. The actual requirements for the design of outlet protection measures at any particular site will depend upon hydraulic, bed erodibility, sediment transport and side slope stability considerations.

Figure C7.22 shows a typical detail for a pipe culvert outlet with nominal scour protection. Culverts with drop outlets have the potential to cause significant erosion at the toe of the culvert headwall, and this can be rectified by a splash pad or stilling basin. For medium-sized culverts, and in situations where downstream materials are susceptible to scour, the requirements for additional protection can usually be assessed using hydrological and engineering geological judgement. It may prove necessary to introduce splash pads, stilling basins, checkdams and cascades into the downstream protection works. Figure C7.23 shows a generic box culvert outlet detail comprising a masonry or concrete cascade, a stilling basin and a riprap or gabion apron. If used, this detail would need to be adapted to suit the actual topography and ground conditions beneath each culvert. For example, Figures C7.24 and C7.25 illustrate a site-specific design constructed beneath a culvert outlet in Laos. In this particular case, the slope on the left-hand side of the photograph (Fig. C7.24) was failing as a result of incision in the stream channel beneath the culvert and the erosion protection measures shown were required to halt this incision in order to reduce ongoing slope movements.

In some instances scour protection measures may be required to extend over considerable distances downstream and comprise a combination of the configurations shown in Figures C7.20, C7.24 & C7.25.

Figure C7.26 shows outlet protection works beneath an arch culvert in Ethiopia. Erosion is continuing due primarily to seepage beneath erosion control structures, leading to under-scour on the downstream side. Additional measures in this case might include masonry or concrete underpinning with flexible gabion mattress protection below. The example illustrates the need to combine the hydraulic and engineering geological considerations of a site before designing outlet protection works.

On very steep gradients in erodible soils and highly weathered rock, it may not be feasible to protect stream channels beneath some culvert outlets from scour. In such cases, consideration should be given to diverting runoff to other culverts or to establishing deeply founded culvert headwalls and adjacent road fill retaining walls that are either supported on bedrock or are well below the level of anticipated scour. In the latter case, scour is allowed to continue without affecting the stability of the road.

For large culverts and bridge crossings it will be necessary to predict velocities and scour depths on a site-by-site basis.

Gabion Cascade Type A - short outfall protection

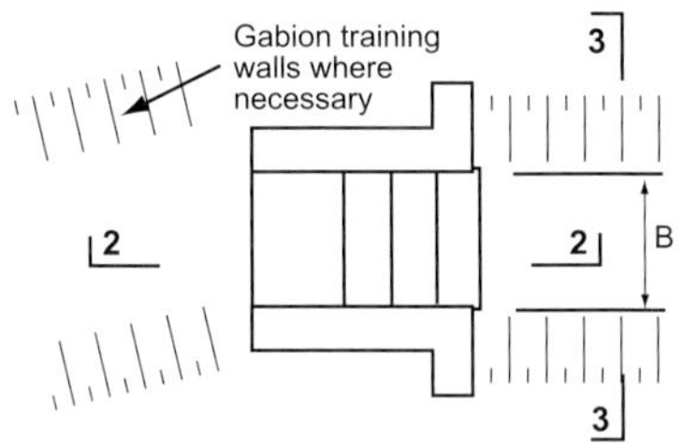

Channel width	B	Drop H									
		3		4		5		6		7	
		D_{sh}	L_{sh}	D_{sh}	L_{sh}	D_{sh}	L_{sh}	D_{sh}	L_{sh}	D_{sh}	L_{sh}
0.6	2.0	1.0	2.3	1.0	2.4	1.0	2.5	1.0	2.6	1.0	2.7
0.9	3.0	1.0	2.9	1.0	3.1	1.0	3.3	1.5	3.4	1.5	3.6
1.2	4.0	1.5	3.6	1.5	4.0	1.5	4.2	2.0	4.4	2.0	4.6

H, L_{sh}, and D_{sh} are shown on Section 2-2

Section 2-2 **Section 3-3**

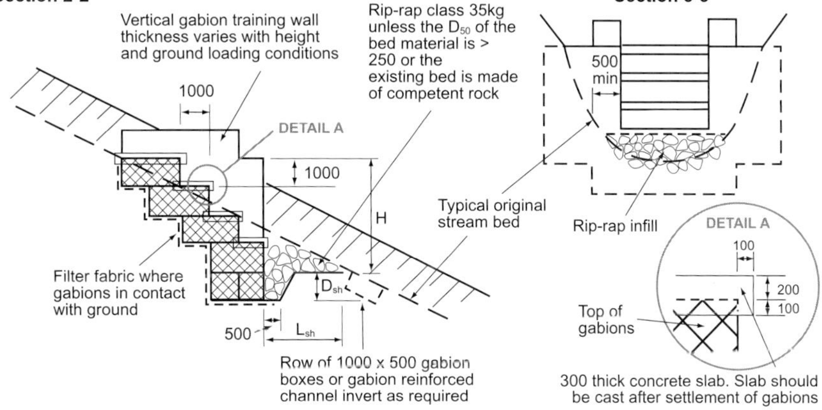

Gabion Cascade - Type B - long outfall protection

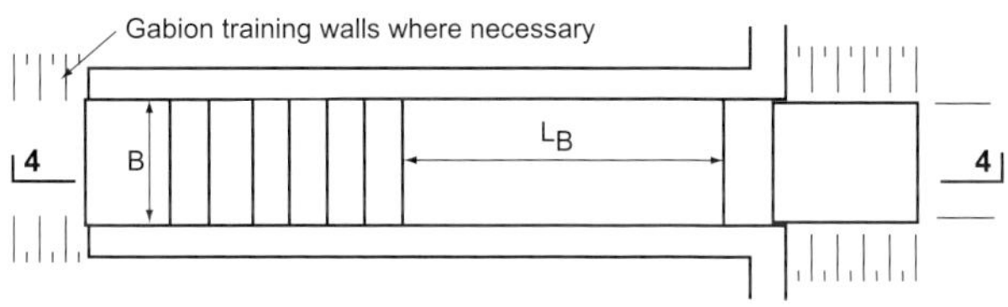

Channel width	B	L_B
0.6	2.0	4.5
0.9	3.0	6.5
1.2	4.0	8.5

Section 4-4

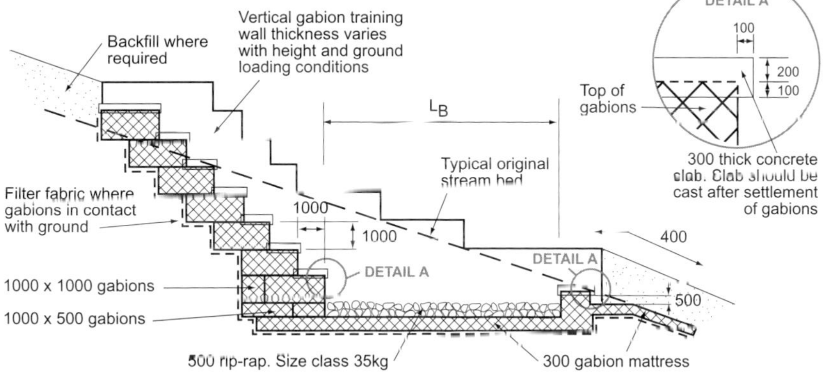

All dimensions in mm, with exception of tables which are in metres
Detail A optional depending upon sediment load and flow velocity

Fig. C7.20. Typical details for a gabion cascade.

Fig. C7.21. Rigid channel lining dislocated by ground movement.

Velocity can be calculated from the Manning (1891) formula using roughness coefficients listed, for example, in Chow (1964). However, it should be borne in mind that maximum scour can be twice the normal scour depth calculated from regime theory (Kellerhals 1967). Furthermore, observed scour depths following the passage of a flood may be misleading due to the infilling of scour holes by deposition during flood recession. Further discussion on scour prediction and hydraulic design for stream crossings on mountain roads is given in TRL (1997). Hoffmans & Verheij (1997), CIRIA (2002) and Neil (2004) provide guidance on scour prediction and protection in the design of hydraulic structures.

C7.5 Slope toe protection and river training

Where roads are located on the slopes above actively eroding rivers and streams, toe erosion can cause instability leading to significant road damage or loss. Scour protection measures may be required to combat this hazard. However, these measures can be extremely difficult to maintain (Fig. C7.27) in high-energy rivers using low to medium cost designs (riprap, revetments and groynes). River training works can assist in diverting flood waters away from vulnerable river banks but, in due course, most high-energy mountain rivers will revert to their original flow pattern during flood flows, destroying under-designed control structures in the process. Gabion structures are often used in these situations because of their ability to withstand a degree of settlement due to under-scour without loss of structural integrity. However, only a certain degree of movement can be tolerated and gabion wire will be prone to abrasion in debris-laden rivers unless protected. Again, this emphasizes the need to combine hydraulic analysis with engineering geological and (in particular) geomorphological assessment in order to derive satisfactory design solutions. The choice of cross-section for new road alignments can be critical to carriageway stability and the maintenance of road access in such situations (Section C2.1).

To illustrate, Figure C7.28 shows a section of road destroyed by river flooding in Nepal. The upper photograph was taken following major flood damage in 1984. This included the destruction of several hundred metres of road and the undermining and failure of a short-span bridge that had been constructed across the debris fan of a tributary river. The road was reconstructed after the 1984 flood and a series of gabion groynes were built in an attempt to divert the river away from the road in this area. The lower photograph shows approximately the same location in

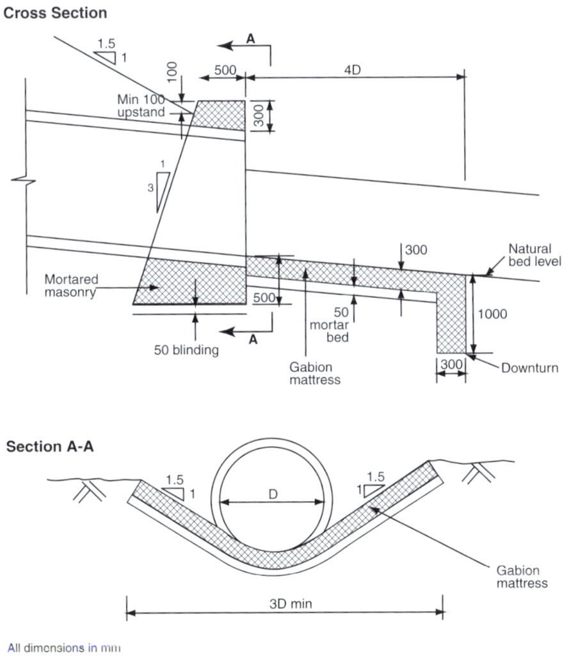

Fig. C7.22. Example of typical pipe culvert outlet scour protection for a low-cost road.

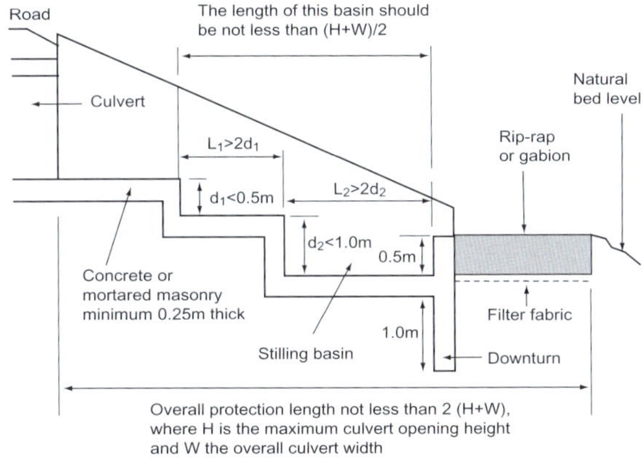

Fig. C7.23. Example of typical box culvert outlet protection works.

2003, where the groynes have been entirely removed by river scour and the road destroyed once more.

For a road to survive in such environments it will be necessary to use protection and retaining structures founded beneath maximum likely scour depth and constructed from materials capable of withstanding the abrasion of debris-laden floodwaters. Where it is not feasible to construct foundations deeply enough comprehensive scour protection will be required, using riprap, aprons and mattresses for example, to prevent the foundation from being exposed. Figure C7.29 shows a road in a riverside location in Laos. It has been constructed on reinforced earth fill protected to full height by gabion mattresses. The embankment is supported by a gabion toe wall constructed on a reinforced concrete scour buttress.

In less erosive riverside locations, the structures detailed in Figure C7.30 may be appropriate and a decision will need to be made as to their required upstream and downstream extent. In some cases this may be obvious, for example at the toe of an unstable slope or on a river meander bend. However, river meanders tend to migrate downstream, sometimes significantly over short periods of time (metres per year), and this may need to be considered. If it is necessary for a road to cross a meandering river then bank protection works will need to extend upstream and downstream of the river crossing sufficiently far to counter future meander movement and to ensure that outflanking cannot occur. The precise details for each river crossing should be considered on a site-by-site basis, although the following general guidance can be given. In order to avoid potential local scour effects, bank protection works should be smoothly aligned and extend a minimum of 0.75 times the flood width of the river upstream and 0.25 times that width downstream of the bridge. It is preferable to key all erosion protection works to an erosion-resistant point on the river bank, such as a rock outcrop. If this is not possible, then consideration should be given to extending the protection further.

Decisions over the type, depth and extent of scour protection works should be made by a hydrologist taking catchment size, runoff regime, river channel cross-section, floodplain geomorphology and bed and bank erodibility into consideration. This assessment should be carried out in conjunction with geological and geomorphological advice.

Fig. C7.24. Example of culvert outlet protection works.

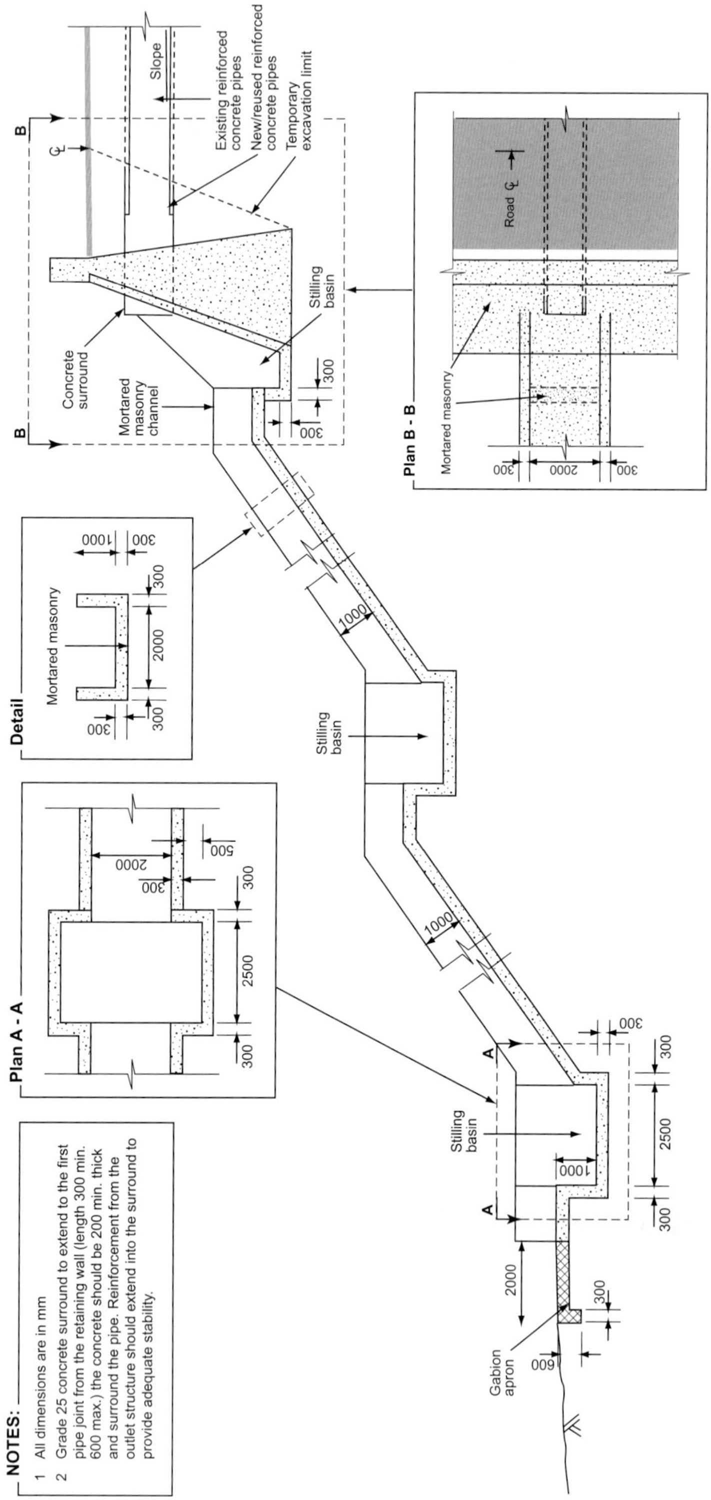

Fig. C7.25. Details of culvert outlet protection shown in Figure C7.24.

Fig. C7.26. Damaged outlet protection.

Fig. C7.28. Flood damage in Nepal: upper 1984; lower 2003.

Fig. C7.27. Undermining of roadside structures due to river scour.

Fig. C7.29. Riverside scour protection in a highly erosive environment.

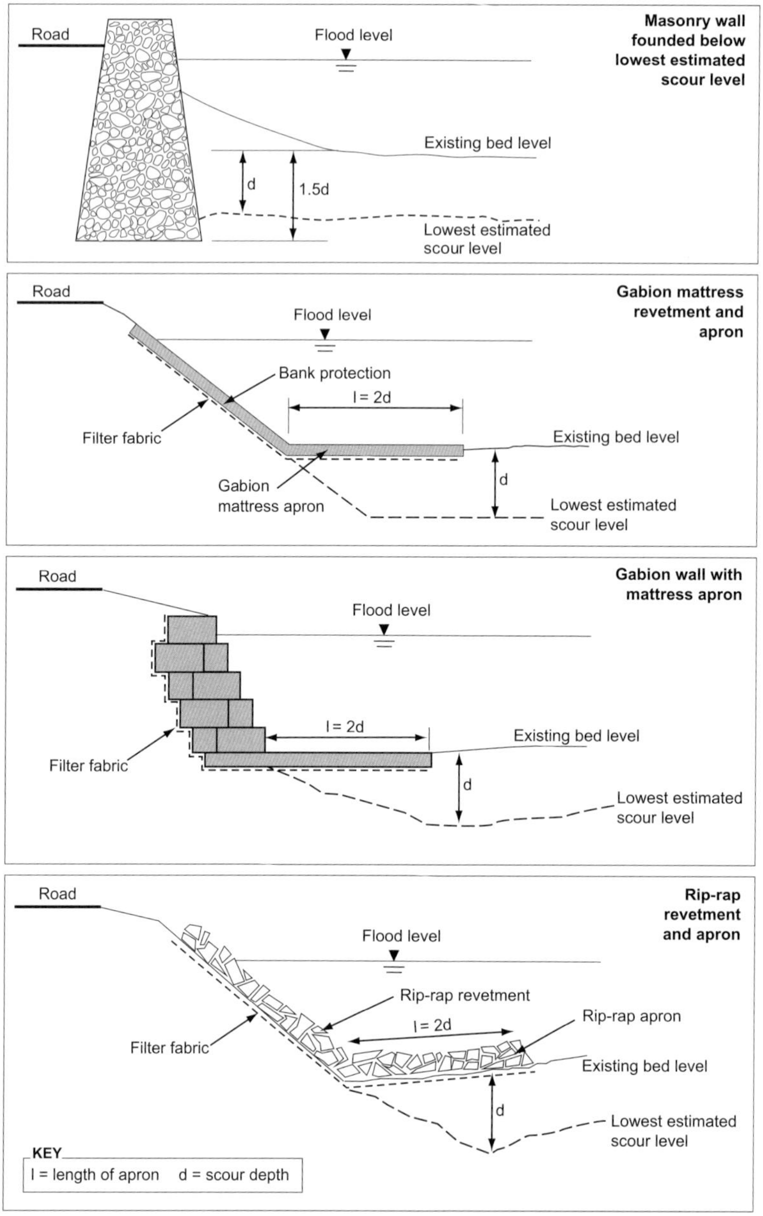

Fig. C7.30. Typical protection in a riverside location of moderate scour potential.

C7.6 Debris flow control and debris fan crossings

As described in Sections A3.4 and C1.2 debris flows and debris fans can pose considerable hazards to mountain roads. Debris flows can be extremely destructive and very difficult to control while debris fans can undergo both scour and deposition during flood events, posing significant problems for road crossings. Figure C7.31 illustrates some of the methods that might be considered when attempting to control debris flows and cross active debris fans. Clearly, the most appropriate method will be dependent on factors such as size and frequency of debris flows, rates of scour

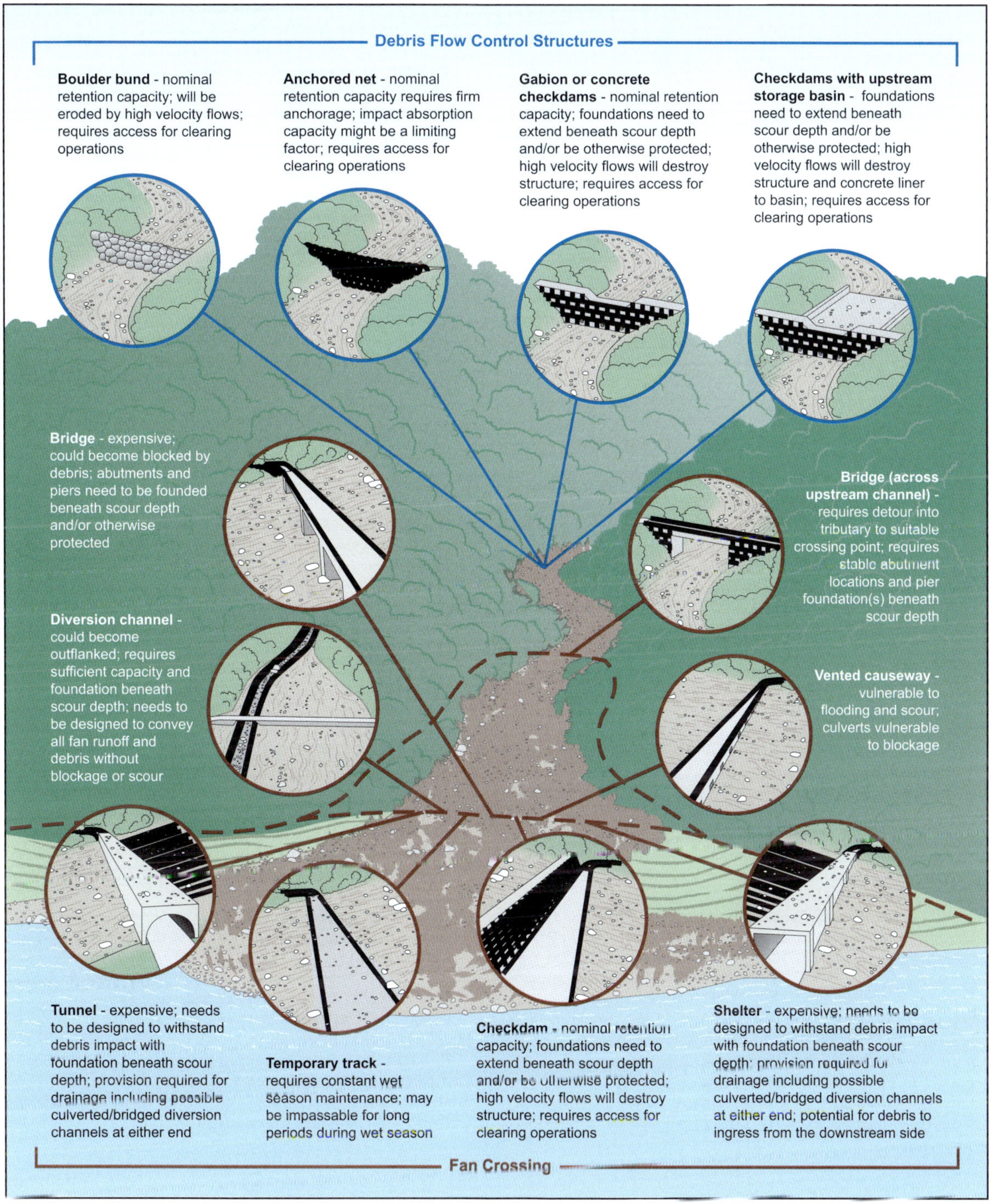

Fig. C7.31. Some options for debris flow control and fan crossings.

and aggradation on fans, traffic volumes and available budgets. Generally, however, the more costly measures, if designed correctly, are likely to prove the most cost-effective in the longer term.

References

BARKER, D. H., WATSON, A. J., SOMBATPANIT, S., NORTHCUTT, B. & MAGLINAO, A. R. (eds) 2004. *Ground and Water Bioengineering for Erosion Control and Slope Stabilization*. Science Publishers, Inc., Enfield.

CAMPOS, S. J. A. M., GALLARDO, A. L. C. F. *ET AL.* 2010. Technological alternatives for erosion control in road construction. *In*: WILLIAMS, A. L., PINCHES, G. M., CHIN, C. Y., MCMORRAN, T. J. & MASSEY, C. I. (eds) *Geologically Active. Proceedings of the 11th IAEG Congress, Auckland, New Zealand, 5–10 September, 2010*, Taylor and Francis Group, London, 4005–4012.

CHOW, V. T. 1964. *Handbook of Applied Hydrology*. McGraw-Hill, New York.

CIRIA. 2002. *Manual on Scour at Bridges and other Hydraulic Structures*. Construction Industry Research and Information Association, Report **C551**, London. Prepared by MAY, R. W. P., ACKERS, J. C., KIRBY, A. M.

CUTLER, D. F., GASSON, P. E. & FARMER, M. C. 1990. The wind-blown tree survey: analysis of results. *Arboricultural Journal*, **14**, 265–286.

FRASER, A. I. & GARDINER, J. B. H. 1967. *Rooting and Stability in Sitka Spruce*. Forestry Commission Bulletin No 40, HMSO, London.

GRAY, D. H. & SOTIR, R. B. 1996. *Biotechnical and Soil Engineering Slope Stabilization. A Practical Guide for Erosion Control*. John Wiley and Sons Inc, New York.

GREENWOOD, J. R. 2006. SLIP4EX – program for routine slope stability analysis to include the effects of vegetation, reinforcement and hydrological changes. *Journal of Geotechnical and Geological Engineering*, **24**, 449–465.

HART, J., HEARN, G. & CHANT, C. 2002. Engineering on the precipice: mountain road rehabilitation in the Philippines. *Quarterly Journal of Engineering Geology and Hydrogeology*, **35**, 223–231.

HEARN, G. J. 1987. *An Evaluation of Geomorphological Contributions to Mountain Highway Design with Particular Reference to the Lower Himalaya*. PhD thesis University of London.

HOFFMANS, G. J. C. M. & VERHEIJ, H. J. 1997. *Scour Manual*. Balkema, Netherlands.

HOWELL, J. H. 1999. *Roadside Bio-engineering*. Department of Roads, Kathmandu.

HUNT, T., HEARN, G. & HOWELL, J. 2010. Slope stabilisation trials in Laos. *In*: WILLIAMS, A. L., PINCHES, G. M., CHIN, C. Y., MCMORRAN, T. J. & MASSEY, C. I. (eds) *Geologically Active. Proceedings of the 11th IAEG Congress, Auckland, New Zealand, 5–10 September, 2010*, Taylor and Francis Group, London, 4207–4216.

KELLERHALS, R. 1967. Stable channels with gravel-paved beds. *Journal of Waterways and Harbors Division*, American Society of Civil Engineers, 63–84.

KOJAN, E. 1978. *Engineering Geological Evaluation of Landslide Problems, Western Hills Road Project: Godavari to Dadeldhura, Nepal*. Department of State Agency for International Development, Washington DC.

KULKARNI, T. 2009. Rock fall and debris flow mitigation techniques. *International Symposium on Geotechnical Engineering, Ground Improvement and Geosynthetics for Sustainable Mitigation and Adaption to Climate Change, Including Global Warming*, 3–4 December 2009, Bangkok, Thailand, 129–135.

MANNING, R. 1891. On the flow of water in open channels and pipes. *Transactions of the Institution of Civil Engineers of Ireland*, **20**, 161–207.

MORGAN, R. P. C. & RICKSON, R. J. (eds) 1995. *Slope Stabilization and Erosion Control: A Bioengineering Approach*. E & F N Spon, London.

National Research Council. 1993. *Vetiver Grass: A Thin Green Line Against Erosion*. National Academy Press, Washington DC.

NEIL, C. R. (ed.) 2004. *Guide to Bridge Hydraulics*. 2nd edn. Transportation Association of Canada/Thomas Telford Ltd, London.

NORRIS, J. E. & GREENWOOD, J. R. 2003. Root reinforcement on unstable slopes in Northern Greece and Central Italy. *Proceedings of an International Conference on Problematic Soils*. Thomas Telford, London, 411–418.

NORRIS, J. E. & GREENWOOD, J. R. 2010. Site evaluation and slope stability analysis for bioengineering. *In*: WILLIAMS, A. L., PINCHES, G. M., CHIN, C. Y., MCMORRAN, T. J. & MASSEY, C. I. (eds) *Geologically Active. Proceedings of the 11th IAEG Congress*, Auckland, New Zealand, 5–10 September, 2010, Taylor and Francis Group, London, 4393–4400.

PHILLIPS, C. J. & WATSON, A. J. 1994. *Structural Tree Root Research in New Zealand: a Review*. Landcare Research Science Series No 7. Landcare, New Zealand.

RAY, D. & NICOLL, B. C. 1998. The effect of soil water-table depth on root-plate development and stability in Sitka spruce. *Forestry*, **71**, 169–182.

SAATHOFF, F. 2003. Geosynthetics in geotechnical and hydraulic engineering. *In*: *Geotechnical Engineering Handbook*, Volume **2**, Ernst & Sohn, Berlin, 507–597.

STHAPIT, K. M. 1996. *Windthrown Tree Roots Survey: Eastern and Central Terai, Nepal*. Forest Research and Survey Centre, Kathmandu.

TRL 1997. *Principles of Low Cost Road Engineering in Mountainous Regions*. Overseas Road Note 16, Transport Research Laboratory, Crowthorne.

WU, T. H. 1995. Slope stabilization. *In*: MORGAN, R. P. C. & RICKSON, R. J. (eds) *Slope Stabilization and Erosion Control: A Bioengineering Approach*. E & F N Spon, London.

ZURICK, D. & KARAN, P. P. 1999. *Himalaya: Life on the Edge of the World*. John Hopkins University Press, Baltimore.

Part D: Slope Management

D1 Slope management

G. J. Hearn* & T. Hunt

URS Scott Wilson Ltd, Scott House, Alençon Link, Basingstoke, Hampshire RG21 7PP, UK
**Corresponding author (e-mail: gareth.hearn@scottwilson.com; garethhearn@talktalk.net)*

D1.1 Components of slope management

Slope management is a critical activity for most road authorities, but particularly those with a road network that traverses hilly or mountainous terrain. Computerized road management systems are now commonplace on high-volume roads (e.g. Paige-Green 1997; Robinson *et al.* 1998; McPherson & Bennett 2005; Russell *et al.* 2008) with in-built deterioration and cost-benefit models to inform the user as to what length of road needs to be upgraded, by how much and when. Although slope management systems have been developed (e.g. Heath *et al.* 1995; Heath & McKinnon 1996; Bujang & Jamaludin 2005; Lee *et al.* 2006; Kwon & Baek 2010; Leyland 2010), these are relatively uncommon on low-volume roads. A significant amount of asset data are required to be collected in the development of road and slope management systems; on low-volume roads it is likely to be most prudent to focus on the difficult areas first, that is, those where road assets are most at risk from slope instability.

Slope management encompasses all factors that affect the stability of a slope, whether natural or man-made, and therefore also includes road and slope drainage, erosion protection, retaining walls and river training works where appropriate. Consequently, a broad range of experience is required for effective slope management with skills in the following areas:

- landslide recognition and engineering geology;
- risk assessment for prioritization of sites;
- hazard assessment for selection of options;
- ground investigation and slope monitoring;
- design of slope stabilization and protection measures;
- design of slope drainage systems;
- design of river training and scour protection works; and
- quality control during the construction and maintenance of stabilization works.

Ideally, slope management along mountain roads should be based entirely on a proactive approach, whereby every potential or developing slope problem is anticipated and dealt with in advance of significant movement and road damage. This approach rarely happens, however, because of three main factors:

- limited manpower resources and maintenance budgets;
- uncertainty over ground conditions and the unpredictability of damage arising from major rainstorms and earthquakes; and
- weaknesses in most road management systems, which largely ignore slope maintenance.

As a consequence, reactive slope management is usually unavoidable in the maintenance of access along low-cost roads. In the extreme (though not uncommon) case of widespread damage during a single rainstorm or an exceptionally heavy wet season, even timely reactive maintenance management can be rendered impracticable and unaffordable due to the need for emergency response on multiple fronts with limited resources.

D1.2 Planning slope maintenance

Even though maintenance funding may be inadequate, a formal plan of slope maintenance for a road network should always be prepared. This helps to justify budget allocations for slope management works and enables problems to be identified and addressed in a timely way. A plan of this nature might typically follow the structure of, and be closely connected to, a standard maintenance management system for the road itself. It would define:

- the assets that need to be maintained (a schedule of slopes and structures; as-built drawings and location plans);
- the anticipated maintenance schedule;
- the inspection schedule (the frequency of inspections required, linked to seasonal changes, particularly the wet season);
- the decision-making procedure (particularly the linkages between site inspectors and managers regarding slope problems that require specialist assessment or detailed investigation);
- the prioritization procedure (the process for determining the order of response in both time and budget allocations);

From: HEARN, G. J. (ed.) *Slope Engineering for Mountain Roads*. Geological Society, London, Engineering Geology Special Publications, **24**, 269–284.
DOI: 10.1144/EGSP24.17 0267-9914/11/$15.00

- the programming system (the timing during which the prioritized response is implemented);
- the responsibilities and resource allocations within the programme (the intervention frequency for different problems and the division of responsibilities between different units e.g. lengthworkers, maintenance contractors and specialist contractors); and
- the quality assurance procedure (the mechanism for monitoring the implementation of the programme to ensure that it is completed satisfactorily and according to specification).

The main problem with slope maintenance, compared to that for a road pavement or for drainage structures, is that slope instability is inherently less predictable than the failure of engineering materials and manufactured assets that have known deterioration rates and prescribed design lives. The resources required to maintain slopes can therefore vary considerably from year to year depending on natural factors, particularly rainfall.

It may remain the case that fully structured maintenance strategies for slopes will tend to be more frequently applied to roads with high traffic volumes than to low-volume mountain roads. Nevertheless, some elements of a formally planned maintenance system can be included at low cost and can contribute significantly to the effective management of slopes.

The sections below describe procedures that can either be adopted as part of a formal slope maintenance plan or can be used in a less structured approach. What is most important is that, for all low-volume roads, the road agency should have an annual programming and budgeting system that allows a high degree of flexibility in the allocation of resources in response to natural events. This is particularly important for emergencies, but is also necessary for situations where unusual rainfall events activate slope movements that are not yet high risk but will get worse if ignored.

D1.3 Categories of slope maintenance

Slope maintenance can be subdivided into four categories:

- *routine* maintenance, which should be carried out throughout the year irrespective of the condition of the slopes;
- *preventative* maintenance, which may be required at certain locations where underlying slope conditions (Section A3.1) indicate a potential for slope failure or where signs of incipient ground movement are detected;
- *emergency* maintenance, which is necessary to clear road blockages and/or to make safe a slope, section of road or retaining wall on a temporary basis until remedial maintenance can be carried out, and
- *remedial* maintenance, which is necessary to rehabilitate a slope, section of road or retaining wall that has already failed.

Special consideration needs to be given to slope movement monitoring (Section B5). Monitoring should form an important part of all four categories of maintenance listed above, and should include:

- visual monitoring of all existing slopes and structures in order to be aware of any changes that might give cause for concern;
- visual monitoring of new works undertaken as part of preventative, emergency and remedial maintenance in order to be certain that the works are performing as designed; and
- measurement monitoring of specific sites of known, suspected or anticipated ground movement, as discussed in Section B5.

D1.3.1 Routine slope maintenance

Table D1.1 lists common routine maintenance works for slopes and retaining walls.

Routine maintenance typically comprises:

- clearing and repair of roadside drains;
- clearing and repair of surface drainage structures on slopes (e.g. cascades, chutes and bench drains on cut slopes);
- clearing of culverts;
- minor repairs to culverts;
- minor repair of erosion damage;
- minor repairs to retaining walls; and
- trimming of vegetation on roadside slopes.

The need for routine maintenance cannot be overemphasized. Timely repairs and clearance of blockages to roadside drains and culverts may prevent uncontrolled rainfall runoff from creating instability below a road, ultimately resulting in slope failure and costly emergency and remedial works. Routine slope maintenance inspections may reveal early warning signs of instability and the need for preventative slope maintenance.

Local landowners and residents will sometimes deliberately block roadside drainage structures in order to facilitate irrigation, prevent road runoff from discharging onto agricultural land or to provide vehicular access to houses and private land next to the road. These practices will often lead to the uncontrolled flow of water onto adjacent slopes, creating the problems of erosion and instability mentioned above. Where anticipated, provision can be made for these drainage modifications during design and construction, but it will be necessary to carry out inspections during routine maintenance in order to ensure that such drainage interventions are managed safely. It is important to remember that land use and land use practices adjacent to roads can frequently change, and that alterations which can affect slopes and road structures are inevitable.

The timing of routine maintenance of slopes must necessarily be devised on a site-specific basis. Figure D1.1 gives two examples of general schedules that would be suitable for slope maintenance in the high-rainfall areas of South

Table D1.1. *Typical routine maintenance works for slopes and retaining walls (modified from GEO 2003)*

Road asset or structure	*Typical maintenance works required*	*Guidance*
Surface drainage channels and catchpits	• Clear debris, undesirable vegetation and other obstructions. • Repair minor cracks with cement mortar or flexible sealing compound. • Rebuild severely cracked channels. • Replace missing or deteriorated joint fillers and sealant.	• Works may be required outside the right of way to prevent debris from blocking the drainage system. • Where large tree roots have damaged drainage channels, appropriate portions of the roots should be removed, taking care not to jeopardize the stability of the tree. Alternatively, the channels may be realigned.
Weepholes and drainage pipes	• Clear obstructions (e.g. weeds and debris) in weepholes and pipe ends. • Probe with rods for deeper obstructions.	• Drainage pipes are prone to being blocked. Where pipes used on slopes leak or are severely blocked, they should be replaced with drainage channels where possible.
Impermeable surface cover (e.g. chunam and shotcrete)	• Remove undesirable vegetation growth. • Repair cracks or spalling. • Re-grade and repair eroded areas. • Replace surface cover that has separated from underlying slope. • Replace missing or deteriorated joint fillers and sealant. • Remove dead, decaying or unstable trees.	• Cracked impermeable surface cover should be repaired by cutting a chase along the line of the crack, and filled with a similar slope cover material or a flexible sealant. • Where large tree roots have damaged the surface cover, the cover should be replaced and the tree roots removed. • Specialist advice may be sought in treating trees. Tree-felling permission should be obtained from relevant authority where necessary.
Vegetated surface cover	• Regrade eroded areas with compacted soil followed by replanting • Replant vegetation in areas where it has died. • Trim vegetation if overgrown. • Remove dead, decaying or unstable trees.	• Where erosion is shallow and does not affect the performance of existing surface drainage channels, the eroded area may be regraded by trimming (without backfilling). • Surface erosion may indicate possible inadequacy of the drainage system. The source of concentrated flow should be identified and rectified. • Specialist advice may be sought on types of cover or species in areas where there is insufficient sunlight to support vegetation growth or where specific species and planting techniques are required for bio-engineering purposes (Section C7.2.4).
Rock slopes and boulders	• Repair cracked or spalled concrete surfaces and support. • Remove loose rock debris. • Remove undesirable vegetation growth.	• It is not advisable to remove all vegetation indiscriminately, but tree roots causing prising action in rock joints should be removed.
Facing	• Re-point deteriorated mortar joints on masonry faces. • Repair cracking or spalling of concrete surface and replace missing or deteriorated joint fillers and sealant.	• Monitor for any continuing movement or deterioration.

South Asia

Routine activity	Jan	Feb	Mar	Apr	May	Jun	Jul	Aug	Sep	Oct	Nov	Dec
Slope earthworks												
Slope trimming		✓	✓	✓								
Small slip clearance						✓	✓	✓	✓	✓		
Erosion repair work									✓	✓	✓	✓
Slope structures												
Masonry wall repairs		✓	✓	✓								
Gabion wall repairs		✓	✓	✓								
Drain cleaning						✓	✓	✓	✓	✓		
Drain repairs							✓	✓	✓	✓	✓	✓
Vegetation works												
Protection from grazing and theft	✓	✓	✓	✓	✓	✓	✓	✓	✓	✓	✓	✓
Extra protection from fire	✓	✓	✓	✓								
Enrichment and replacement planting						✓	✓	✓				
Mulching new plants											✓	✓
Grass cutting							✓	✓	✓	✓	✓	
Pruning and thinning shrubs	✓	✓										✓
Inspections					✓	✓	✓	✓	✓	✓		

Routine activity	Jan	Feb	Mar	Apr	May	Jun	Jul	Aug	Sep	Oct	Nov	Dec
Slope earthworks												
Slope trimming	✓	✓										✓
Small slip clearance							✓	✓	✓	✓	✓	
Erosion repair work										✓	✓	✓
Slope structures												
Masonry wall repairs	✓	✓										✓
Gabion wall repairs	✓	✓										✓
Drain cleaning						✓	✓	✓	✓	✓		
Drain repairs	✓	✓					✓	✓	✓	✓	✓	✓
Vegetation works												
Protection from grazing and theft	✓	✓	✓	✓	✓	✓	✓	✓	✓		✓	✓
Extra protection from fire	✓	✓	✓									
Enrichment and replacement planting				✓	✓	✓						
Mulching new plants										✓	✓	
Grass cutting				✓	✓	✓	✓	✓	✓	✓	✓	
Pruning and thinning shrubs	✓											✓
Inspections			✓	✓	✓	✓	✓	✓	✓	✓	✓	

Fig. D1.1. Indicative schedules of routine slope maintenance activities for two geographical areas (wet season shown as shaded cells). Information provided by J. Howell.

Asia and West Africa, and indicates how interventions need to be timed in relation to the wet season.

D1.3.2 Preventative slope maintenance

Preventative slope maintenance is the one maintenance activity that is likely to be ignored in the context of low-volume roads, usually because the overall maintenance funding is inadequate and any funds spent on early preventative maintenance may jeopardize later expenditure required for emergency and remedial maintenance. Furthermore, expenditure on preventative maintenance may, in some situations, be unnecessary if observations and predictions of future slope deterioration and ground movement prove to be incorrect.

Preventative slope maintenance should be undertaken where:

- underlying ground conditions (conditioning factors) or external processes such as toe erosion, earthworks or irrigation practices (triggering factors) create a recognized potential for slope failure (Section A3.3); or
- there are signs of incipient failure of slopes and structures.

Signs of incipient failure might include:

- curvilinear cracking or depression of the road at the crest of a fill slope or behind a retaining wall;
- cracking to retaining walls and other structures; or
- tension cracks in the ground surface at the crest of a cut slope or in the natural hillside above (Section B3.4).

Options for increasing the stability of slopes are given in Tables C3.7 and C4.2. The timing of preventative slope maintenance is important and ideally takes place during the dry season.

D1.3.3 Emergency slope maintenance

Emergency inspections should be carried out during or immediately after prolonged periods of heavy rain, a significant seismic event or whenever a slope failure is reported.

Emergency maintenance typically comprises:

- the installation of warning signs and other road safety measures to ensure the safety of road users (where appropriate, the police and other civil authorities should be given details of the actual danger along with warnings to inhabitants of occupied buildings that may be at risk);
- the temporary reinstatement of sections of road that have failed due to slope instability or wall failure below;
- clearance of landslide debris from the road carriageway, roadside drains and other drainage channels;
- the provision of temporary surface drainage to divert surface water flows away from the area of instability;
- the use of temporary retaining structures to support slopes that have failed, either above or below the road, using gabions or sheet piling for example;
- repair of culverts, which may include replacing and backfilling of damaged sections or replacement with larger structures if necessary;
- repair to damaged scour protection works or the construction of additional scour protection that may be necessary as an emergency measure; and
- repair of retaining walls, which may include any works necessary to temporarily stabilize a wall prior to replacement.

Emergency works require careful supervision. Strong pressure to re-open a road can lead to unplanned actions that can have serious long-term consequences. A common example is shown in Figure D1.2, where debris cleared from a small slip onto a road has been tipped over a newly stabilized slope, creating the potential for further instability. Guidance on the safe disposal of debris is given in Section C2.6 and it is recommended that suitable spoil disposal areas are identified and documented in advance of emergency situations. More substantial problems are sometimes caused during unplanned emergency works by altering drainage channels, either by using them for spoil disposal or by installing badly designed temporary drainage diversion measures.

D1.3.4 Remedial slope maintenance

Remedial slope maintenance is necessary where a slope, section of road or retaining wall has failed, and where the emergency maintenance carried out to safeguard road access during the wet season needs to be replaced or strengthened as part of the permanent works.

Remedial slope maintenance will usually need to be delayed until the onset of the dry season and should follow a programme of investigation and design as described in Parts B and C. Typical engineering management options for remedial works are given in Tables C3.6, C3.7 and C4.2.

Fig. D1.2. Landslide debris tipped onto a newly stabilized slope.

D1.4 Inspection

Although most road management systems will include a database containing basic information on the pavement, retaining structures, side drains, culverts and bridges within the road network, as noted earlier they are very unlikely to contain any information on slopes. However, there are some computerized slope management systems available that are capable of holding a detailed inventory of slopes and retaining walls on a road network. As a minimum, these record slope or wall location, areal extent or size and condition, and are able to be updated. They are usually structured as a database with drop-down menus of descriptor categories and classifications, and are usually GIS-based (Section B2.7).

In the absence of such a system, a spreadsheet inventory (preferably backed up with digital photography) could be used to record the main attributes of each slope or retaining wall and a simple example of this is given in Figure D1.3. Routine and detailed slope inspections will be necessary to update the inventory and provide additional information on the maintenance strategies to be followed.

D1.4.1 Routine inspection

Slope instability is most likely to occur after periods of heavy or prolonged rainfall. In countries with a pronounced wet season, routine inspections of roadside slopes, retaining walls and drainage structures are recommended at the following intervals:

- shortly before the onset of the wet season, to check that the dry season maintenance has been carried out and that all drains are clear of debris;

FEATURE INVENTORY										
Road No:	13	Section from:		Luang Prabang			To	Kasi		
Location	Type of feature						Comments	Date	Action	
	Above road			Below road						
	Slope	Wall	Drain	Slope	Wall	Culvert			Maint Team	Detailed inspection
65+280/380			100L M				Drain blocked at 65+355	03.05.09	✓	
65+350/375	15x15 C						Erosion of cut slope: existing slope drainage ineffective	03.05.09	✓	
65+380						1P	Culvert outlet showing signs of movement	03.05.09		
65+380/700			320L M				Good condition	03.05.09		
65+430/450				20x30 F			Some large tension cracks at top of slope	03.05.09		✓
65+600/615					4x15 M		Evidence of scour at lower end of wall	03.05.09	✓	
65+650/670		3x20 G					Gabion wires corroding	03.05.09		
65+650/670	20x30 C						Remedial works last carried out in 2005	03.05.09		

LEGEND	
Slope	5x10 = length (perpendicular to road) and width (parallel to road) of feature in metres C = Cut slope F = Fill slope N = Natural ground
Wall	6x15 = height and length (parallel to road) in metres M = Masonry G = Gabion RC = Reinforced concrete
Drain	50L/R = length (parallel to road) on Left or Right side looking up chainage, in metres U = Unlined M = Masonry RC = Reinforced concrete
Culvert	1 = culvert diameter or width in metres P = Pipe B = Box S = Slab

Fig. D1.3. Slope, wall and culvert inventory.

- regularly during the wet season, to check that the drainage structures are functioning correctly and to identify any incipient stability concerns so that they can be dealt with in a timely manner as preventative maintenance; and
- immediately after the wet season, to ascertain the extent of any damage and plan the remedial works for the remainder of the dry season.

A routine inspection should yield one of the following outcomes:

- continued monitoring;
- continued routine maintenance; or
- identification of the need for a detailed inspection for possible preventative or remedial action.

Routine inspections should include structures that are not otherwise easily seen from a passing vehicle, such as culvert inlets and outlets, below-road retaining walls and drainage structures above the road. Figure D1.3 can be used as the basis for recording the outcome of a routine inspection. The right-hand columns indicate the need for routine maintenance or detailed inspection prior to carrying out preventative, emergency or remedial maintenance.

D1.4.2 Detailed inspection

In order to ascertain the scope of preventative, emergency or remedial works a detailed inspection is required. The results of this inspection should be entered into an Instability Report such as that illustrated in Figure D1.4. Guidance on the completion of this Instability Report is given in Table D1.2. The report should be accompanied by a sketch map and cross-section, together with digital photographs. It should be of sufficient detail to enable the approximate scope and cost of temporary (in the case of emergency) works or permanent (in the case of remedial) works to be ascertained. Figure D1.5 illustrates a fairly common outcome where the full extent of ground movement is not properly assessed and recorded by detailed inspection prior to the scheduling of remedial works.

D1.5 Risk management

The recommended flow path for risk management decision-making that commences with inspection and problem recognition and progresses through risk assessment and prioritization to works implementation is given in Figure D1.6. This should be used to judge the urgency for action and to prioritize preventative, emergency or remedial works.

For many low cost roads it is common to find road authorities adopting a 'wait and see' approach, as discussed in Section A.4.3.3, with respect to risk management. Where the risk posed by ground movement is judged to be moderate or low (risk ranking 3 and 4 in Section D1.6), the use of low-cost and temporary measures might prove the most cost-effective in the short to medium term until:

- a greater knowledge of the extent and rate of ground movements can be ascertained through monitoring and observation, thus allowing more informed decision-making; and
- the slope movements reduce under the process of self-stabilization, thus negating the need for costly remedial measures.

In order to be able to make these judgements, a detailed inspection, along the lines described in Section D1.4.2, should be carried out.

D1.6 Prioritization

Given the usual conflict between the need to maintain, repair or reinstate a large number of slopes and structures on the one hand and limited resources on the other, some form of prioritization is required. Section A4 illustrates this in relation to the categorization of hazard and the value and vulnerability of road assets at risk. However, the prioritization should also take into account:

- the consequences to life – to people either occupying buildings or travelling in passing vehicles;
- the extent to which traffic can still use the road and, if not, the period required to make the road trafficable;
- whether or not the situation is likely to get worse (more dangerous and more expensive to rectify) if remedial works are delayed.

To assist in decision-making, these factors should be quantified wherever possible in terms of risk, that is, the economic and social losses that will occur if no action is taken compared to the cost of taking action. As discussed in Section A4, the calculation of risk is made difficult because of uncertainties over probability and vulnerability when event data are unavailable. Consequently, some of the risk calculations described by Lee & Jones (2004), Chen *et al.* (2010) and Jaiswal *et al.* (2010), for example, can be difficult to apply. Instead, a more workable approach is required for low-cost road maintenance purposes. Winter *et al.* (2010) devised a method of risk assessment for landslides affecting Scottish hill roads, taking both disruption and potential injury and fatality into consideration for roads classified according to AADT. The assessment was based on a scoring system in the absence of event data. GEO (2009) provides a system for priority ranking of man-made slopes and retaining walls in Hong Kong using instability scores, consequence scores, AADT and other factors. The instability score includes consideration of geology and hydrogeology, signs of distress, age of cutting and level of geotechnical input to its original design, the consequence score includes slope height, land use facility above and below the cut slope and its location and vulnerability with respect to potential failures. A review of risk assessment systems for highway slopes is given by Pantelidis (2011).

INSTABILITY REPORT

1: LOCATION

Road designation [ ] Distance [ ] Reporter [ ] Digital photo Nos [ ] Date of report [ ]

Roadside affected: Left [ ] Right [ ] Both [ ] GPS coordinates [ ]

2: HISTORY

New failure [ ]

Movement/extension of existing failure [ ]

3: SITE DESCRIPTION

[ ]

4: TYPE(S) OF INSTABILITY

Failure of cut slope [ ] Entire hillside [ ]
Failure of fill slope [ ] Erosion of cut slope [ ]
Hillside above road [ ] Erosion of fill slope [ ]
Hillside below road [ ] Erosion of hill slope [ ]

5: MECHANISM AND CAUSE OF INSTABILITY

Plane [ ] Circular [ ] Wedge [ ] Fall/topple [ ] Other [ ]

6: MATERIAL TYPE INVOLVED IN INSTABILITY

Mostly rock [ ] Mostly soil [ ] Both [ ]

7: TOPOGRAPHY

Original hillside angle [ ] Angle of failed material [ ] Angle of original cut slope [ ] Orientation of slope *(°N)* [ ]

8: EXTENT OF UNSTABLE AREA DIRECTLY AFFECTING ROAD

Length *(m perpendicular to road)* [ ] Width *(m parallel to road)* [ ] Ave. Depth *(m estimated)* [ ] Volume estimated *(LxWxD)* [ ]

9: MATERIALS DESCRIPTION

Rock [ ] *In situ* weathered soil [ ] Taluvium [ ] Colluvium [ ] Other [ ]

Rock type [ ]
Weathering grade [ ]
Field description of rock strength [ ]

10: ROCK STRUCTURE

Bedding/foliation dip/dip direction [ ] Other joints, dip/dip direction and spacing [ ]

11: SLOPE DRAINAGE

Dry [ ] Seepage [ ] Water table at surface [ ]

Description of natural/constructed drainage: [ ]

12: VEGETATION AND LAND USE

On slope [ ]
Above slope [ ]

13: IMPACT ON ROAD

Road partially blocked [ ] Road in danger of failure [ ]
Road totally blocked [ ] Road partially failed [ ]
Road in danger of blockage [ ] Road totally failed [ ]

14: IMPACT ON ROAD DRAINAGE

		Damaged	Blocked
Slope drains:	Left	[ ]	[ ]
	Right	[ ]	[ ]
Roadside drain:	Left	[ ]	[ ]
	Right	[ ]	[ ]

Masonry [ ] Unlined [ ] Concrete [ ]

		Damaged	Blocked
Culvert:	Box	[ ]	[ ]
	Pipe	[ ]	[ ]

Pipe dia(m) [ ] Box span(m) [ ]

15: IMPACT ON STRUCTURES

Revetment: Height (m) [ ]

	Above road	Below road
Length affected (m)	[ ]	[ ]
Repair (m)	[ ]	[ ]
Replace (m)	[ ]	[ ]

Retaining wall: Height (m) [ ]

	Above road	Below road
Length affected (m)	[ ]	[ ]
Repair (m)	[ ]	[ ]
Replace (m)	[ ]	[ ]

Cause of damage: Overturning [ ] Foundation failure [ ] Sliding failure [ ]

Impact on road of wall failure if not repaired or replaced: None [ ] Debris onto road [ ] Road blockage [ ] Partial or complete loss of carriageway [ ]

16: IMPACT ON OCCUPIED BUILDINGS (NUMBER)

Damage [ ] Collapse [ ] At risk [ ]

17: PROPOSED REMEDIAL MEASURES

[ ]

On a separate sheet provide a description of the geomorphology of the site and a geomorphological sketch map and sketch cross section. (Include outline of unstable area in relation to road, back scarp, failed material (including failed rock masses and landslide debris), rock outcrop (unfailed), springs and seepages, wet ground, tension cracks and ground cracking, evidence of ground heave, road and structural damage).

Fig. D1.4. Instability report.

Table D1.2. *Checklist for completing the Instability Report (Fig. D1.4)*

No	Descriptor	Notes
1	Location	Road, chainage/kilometre on road, LHS/RHS (looking up-chainage), GPS for GIS entry
2	History	a) New failure/first-time failure; b) Previous failure location/renewed or extended movement
3	Site Description	General description of the site and its surroundings, nature of ground movements and overall relationship to the road. Any specific observations that need to be recorded, such as engineering or land use considerations and effects
4	Type of instability	Many unstable slopes are reactivations of existing landslides. It is important to define the entire extent of ground movement as defined by lateral and back scarps, i.e. not just the most obvious recent/current failure or movement. This identification can often be assisted by changes in land use and slope drainage across landslide boundaries. These details should be drawn on a geomorphological map. Differentiate between slope failures above and below the road, cut and fill slope failures and slope erosion.
5	Mechanism and cause of instability	The classification of mechanism can be made by reference to the plan shape of the landslide or movement, the height and steepness of the back scarp, the apparent depth of failure and the presence of exposed bedding, foliation and joint orientations that may form all or part of the failure surface. Usually, a failed slope angle that is similar and approximately parallel to the slope of the surrounding ground indicates planar failure A failed slope angle that is much less or is compound may indicate circular failure. Wedge failures and falls/topples should be straight forward to define in the field as a result of the configuration of jointing and the materials comprising failure deposits. **Some conditioning factors** (see Table A3.2): a) Adverse jointing; b) Low strength soils; c) Steep slope; d) Other (specify) **Some triggering factors** (see Table A3.2): e) Groundwater table rise; f) River or stream erosion; g) Failure of engineering structure; h) Runoff/drainage from adjacent land use; i) Other (specify)
6	Material types involved in instability	Many slope movements/failures occur along the boundary between overlying soil and underlying weathered rock. In cut slopes these are fairly easy to identify but on natural slopes and where the extent of movement may be only minor, this may not be so easy to assess. It will be necessary to make a judgement based on the area of failure and an assessment of soil depth from surrounding exposures and geomorphology. Failure in rock only should be fairly easy to distinguish from the scarp and deposit.
7	Topography	This information can contribute to an assessment of the maximum slope angles at which slope materials can stand. It can also help in the assessment of mechanisms (5 above). Slope orientation with respect to magnetic North is important if underlying rock discontinuity orientations play a role in conditioning the slope to failure.
8	Extent of unstable area	The areal extent of slope failure or movement can usually be defined by reference to morphological features. A geomorphological map helps in the assessment of landslide/movement depth and mechanism and allows the extent of required remedial measures to be determined. Without ground investigation and in the absence of slip surface evidence in landslide scarps, the assessment of depth can be difficult and inexact. The morphology of the slope can assist in the assessment of depth. The survey of a slope section from back scarp to toe of slope can often help in determining credible landslide depth, especially where landslide mechanism is known.
9	Material description	The definition and description of soil and rock materials allows the likely range of applicable strength parameters to be determined. Standard descriptions and classifications should be used.
10	Rock structure	This can only be described where bedding and other joints are sufficiently well exposed in the landslide scarp or in neighbouring cut slopes and natural exposures. Conventional engineering geological methods require measurement of the maximum angle of dip of each joint plane, the orientation of the maximum angle of dip relative to magnetic North, and an assessment of joint persistence (extent), smoothness and spacing. If there is any infill between the joints then this needs to be described as well.
11	Slope drainage	Slope drainage features including springs, seepages and areas of waterlogged ground should be described. Quite often springs can be seen to emerge on the slopes beneath below-road retaining walls and at the junction between soil and underlying rock. Seepages may also be associated with the locations on the slope where the slip surface 'daylights'. Include reference to any constructed drainage works and their performance.
12	Vegetation and land use	Vegetation patterns can indicate failed ground. Patches of jungle amid agricultural land can indicate unstable ground unsuitable for cultivation. Certain vegetation types indicate high water tables that might be associated with failed ground. Deflection of tree trunks can result from active slope movement as might fallen trees and dead trees. The land use above a landslide area can help assess the effects of runoff and irrigation on slope stability and hence whether drainage or a change in land use is required.
13	Impact on road	An assessment of a) the impact of observed slope movement/failure on the road and b) future impact if no action is taken is important in the assessment of risk and prioritising remedial works. Impact should be carefully recorded because this can also contribute to hazard monitoring. Assessing future outcomes is more of a judgemental exercise, and can only really be undertaken with any confidence when the details described above are recorded and interpreted.
14	Impact on road drainage	This requires factual description in order to assess the extent of required repairs and also to determine the role that both controlled and uncontrolled drainage might be playing in current and possible future slope movement.
15	Impact on structures	This requires inspection of all revetments and retaining walls affected by slope movement. Failure or movement of a retaining wall below the road, for example, might actually be the cause of slope failure and damage to the road. Therefore, it is important to ascertain the relationship between wall stability and the stability of the road and adjacent slopes. It is common for wall movements below the road to be caused by deeper-seated slope movements rather than bearing capacity failure or overturning.
16	Occupied bldgs affected	Includes evidence of cracking caused by structural movement. Care is required not to confuse this with damage due to differential foundation settlement, poor construction or heave due to expansive soils. Includes assessment of potential for slope failure to impact occupied buildings in the future (at risk condition).
17	Remedial measures	Based on the above what combination and extent of remedial measures might be envisaged? Consider earthworks, drainage, retaining structures and erosion protection.

Fig. D1.5. Inadequate appreciation of the full extent of failure prior to scheduling remedial works.

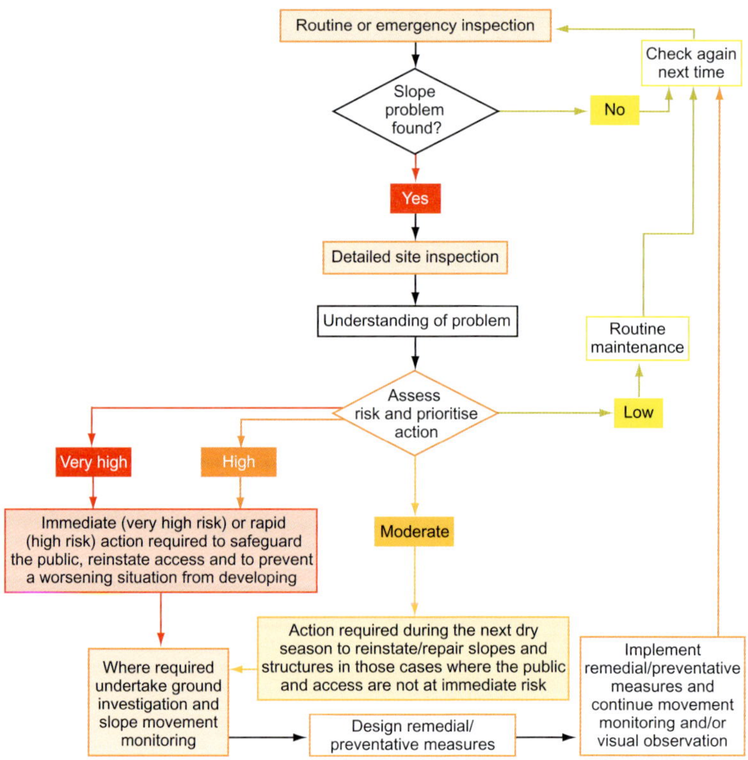

Fig. D1.6. Decision-making process for slope inspection and maintenance. Note that the risk ranking categories are defined in Tables D1.3 and D1.6.

Table D1.3. *Risk-based prioritization matrix for preventative works*

Asset Value or Potential Economic/Social Loss

AADT	*Strategic Importance*	*Occupied buildings at risk?*	*Loss of access time*	*AADT*	*Strategic Importance*	*Occupied buildings at risk?*	*Loss of access time*	*AADT*	*Strategic Importance*	*Occupied buildings at risk?*	*Loss of access time*
>800	Military, access to hospitals, towns	Within hazard area	>3 days	200-800	Access to medical facilities, villages	On margin of hazard area	1-3 days	<200	Road network only	Outside of hazard area	<1 day

Hazard				*Vulnerability of Road, its Structures, Access, Travelling Public and Occupied Buildings if Hazard Occurs (Table D1.4)*								
				High	Moderate	Low	High	Moderate	Low	High	Moderate	Low
				Loss Value Should the Hazard Occur (Value x Vulnerability)								
				5	4	3	4	3	2	3	2	1
Type	Probability			Risk Number (Loss Value x Probability)								
	Descriptor (Section B3.4)	Level										
Landslides	Adverse geological structure, steep soil slopes, seepages, toe erosion and/or fresh scarps, bulges, cracks	High	5	25	20	15	20	15	10	15	10	5
Debris flows	Landslide sources, fresh deposits and/or frequent occurrence (at least every 2 years)											
Fill slope and RW failures	Steep slopes, weak/unstable foundations and/or fresh cracking and settlement											
Landslides	Landslide morphology but no visible active movement	Moderate	3	15	12	9	12	9	6	9	6	3
Debris flows	Potential landslide sources, previous deposits and/or periodic occurrence (2-5 years											
Fill slope and RW failures	Evidence of previous movement											
Landslides	No evidence of ground movements	Low	1	5	4	3	4	3	2	3	2	1
Debris flows	No apparent sources and no evidence of previous flows											
Fill slope and RW failures	Stable founding conditions, no evidence of movement or damage											

Risk Number Range	*Risk Rank*	*Action*
20 and 25	1. Very High	Implementation of immediate preventative measures
10-15	2. High	Implementation of preventative measures during the following dry season
5-9	3. Moderate	Implementation of monitoring measures to detect any movements
1-4	4. Low	None

Risk assessment for works prioritization can be applied at two main stages of engineering intervention:

- preventative intervention; and
- emergency and remedial intervention.

D1.6.1 Preventative intervention

In the case of preventative maintenance on slopes that have not yet failed, it will be necessary to carry out an assessment of the following parameters in order to yield a risk ranking:

- the value of assets at risk (engineering structures, traffic volumes, strategic importance of access, presence of occupied buildings);
- the vulnerability of these assets to total or partial loss or damage should the hazard occur; and
- the relative probability of the hazard occurring during a given period.

Table D1.3 illustrates a matrix devised to take account of these various parameters. The hazards that pose potential risk have been split into landslides, debris flows and fill slope and retaining wall failures. A risk ranking is developed based on asset classification, vulnerability and relative probability of hazard occurrence. *Hazard probability* has been assigned the same level of importance in the computation as the combined *value* and *vulnerability* parameters. Table D1.4 illustrates how vulnerability might be assessed for the various engineering, land use and road user assets at risk. Risk numbers have been consolidated in Table D1.3 into a four-fold risk ranking for prioritization, and recommended actions have been assigned to each. It is emphasized that this matrix is indicative only and will need to be modified to suit particular road networks and sections of road according to national and regional conditions. Table D1.5 uses three scenarios to illustrate how Table D1.3 is utilized.

In Scenario 1, field observations indicate that there is a *moderate* probability of a landslide occurring on the slopes above a section of road. The road carries more than 800 vehicles per day but does not provide the only access to villages, towns or medical facilities. It is judged that, should a landslide occur, it could block the road for between 1 and 3 days. Given the high traffic volumes and the size of the potential landslide, it is considered quite possible that injury or fatality could occur to the travelling public should the landslide occur, and consequently a *high* vulnerability is assigned (Table D1.4). From Table D1.3, the potential loss value is 5 (on account of the high traffic volume combined with the *high* vulnerability). With a *moderate* probability number of 3 this scenario yields a Risk No of 15, corresponding to a *High* Risk Rank. Preventative measures are therefore scheduled for the following dry season.

In Scenario 2, the hazard posed to the road relates to the potential failure of a retaining wall supporting a section of road. In this instance, although the traffic volume is less than 800 AADT, the road provides access to a hospital and the failure of the wall would prevent access for more than 3 days and would be a significant hazard to drivers until traffic warning and control measures could be implemented. The combined asset value and vulnerability yields a potential loss value of 5. Field investigations indicate that the stability of the wall foundation is being undermined by seepage erosion and it is concluded that there is a *high* probability of

Table D1.4. *Possible vulnerability level definitions for use in Table D1.3*

Vulnerability level	*Elements at Risk*			
	Road assets	Strategic, commercial or social benefit from access	Occupied buildings	Travelling public
	Vulnerability level definitions			
Low	No damage, easily repairable damage	Access could still be maintained by through traffic	No damage, minor repairable damage	Slow moving landslide would pose no risk to vehicles
Moderate	Partial loss	Access could only be maintained by foot or temporary FWD track	Partial damage	Small cut slope failure could damage passing vehicles
High	Complete loss	Access could not be maintained until the blockage was cleared or the road reinstated	Partial or total structural failure	Rockfalls and large slope failures could cause injury or fatality to travellers in vehicles and pedestrians. Sudden partial or complete road loss could cause drivers to lose control of their vehicles

Table D1.5. *Hypothetical scenarios to illustrate the use of Table D1.3*

Scenario No	Hazard type	Asset Value/Economic/Social Loss Potential				Vulnerability	Loss value	Probability	Risk No.	Risk rank
		AADT	Strategic importance of road	Occupied buildings at risk?	Loss of access					
1	Landslide	>800	Road network only	No, outside hazard area	1–3 days	High	5	Moderate (3)	15	High
2	Retaining wall failure	200-800	Access to hospital	No, outside hazard area	>3 days	High	5	High (5)	25	Very High
3	Debris flow	>800	Road network only	Yes, within hazard area	<1 day	High	5	Low (1)	5	Low

failure. The overall Risk No is therefore 25, corresponding to a *Very High* Risk Rank, and preventative measures are required to be undertaken immediately.

In Scenario 3, a high traffic volume and the presence of occupied buildings within the debris flow hazard area combine with a *high* vulnerability to yield a potential loss value of 5, but a *low* expected probability resulted in a Risk Number of 5, corresponding to a *Moderate* Risk Rank and a recommendation for monitoring rather than the carrying out of any preventative measures.

Das *et al.* (2010) describe the use of methods based on landslide susceptibility (Sections B2.6 & B3.3) and on rock mass analysis (see also Section C4.2 and Marinos *et al.* 2005 for example) to identify slopes most vulnerable to failure. One of the observations drawn from the Das *et al.* (2010) study was that the generalizing process used in the susceptibility-based methods causes some slopes that are geotechnically critical to be missed. It is therefore important to ensure that slopes along mountain roads are assessed on a case-by-case basis using engineering geological mapping (Section B3.4) and classification techniques (Section C4.2.2) for rock slopes. Koo *et al.* (2010) and Liu *et al.* (2010) describe cut slope stability ratings that could be used to identify priority locations for preventative

Table D1.6. *Risk-based prioritization matrix for emergency and remedial works*

Actual consequence arising from failure (Emergency and Remedial maintenance)	*Risk Ranking*			
	1. V High	*2. High*	*3. Mod*	*4. Low*
Occupied buildings damaged or destroyed	✓			
Road completely lost	✓			
Road partially lost		✓		
Below-road retaining structure collapse		✓		
Road subsidence greater than 0.5m		✓		
Road completely blocked		✓		
Above-road retaining structure collapse			✓	
Road partially blocked			✓	
Retaining structure damage or movement			✓	
Road subsidence less than 0.5m			✓	
Rockfall onto road				✓
Roadside drain damaged or blocked				✓

Table D1.7. *Road category*

Road attribute	*Category*		
	A	*B*	*C*
Strategic importance	High	Medium	Low
AADT	>800	200–800	<200

maintenance, based on factors such as rock type, structural orientations and weathering condition, water table, slope height and slope angle. It should be reiterated (Section C3.1) that, when carrying out any form of slope stability assessment (descriptive, rating-based or analytical), it is imperative to differentiate between slopes that are anticipated to behave principally as rock or soil and to ensure that the factors that control stability are adequately considered (Leyland 2010).

D1.6.2 Emergency and remedial intervention

In the case of emergency and remedial intervention, a landslide has already occurred or a section of road has already failed and the degree of damage can be observed. Consequently, neither probability nor vulnerability (degree of loss) needs to be predicted as part of the works prioritization. Table D1.6 provides a simple four-fold risk ranking which can be applied based on observed damage:

- Very High risk ranking poses a constant and unacceptable level of hazard to the road, road users or adjacent occupied buildings, thus requiring immediate engineering intervention;
- High risk ranking implies that although the road will continue to be trafficable once landslide clearance has taken place and access has been reinstated, there is a very real threat to road traffic safety and failure could continue to develop into a more significant hazard, thus requiring rapid engineering intervention;
- Moderate risk ranking implies that the slope or structure could fail or continue to fail with a level of hazard to the road that is tolerable in the short term, thus avoiding the need for immediate or rapid engineering intervention;
- Low risk ranking implies maintenance that needs to be carried out in the medium term, or a location that will self-stabilize over a given time period (perhaps 5 years is acceptable in the context of low-cost, low-volume road maintenance) with a reducing and acceptable level of hazard, thus avoiding the need for engineering intervention.

This risk ranking may need to be varied from one country to another and possibly from one region to another.

The ranking can be adjusted to take into account the relative importance of a particular road (or section of road) compared to another, in a similar manner as shown in Table D1.3 for the prioritization of preventative works. The rankings given in Table D1.6 might apply to a category B road (Table D1.7). The risk rankings would increase by one rank for a category A road and decrease by one rank for a category C road. The term 'strategic importance' refers to the need to maintain access irrespective of the AADT; for instance if the road connects to a hydropower scheme or a hospital. Only the highest category of the two attributes would apply to a particular road or length of road. Table D1.7 may need to be modified to suit regional circumstances.

D1.7 Maintenance procurement

Traditionally, road authorities have been responsible for routine and emergency maintenance. Sometimes lengthworkers who live locally are appointed to carry out routine inspection and maintenance on a specific length of road, and undertake activities such as grass cutting and roadside drain clearing.

Text box D1.1. Performance-based contracts

In one southeast Asian country, routine and emergency maintenance was increasingly being carried out under performance-based contracts where a contractor was appointed to maintain a section of road up to 50 km in length. This worked reasonably well where roads were located in level or rolling terrain, but became increasingly problematic in mountainous areas where a slope failure, for instance, might vary from a few cubic metres of soil and weathered rock slumping onto the road to the road being severed by much more extensive movement. Originally, only debris clearances over a certain volume were paid on a re-measurement basis, but this was eventually abandoned when it was realized that contractors would then wait until individual landslides exceeded that volume, rather than commence timely clearance. This was replaced by a clause which stated that only in exceptional cases, where the event causing major landslides was categorized as a National Disaster by the government, would the contractor be entitled to negotiate additional payment for the remedial works on a re-measurement basis. Contractors maintaining the mountainous sections of roads on this basis were exposed to major financial risks.

Emergency maintenance is usually carried out by the road authority itself or by a standby contractor. In the latter case, this appointment is normally restricted to high-volume roads where the rapid clearance of landslides is both an economic and political necessity. For low-volume roads, a delay in reopening the road is usually less critical.

Preventative and remedial maintenance is usually let as a separate contract to national or international contractors, depending on the complexity and extent of the work to be undertaken. It is important to ensure that the contract is sufficiently flexible to allow for variations in quantities, since slope stabilization works are, by necessity, based on estimated quantities and assumed subsurface conditions. Sometimes these variations can be significant and the road authority should always be aware of this possibility. Such variations can sometimes create time-consuming contract administration problems, where permission for major revisions to the quantities has to be approved at high level. Trial pit and borehole ground investigations can be used to reduce some of this uncertainty (Section B4).

Road maintenance may also be carried out by a maintenance contractor appointed by the road authority on a *term* or *performance-based* contract, and this can also include the maintenance of roadside slopes.

Under a *term* contract, the contractor is usually instructed on what resources he should deploy and where. Payment to the contractor is based directly on the resources utilized and the road authority bears all the financial risk of maintaining the road. An emergency maintenance standby contractor may be appointed on a term basis.

Under a *performance-based* contract, the contractor decides his own deployment in order to provide a specified level of serviceability. On high-volume roads this serviceability level might be gauged by reference to pavement roughness, amongst other things, while on low-volume roads the number of days the road remains open might apply. Payment is based on the satisfactory serviceability of the road and the contractor bears all the financial risk of its maintenance. In respect of slope works, the contractor may consequently seek to minimize his costs by avoiding certain activities if he is not going to be paid specifically for them, even though they may be required under the specification. This might lead, for example, to the dumping of landslide debris over the road edge rather than transporting it to an approved disposal site, or the construction of a retaining wall foundation as a continuous excavation rather than in bays. Performance-based contracts are becoming increasingly popular, often with very optimistic performance targets. Although performance-based contracts for roads should include routine slope maintenance, they are much less suited to preventative, emergency or remedial works because of the inherent uncertainties in the scale of the works that might be required from one year to the next (Text box D1.1). It is recommended that preventative, emergency and remedial slope maintenance be included only on a re-measurement basis where the contractor is paid for the actual resources he has to deploy (Section A2).

References

Bujang, B. K. H. & Jamaludin, S. 2005. Evaluation of slope assessment system in predicting landslides along roads underlain by granitic formation. *American Journal of Environmental Science*, **1**, 90–96.

Chen, L. X., Yin, K. L. & Wang, Y. 2010. Risk analysis for landslide-related road-traffic fatalities. *In*: Williams, A. L., Pinches, G. M., Chin, C. Y., McMorran, T. J. & Massey, C. I. (eds) *Geologically Active. Proceedings of the 11th IAEG Congress, Auckland, New Zealand, 5–10 September 2010*, Taylor and Francis Group, London, 1015–1021.

Das, I., Sahoo, S., van Westen, C., Stein, A. & Hack, R. 2010. Landslide susceptibility assessment using logistic regression and its comparison with a rock mass classification system along a road section in the northern Himalayas (India). *Geomorphology*, **114**, 627–637.

GEO 2003. *Layman's Guide to Slope Maintenance*. Geotechnical Engineering Office, Civil Engineering and Development Department, Hong Kong.

GEO 2009. The new priority ranking systems for man-made slopes and retaining walls. *Special Project Report* **SPR 4/2009**. Cheng, P. F. K. Geotechnical Engineering Office, Hong Kong.

Heath, W. & McKinnon, B. 1996. Highway earthwork management; a need for strategies to control slope problems. *Highways into the Next Century Conference*, Kowloon, Shangri-la, Hong Kong, 24–26 November 1996.

Heath, W., Hearn, G. & McKinnon, B. 1995. Comparisons between aerial and ground-based assessments of road earthworks in Nepal. *Proceedings of the 8th Road Engineering Association of Asia and Australasia (REAAA) Conference*, Taipei, April 1995.

Jaiswal, P., van Westen, C. J. & Jetten, V. 2010. Quantitative assessment of landslide risk along transportation lines in southern India. *In*: Williams, A. L., Pinches, G. M., Chin, C. Y., McMorran, T. J. & Massey, C. I. (eds) *Geologically Active. Proceedings of the 11th IAEG Congress, Auckland, New Zealand, 5–10 September, 2010*, Taylor and Francis Group, London, 1185–1193.

Jamaludin, S. & Hussein, A. N. 2006. Landslide hazard and risk assessment: the Malaysian experience. *In*: Pankhurst, B. (ed.) *Engineering Geology for Tomorrow's Cities. International Association of Engineering Geology 10th International Congress*, Nottingham, UK, Paper No 455.

Koo, H. B., Kim, S. B., Bautista, F. E., Park, Y. E. & Hong, N. K. 2010. Scientific and systematic cut slope management of Korea. *In*: Williams, A. L., Pinches, G. M., Chin, C. Y., McMorran, T. J. & Massey, C. I. (eds) *Geologically Active. Proceedings of the 11th IAEG Congress, Auckland, New Zealand, 5–10 September, 2010*, Taylor and Francis Group, London, 1217–1224.

Kwon, O. I. & Baek, Y. 2010. Development of advanced digital investigation equipment for cut slope management. *In*: Williams, A. L., Pinches, G. M., Chin, C. Y., McMorran, T. J. & Massey, C. I. (eds) *Geologically Active. Proceedings of the 11th IAEG Congress, Auckland, New Zealand, 5–10 September 2010*, Taylor and Francis Group, London, 2037–2042.

LEE, E. M. & JONES, D. K. C. 2004. *Landslide Risk Assessment*. Thomas Telford, London.

LEE, W. F., LIAO, H. J., CHEN, R. H., WEIL, C. Y. & HUANG, Y. M. 2006. The development and application of the slope management system. *In*: DEGROOT, D. J., DEJONG, J. T., FROST, D. & BAISE, L. G. (eds) *Geotechnical Engineering in the Information Technology Age*. American Society of Civil Engineers, Feb 26–Mar 1, 2006, Atlanta, GA, 1–6.

LEYLAND, R. C. 2010. The development of a strategic slope management system for use in South Africa. *In*: WILLIAMS, A. L., PINCHES, G. M., CHIN, C. Y., MCMORRAN, T. J. & MASSEY, C. I. (eds) *Geologically Active. Proceedings of the 11th IAEG Congress, Auckland, New Zealand, 5–10 September 2010*, Taylor and Francis Group, London, 1265–1277.

LIU, C., YIN, K. & DU, J. 2010. Cut slope susceptibility assessment at Badong County in the Three Gorges Reservoir. *In*: WILLIAMS, A. L., PINCHES, G. M., CHIN, C. Y., MCMORRAN, T. J. & MASSEY, C. I. (eds) *Geologically Active. Proceedings of the 11th IAEG Congress, Auckland, New Zealand*. Taylor and Francis Group, London, 1295–1302.

MARINOS, V., MARINOS, P. & HOEK, E. 2005. The geological strength index: applications and limitations. *Bulletin of Engineering Geology and the Environment*, **64**, 55–65.

MCPHERSON, K. & BENNETT, C. R. 2005. *Success Factors for Road Management Systems*. World Bank, Washington DC.

PAIGE-GREEN, P. 1997. A slope management system for transportation routes. *Technical Report* No **TR-97/005** CSIR Division of Roads and Transport Technology, Pretoria, South Africa.

PANTELIDIS, L. 2011. A critical review of highway slope instability risk assessment systems. *Bulletin of Engineering Geology and the Environment*, **70**, 395–400.

ROBINSON, R., DANIELSON, U. & SNAITH, M. 1998. *Road Maintenance Management – Concepts and Systems*. Macmillan Press, Basingstoke, UK.

RUSSELL, C. P., SANTI, P. & HUMPHREY, J. D. 2008. Modification and statistical analysis of the Colorado rockfall Hazard Rating System. Colorado Department of Transportation Research Report, No **CDOT** – 2008–7.

WINTER, M. G., MACGREGOR, F., HARRISON, M. & SHACKMAN, L. 2010. Scottish road network landslide study: hazard assessment and ranking. *In*: WILLIAMS, A. L., PINCHES, G. M., CHIN, C. Y., MCMORRAN, T. J. & MASSEY, C. I. (eds) *Geologically Active. Proceedings of the 11th IAEG Congress, Auckland, New Zealand, 5–10 September 2010*, Taylor and Francis Group, London, 1515–1523.

Glossary of terms

The terms defined below are generally of a geological or geomorphological nature. Other terms are defined in the body of the text. Standard engineering terms are not defined here, as it is assumed that the readership is already familiar with those.

AADT: Annual average daily traffic, usually motorised vehicles.

Active pressure: the lateral pressure exerted by a soil on a retaining structure that is reached when the structure yields and allows the soil to expand in a horizontal direction.

Adverse jointing: joints or discontinuities in the rock mass are orientated (i.e. they dip) in a direction that is towards the slope, thereby forming planes along which slope failure might take place.

Allophane: amorphous glassy aluminosilicate clay mineral often derived from weathering of volcanic rocks.

Allowable bearing pressure: the maximum safe stress that can be applied to a foundation for an assumed acceptable settlement.

Arcuate: a landform (see below) that is arc-shaped (crescentic/semi-circular) in plan-form.

Automatic classification: The use of digital remote sensing imagery to classify and map attributes on the ground, such as land use, soils, wet ground and erosion areas based on spectral reflectance bands.

Back analysis: numerical analysis of a failed slope, assuming a factor of safety (see below) at or close to 1.0, in order to determine material strength parameters and other variables at the time of failure.

Back scarp: Slope or slopes from which slope failures detach. Back scarps are usually steeper than the surrounding topography and are usually arcuate in plan in the case of circular/rotational failures or linear in the case of planar failures. A landslide scar is the entire landslide outline: source area and deposit. The landslide scarp is the landform that forms the source area detachment slopes or release surfaces and the back scarp is the uppermost portion of this.

Band ratios: One of the most commonly used satellite image interpretation techniques; band ratios are created by dividing one spectral band against another, as this helps to highlight differences in the spectral signatures for each band.

Bearing capacity: the maximum stress that can be applied to a given foundation without inducing shear failure.

Bench: Platform on a cut slope that has been excavated as part of a series of steps rather than as a continuous slope. Also a level area in the landscape formed by erosion, river deposition or geological structure control.

Berm: Platform on a fill slope that has been constructed as a part of one or more steps rather than as a continuous slope. Also stabilizing embankment at toe of slope.

Bio-engineering, bio-engineer: the use of plants to provide soil reinforcement and erosion control for engineering purposes, usually associated with techniques designed to speed up the vegetation of road cuts and fills slopes. A bio-engineer is often trained as a forester and may have a post-graduate qualification in soil science or a related discipline.

Black cotton soil: smectite (see below) clay-rich soil that undergoes large volume change upon wetting and drying. These soils develop on basalt rocks especially, though not exclusively.

Brush layering: layers of vegetation in shallow trenches with the top of the vegetation layer protruding above the slope surface to form an immediate barrier to surface runoff erosion.

Catchment area: drainage basin or watershed that contributes runoff to single stream or river flow.

Chunam: lime-based screed which is manually applied to a soil or weathered rock slope. The technique was used extensively in Hong Kong before the wider use of shotcrete.

Clast: gravel, cobble or boulder-sized particle within a soil mass. A clast-dominated soil is one where this sized material makes up the majority of the soil composition.

Cleavage: tendency of a rock to split along closely-spaced parallel planes developed under pressure metamorphism.

Coarse-grained: soil with particle grain size >2 mm.

Colluvium: a fine grained slope deposit.

Colour composite image: colour image prepared by projecting individual black-and-white multispectral images, each through a different colour filter. When the projected images are superimposed, a colour composite image results.

Compound slope profile: slope cross-section made up of two or more segments of different angle.

Compression ridge: linear raised ground caused by convergence of slope materials at toe of slope movement or along margins.

Concave break in slope: slope angle reduces in downslope direction.

Concavo-convex slope: slope reduces and then increases in a downslope direction.

Conditioning factors: attributes of a slope or a material that help determine its susceptibility to failure.

Continuous flight augering: a method for pile construction or ground investigation by injecting concrete through the hollow centre of the auger as it is withdrawn. Soil samples can be taken through the hollow stem.

Convex break in slope: Slope steepens in downslope direction.

Corestones: undecomposed rock surrounded by decomposed material, often rounded (depending on weathering) and found in jointed igneous (and some jointed sedimentary) rocks.

Creep: in this context, progressive failure due to 'plastic' flow of soil or closely jointed rock without the distinctive morphology of a slide with a basal shear surface.

Daylight: the location on a slope where a bedding plane, foliation plane or other joint set or stratum is exposed. If a joint is steeper in angle than the average angle of the slope then it is said not to 'daylight' and it cannot form a single plane along which sliding is able to take place. The term 'daylight' is also used to describe where a cutting angle or a joint set or other plane will become exposed on the slope above.

Debris (size): there is no accepted grain-size definition for debris, but in geomorphological terms debris is usually regarded as being coarse-grained.

Deep-seated: In the context of slope failure deep-seated is usually taken to mean involving movement through rock; i.e. the failure plane passes beneath the soil and weathered rock into the underlying bedrock. This definition might not be as easy to apply in the case of circular failures or planar failures occurring on a slope formed in unweathered rock. Consequently, deep-seated is often used to infer actual depth. In the context of low-cost mountain roads the term deep-seated might be applied to slope failures deeper than 5 m, or perhaps 10 m.

Dendritic: multiple-branching drainage network, similar to the veins on a leaf, and usually formed on uniform lithology without underlying structural control.

Desiccate: cracking caused by shrinkage in clay soil upon drying.

Digital Elevation Model (DEM): Electronic, three-dimensional representation of a part of the Earth's surface based on multiple x, y and z co-ordinates or digital contour data.

Digital Terrain Model (DTM): Electronic, three-dimensional representation of a part of the Earth's surface based on multiple x, y and z co-ordinates or digital contour data.

Dilation: expansion or widening of joints or fissures in rock due to mechanical or chemical weathering and stress release caused by erosion and mass movements. In geotechnical terms dilation is used to describe the expansion of a soil mass as it becomes sheared. Particle rearrangement leads to an increased volume (dilation) and a significant increase in the angle of friction above that for inter-particle friction alone.

Dilation angle: in dense sands describes the rate of increase in volume as a result of shearing.

Discontinuities: joints, fissures and other breaks in a rock mass.

Double case/tube core barrel: comprises an outer barrel and an inner tube used for extracting core during rotary drilling. The sample is recovered in the inner tube.

Draping: whereby a digital satellite image is moulded onto a digital elevation model of the same area to produce a 3D visualisation of the landscape.

Duricrust: hardened horizon in a soil profile formed by precipitation of various compounds from solution, for example iron-rich horizon (ferricrete) and limestone-rich horizon (calcrete).

Effective stress: the difference between a) the total stress acting on a slope as a result of the weight of soil and any overlying materials and structures and b) the pressure exerted by water in the soil pore spaces.

Embayment: broadly rounded indentation into a hillside or valley side.

Empirical approach: in this context this approach is used to derive relationships based on observations. In the case of debris flow runout distances, for example, the approach combines observations of previous failure volumes with angle of reach of total distance travelled.

Engineering geology, Engineering Geologist: the practice of geology in relation to engineering, principally the study of the engineering behaviour of soils and rocks, including considerations of structural geology, rock strength and weathering grade, the origin and composition of soils and groundwater considerations. An engineering geologist is usually educated to post-graduate level, usually with an MSc in the subject.

Equatorial: the equatorial zone is within 5° latitude of the equator with a climate controlled by convective rainfall, experiencing 2000–3000 mm of rain each year, temperatures between 25° and 35 °C and humidities of 80–90%. An equatorial climate will not necessarily apply to the higher altitude locations within this zone.

Extensometer: Instrument for measuring changes in linear dimension. In the context of slope monitoring a wire is positioned across a tension crack and movement of the crack is recorded by reference to markers or trip blocks against a scale. Extensometers are also used in boreholes.

False colour composite: An image produced by displaying multiple spectral bands as colours different from the spectral range they were taken in.

Fan: cone-shaped accumulation of sand, gravel and boulders deposited where a stream or river undergoes a sudden decrease in velocity, such as at the mouth of a tributary stream where the bed gradient reduces or where the channel width increases.

Fascine: a bundle of live branches buried by soil in a shallow trench across a slope. Roots are put out and the fascine then forms a strong line of vegetation. Sometimes referred to as contour wattling.

Faulting: rupture within rock strata due to tectonic stresses as a result of which displacement takes place.

Ferricrete: duricrust formed by precipitation of iron oxides.

Filter fabric: usually a non-woven geotextile used in retaining wall and drainage applications. The fabric has an aperture size that is large enough to allow water migration but small enough to prevent passage of fine material.

Fine-grained: soil with particle grain size <0.1 mm.

First-time failure: movement on a slope where it has not occurred before.

Flood recession: declining flow in a river or stream following the passage of the peak stage. Sediment deposition often takes place during flood recession due to decreasing velocities.

Folding: flexure or bending of rock strata due to tectonic stresses.

Foliation: preferred orientation of minerals creating a planar fabric in metamorphic rock.

Formation level geological mapping: rock strata are grouped into 'formations', conventionally on the basis of age. The map legend shows the rock types comprising each formation. The map shows the distributions of these formations but not the rock types within them.

Formation width: total required width for construction of carriageway, shoulders and side drains.

Foundation stability: the resistance of a given founding level for a wall, for instance to failure, either as a result of inadequate bearing capacity or due to deeper-seated slope failure.

Friction angle/angle of internal friction: the angle on the graph (Mohr's Circle) of the shear stress and normal effective stresses at which shear failure occurs. Angle of internal friction can be determined in the laboratory by the Direct Shear Test or the Triaxial Stress Test.

Freeboard: safety margin (elevation level) above design flood level. In the event of the design flood occurring the bridge soffit (underside of bridge) or finished road level on a riverside embankment is above this level by an amount equal to the freeboard.

Furrowed slope: parallel trenches in the slope surface caused by tension arising from slope movement.

Gap-graded: a soil with a discontinuous size grading, i.e. with certain particle sizes unrepresented, for example a soil comprising gravel and clay only.

Geogrid: Flexible plastic grid with open structure to allow soil to soil contact through the grid, usually made from polymer-based interconnected tensile elements and used for earth-reinforcement and erosion control.

Geophysical investigations: use of electrical resistivity, seismic velocity and other properties of soil and rock in order to determine soil depth and soil/rock stratification etc.

Geohazard: strictly geological/gemorphological processes (landslides, earthquakes, volcanic eruptions, solution features etc.) that pose a potential threat to man, infrastructure and environment.

Geology, Geologist: geology is the study of the formation of the Earth and the stratigraphic succession, structure and lithology of rocks that make up the Earth's lithosphere. A general geologist is normally educated to at least degree level in the subject.

Geomorphology, Geomorphologist: geomorphology is a discipline of geology (though historically of geography in the UK) that studies the origin of landforms taking geology and past and present surface processes eg weathering, mass movement and erosional activity into consideration. A geomorphologist is usually qualified to degree level in either geology or physical geography and normally with a post-graduate qualification (usually a PhD) focusing on landforms and earth surface processes and geohazards.

Geo-rectification: usually undertaken for aerial photographs whereby the variable scale of ground detail portrayed on an aerial photograph is made constant on a digital output map using control points of known or surveyed location on the ground that are visible in the aerial photograph.

Geosynthetic: term used to describe polymer-based products including geotextiles, geogrids and geomembranes. Filter fabrics for drainage control fall into this category.

Geotechnical Engineering, Geotechnical Engineer: geotechnical engineering is a discipline of civil engineering that is concerned with the engineering behaviour of soils and rocks, particularly foundations and slope stability. A geotechnical engineer is usually a qualified civil engineer with a post-graduate degree in Soil Mechanics or Foundation Engineering.

Geotextile: is a permeable woven or non-woven needle-punched or heat-bonded synthetic fabric that is used in civil engineering to filter, reinforce, protect and drain a soil.

GIS: Geographical Information System used to collate and analyse layers or maps of spatial (geographical) digital data.

Gravel-sized: 2–60 mm in dimension.

Ground model (terrain/elevation): digital representation of the topography, usually derived from photogrammetry or digital remote sensing (e.g. LiDAR).

Ground model (geology): a representation, usually as a cross-section, showing soil and rock horizons and groundwater and slip surface (s), where present, for purposes of stability analysis.

Ground conditions: general term used to describe surface and sub-surface geology, i.e. depth of soil, soil types and strengths, groundwater levels, rock head level, lithology, weathering, strength and geological structure, i.e. aspects of the ground relevant (in this case) to engineering.

Ground truthing: verification of an interpretation derived from desk study by field observations.

Groyne: in the context of this document, a structure built into the flood plain of a river to divert flood waters away from a riverside structure, such as a road or bridge abutment.

Hazard: a hazard is an event or process that has the potential to cause damage. The hazard is defined as the quantum of a given event and the probability of its occurrence over a given time period, such as the design life of a road. Hazard can also be derived from human action or inaction.

High-grade metamorphic rocks: are rocks that have formed under high temperatures and pressures, such as granite-gneisses and gneisses.

Hill wash: removal of surface soil particles as a result of surface runoff or overland flow during rainfall.

Hornfelsed: rock altered through thermal metamorphism.

Hydrogeology: the study of the interaction between soil and rock permeability and structure and the movement of groundwater and the groundwater table.

Igneous: rocks formed from the solidification of magma (subterranean molten rock). Extrusive igneous rocks are derived from volcanic eruptions.

Inclinometer: devise inserted into drillholes or boreholes for measuring the depth, direction and extent of movement within a soil or rock mass or across a sliding surface.

***In situ*:** slope material, usually rock, that is located in its original position, i.e. it has not failed or been otherwise removed or transported.

***In situ* weathered soils:** soils that have developed *in situ* as a result of the decay of the parent rock. These soils are classified into various weathering grades and their composition and structure is dependent on that of the parent rock.

Intrusive ground investigation: sub-surface exploration using trial pits, probes, drillholes and boreholes but not geophysical techniques, which are not intrusive.

Isotropic: rock that has the same properties in all directions. In this context it does not have dominant planes of weakness in any one orientation that control its stability when exposed on slopes.

Joint-controlled slopes: slopes that are formed along a single set of joints in the underlying rock. These slopes are usually smooth and linear and have normally been formed as a result of original detachment of overlying rock from the joint surface.

Kinematic feasibility/admissibility: the potential for a slope to fail along a single or combination of joints and planes of weakness as determined by the geometry of the planes and the topography of the slope.

Laterite: refers specifically to ferricrete duricrust or plinthite (or latosol), the latter being a soft horizon within the weathering profile made up of hydrated oxides of iron and aluminium which hardens upon exposure (Fookes 1997).

Landform: a topographic feature that has evolved as a result of a specific combination of underlying material type, strength or structure and the processes acting upon it (weathering, erosion, mass movement etc.).

Landsat: series of unmanned earth-orbiting NASA satellites that acquired multispectral images in various visible and IR bands.

Landslide: the movement of a mass of rock, debris or earth down a slope.

Landslide scar: the boundary of a landslide, including its source area (zone of detachment) and its deposit.

Landslide scarp: the source area landform derived from the detachment of a landslide from a slope.

Latosols: soils developed widely in the humid tropics and sub tropics in which chemical weathering and leaching are intense, leading to deep profiles of free-draining, yellow to red acid soils rich in hydrated oxides of iron, aluminium and manganese (ferralitic soils).

Leaching: the process whereby water percolates down through a soil profile removing soluble bases (alkaline constituent) and sesquioxides.

Lengthworker: person employed to maintain a specific section of road by a road authority.

Levee: in the context of debris flow deposits, levees are linear ridges that border either side the central flow path of a debris flow.

LiDAR: Light detection and Ranging. The distance (range) to an object is determined by measuring the time delay between transmission of a pulse and detection of the reflected signal.

Limit equilibrium: This applies to the state where forces promoting slope or wall failure are balanced against those resisting failure.

Limit state: this refers to the condition at failure, i.e. the ultimate stress that the design can accommodate. Eurocode 7 defines 'ultimate limit state' as '... states associated with collapse.....'

Limiting slope angle: geomorphological term to describe the steepest angle at which a given material and landform (such as a talus slope for example) are found to stand.

Linear elastic model: Soil is perfectly elastic at all stress levels with a constant stiffness defined by either elastic modulus E and Poisson's ratio n, or bulk modulus K and shear modulus G.

Liquefaction: the process whereby fine sediments and soils collapse due to sudden loss of cohesion following a loss of shearing resistance. This is usually brought about by a sudden increase in pore water pressure due to seismic loading.

Liquid limit: The water content of a soil at which the soil behaves as a liquid.

Lithology: rock type, mineralogy and grain size.

Litho-relic: remnant of rock fabric or structure within profile of more weathered material.

Lithosphere: crustal portion of the Earth.

Lobate: a deposit of material shaped like a tongue (or lobe)

Low-grade metamorphic rocks: are meta-sedimentary rocks such as phyllite and slate that have formed under low temperatures and pressures.

Mafic: mafic minerals are manganese and iron and are found in rocks such as basalt and gabbro.

Mass movement: the downslope movement of surface materials under gravity, including soil creep, hill wash and landslide processes.

Meander bend: loop-like bend in a mature river.

Metamorphic rock: rock that has been altered from its original state by temperature and pressure through crustal tectonics.

Mohr-Coulomb Failure Criterion: this represents the linear envelope that is obtained from a plot of the shear strength of a material against the applied normal stress.

Monsoonal climate: a climate dominated by an annual change in wind direction, and accompanied by a distinct wet and dry season. The strongest monsoon effects are found in South Asia, South East Asia and China.

Morphology: the study of the shape of the ground, including landforms.

Multi-spectral analysis: study of remotely-sensed data in different spectral bands.

Multi-spectral data: sets of data obtained simultaneously, but each set obtained by sensing a different part of the electromagnetic spectrum.

Net present value: measures the excess or shortfall of cash flows, in current value terms, once financing charges are met.

Open-hole drilling: drilling without casing or tubing.

Palisade: A line of cuttings or seedlings placed across a slope to form a barrier to soil movement when growth takes place.

Partial factors: index of less than 1.0 applied to soil or rock strength parameters to take uncertainty and variability into consideration (Eurocode 7).

Passive pressure: the upper limiting value of lateral pressure, or resistance, reached when a structure yields and causes the soil in front of it to be compressed in a horizontal direction.

Peak strength: maximum strength attained by a soil prior to shear failure.

Perched water: a water table created above the groundwater table due to the presence of an impermeable layer and sub-surface water percolation above it.

Permeability: the rate at which a soil or rock permits water to pass through it.

Pixel: contraction of picture element. Smallest screen element or cell to which attributes can be given, for example light reflection in the case of remote sensing or land use type in the case of GIS.

Plasticity: the range of water contents over which a soil exhibits plastic properties.

Plastic limit: the water content of a soil at which the soil starts to exhibit plastic properties.

Plum concrete: concrete containing boulders.

Pore water pressure: pressure exerted by water contained within the pore spaces of soil or rock. Positive pore water pressure acts to reduce effective stress, while negative pore water pressure creates soil suctions, thus increasing apparent soil cohesion.

Porosity: percentage of pore space within a rock or soil mass that allows water to be transmitted through it.

Pre-split blasting: technique by which blasting is used to create a linear shear in a rock mass prior to bulk blasting.

Principal Components Analysis: mathematical procedure that identifies which variables account for the maximum variance in other variables.

Probe: instrument for testing the relative density/strength of sub-soil.

Progressive failure: the process of slow loss of strength of a slope material over time, owing to progressive softening due to weathering.

Pyroclastic: rock formed from fragmental material derived from explosive volcanic activity

Radar: acronym for radio detection and ranging. Radar is an active form of remote sensing that operates in the microwave and radio wavelength regions.

Raft (of rock): large failed blocks of rock contained within taluvium or landslide debris. These can be misinterpreted as *in situ* rock in drill holes if the drilling depth terminates within them.

Ravelling: gradual process of shallow failure of a steep slope through detachment of soil particles or rock fragments.

Re-entrant: name given to a marked recess or indent in an escarpment, a valley side or an otherwise steep slope, in this instance usually associated with a tributary mountain stream.

Regrading (of a slope): this usually refers to the cutting back of a slope to reduce its angle.

Relic joints: planar weaknesses that remain in weathering grade V materials, inherited from the original rock structure.

Remote sensing: collection of data relating to objects without being in physical contact with them. In this context the term relates to satellite and airborne imagery.

Residual soil: *in situ* soil remaining after full weathering decomposition of the parent rock mass.

Residual shear strength: final (post-peak) shear strength on a rupture plane that a soil maintains at large displacements.

Regressive/retrogressive: this refers to the headward extension of the back scarp of a landslide. The initial failure creates an over-steep back scarp which then fails and the process continues until an equilibrium slope is reached.

Right of Way: in the context of this document, the strip of land acquired by a road authority within which a road is constructed. The strip is usually defined as the width either side of the road centreline and may make provision for future road widening.

Rippable: rock that can be excavated by the claw of a machine.

Risk: probability of economic and social loss due to the occurrence of a hazard.

Rill: shallow erosion channel, incipient to a gully.

River regime: the flow characteristics of a river in response to rainfall in its catchment. Regime is controlled by catchment steepness, catchment shape, permeability and the 'flashiness' of storm rainfall.

Rock head: term used to refer to the boundary between overlying soil and underlying rock. It is used mostly to describe the top of the rock beneath a cover of transported soil (colluvium, taluvium or landslide debris).

Rock fabric: describes the texture and structure of a rock mass, including grain size and orientation.

Rock spall: the failure of individual fragments of rock from cliffs and other steep rock slopes, predominantly due to freeze-thaw effects in high latitudes/high elevations, and also due to blasting effects. Rock spalls also occur locally on the margins of slow moving deep-seated movements in rock.

Route corridor: the linear area or zone within which an alignment is designed. Depending upon the steepness and complexity of the terrain the route corridor can vary from several tens of metres to a kilometre or more in width.

Saprolite: *in situ* weathered soil (less than 30% rock) that is intermediate between residual soil (weathering grade VI) and weathered bedrock (weathering grade III).

Sedimentary: rocks formed by the accumulation of fragments derived from the erosion of pre-existing rocks or from organic sources or chemical sources.

Shearing: the action caused by the application of a tangential stress to a solid material. It results in the formation of a shear zone, shear fault or shear surface.

Shotcrete: sprayed concrete usually applied to a steel mesh fixed to a slope by steel dowels.

Slickensided: polished and scratched planar surface, in this case forming a landslide basal or lateral shear plane.

Slip indicator: a rod that is lowered into a tube inserted into a borehole. Where the tube has been kinked as a result of shear surface displacement in the hole the prevented progress of the rod identifies the level of the shear surface.

Slope engineering: the design, construction, stabilisation and maintenance of earthworks slopes and natural slopes for the purpose (in this case) of establishing and maintaining a stable road alignment.

Smectite: clay mineral group containing montmorillonite and other similarly structured clay minerals that undergo volume change upon wetting and drying.

Soil creep: the imperceptible but continuous or discontinuous movement of soil downslope.

Spatial resolution: accuracy with which ground detail is portrayed on an aerial photograph or satellite image.

Spectral bands: remote sensing sensors record electromagnetic radiation in one or more such separate wavelength ranges (called spectral bands) which provide information about the 'spectral response' of a feature on the ground.

Spoil: excavation material unsuitable or otherwise unused as fill or aggregate and therefore required to be disposed.

SPOT-Systeme Probatoire d'Observation del la Terre: unmanned French remote sensing satellite.

Spurs: sub-ridges that project downwards from main ridges or mountains.

Standpipe piezometer: filter tip attached to a pipe inserted into a borehole to enable groundwater levels to be monitored using a probe lowered into the pipe.

Stereonet: used to plot graphically the declination and orientation of measured rock joints. Can be used in relation to topographic and cut slope angle and orientation to determine their combined feasibility to promote slope failure.

Stereo photographs: overlapping photographs (usually 60%). The overlap, when viewed in a stereoscope, provides a 3D image of the ground surface. Stereo photographs are usually taken vertically from an aircraft, though oblique photographs are also sometimes taken.

Strength parameters: cohesion and friction between particles. In granular soils and jointed rock masses shear strength is considered to be predominantly derived from the friction angle between particles or the friction angle between joints and other planes of weakness.

Subgrade: natural ground upon which road fill and pavement layers are constructed.

Subtropics: zone of the Earth's surface between the latitudes of 40°N and 23°N and 40°S and 23°S.

Susceptibility: the degree to which, in this context, a slope might be prone to movement.

Synthetic-aperture radar (SAR): radar system based on the relative motion between the antenna and its target. High azimuth resolution is achieved by storing and processing data on the Doppler shift of multiple return pulses in such a way as to give the effect of a much longer antenna.

Talus: rockfall accumulation material. A talus slope is the accumulation slope beneath a source of progressive rockfall, usually with a slope equal to the angle of repose of the material.

Taluvium: a slope deposit that comprises a mixture of coarse and fine material, and is often used to describe weathered talus (rockfall deposit) and landslide debris.

Tectonics: the geo-dynamics of the Earth's crust, the processes of which have given rise to mountain-building and metamorphosed, sheared, faulted and joined rock masses.

Tectonism: metamorphosism, folding, faulting and shearing arising from compressional and other geo-dynamic deformational processes acting within the Earth's crust.

Tension crack: crevasse caused by the pulling apart of ground due to slope movement.

Terracette: narrow, horizontal step across a slope, usually forming parallel lines on a steep slope, probably due to soil creep.

Terrain model: classification of the landscape to identify patterns of landforms and groupings of slopes and landforms with broadly similar materials and geological/geomorphological processes.

Thin wall sampling: use of a narrow gauge tube sampler located at the end of a hollow auger to retrieve an undisturbed soil sample. The thin wall enables minimal soil disturbance during sampling.

Thermal imaging: a form of imaging that uses cameras that can detect radiation in the infrared range of the electromagnetic spectrum (roughly 900–14 000 nanometers (1×10^{-9}).

Thermal IR: Infrared region from 3 to 14 μm (1×10^{-6}) that is employed in remote sensing. This spectral region spans the radiant power peak of the Earth.

Tile (or scene): the area covered by a single satellite image. The size of this area varies for different satellites.

Topographic amplification: the effect of topography on the distribution of strong ground motions during an earthquake. Amplification tends to occur at distinct convex changes in slope profile, for example cliffs, scarps and sharp ridges.

Total station: a surveying instrument that enables electronic distance measurement.

Transported soil: soil derived from material that has moved downslope or by a river or glacier etc.

Triggering factors: dynamic processes, such as rise in groundwater, erosion, seismic loading that initiate slope failure.

Trimming: removal of loose soil, debris and rock fragments from a slope, sometimes in combination with a reduction in slope angle.

Triple case/core tube barrel: split tube is mounted in the inner tube of a double tube core barrel in order to recover 100% undisturbed sample.

Tropics: zone of the Earth's surface between 23°N and 23°S.

Truncheon: woody stem cut from a live tree or shrub that is inserted into the ground. Roots develop allowing a new plant to grow.

Tuff: compacted fine to medium-grained pyroclastic material.

Turnout (drainage): discharge point from a side drain, intermediate between culverts to accommodate side drain surcharge.

Unconformity: break in a geological sequence marked by a break in deposition or by tectonic juxtaposition. The lower surface of a sedimentary unconformity may be a surface of denudation.

Variable actions: these are loads or forces applied to a slope or structure as a result of external engineering effects (Eurocode 7).

Vertisols: usually soils containing montmorillonite clay minerals, exhibiting high shrink-swell upon wetting and drying.

Virtual back: in the design of walls such as L-shaped gravity walls (which use the soil weight over the base to resist sliding and overturning), the back of the wall is taken to be a vertical plane running up from the wall heel. This is known as the 'virtual back'. The virtual back is the plane on which the disturbing earth pressures are taken to act. All of the soil in front of the virtual back contributes to the wall resistance.

Void ratio: ratio of the total volume of voids in a soil to the total volume of solid particles.

Warm temperate: mid-latitude climate zone (see Kloppen's climate classification Figure A1.1).

Water table: the upper level of the zone of groundwater saturation in soils and permeable rocks.

Weathering: physical breakdown and/or chemical decay of rock in response to contact with air, water and organisms.

Weathering grade: a classification of the extent of alteration and weakening that has taken place as a result of physical breakdown and chemical decay of a rock mass.

Weathering profile: soil column comprising horizons of soil and weathered rock derived from rock weathering. A fully developed weathering profile in the humid tropics and subtropics comprises surface horizons of residual soil overlying successive horizons of completely weathered, highly weathered, moderately weathered and slightly weathered rock, with or without duricrust development.

Windthrow: uprooting of trees due to the force of wind.

Wire line drilling: an inner tube containing the core is detached from the core barrel assembly via a wire line. The tube and core contained in it are pulled up to the surface by the wire which has been dropped down the string of drill rods. This allows rapid recovery of core and minimises sample disturbance in the process.

Index

Page numbers in *italics* denote figures. Page numbers in **bold** denote tables. Letter 'b' after a page number indicates a text box.